# SIMULATED MOVING BED TECHNOLOGY

# SIMULATED MOVING BED TECHNOLOGY

## Principles, Design and Process Applications

**A.E. RODRIGUES**

**C. PEREIRA**

**M. MINCEVA**

**L.S. PAIS**

**A.M. RIBEIRO**

**A. RIBEIRO**

**M. SILVA**

**N. GRAÇA**

**J.C. SANTOS**

**ELSEVIER**

Amsterdam • Boston • Heidelberg • London • New York • Oxford
Paris • San Diego • San Francisco • Singapore • Sydney • Tokyo
Butterworth-Heinemann is an imprint of Elsevier

Butterworth-Heinemann is an imprint of Elsevier
The Boulevard, Langford Lane, Kidlington, Oxford OX5 1GB, UK
225 Wyman Street, Waltham, MA 02451, USA

**Notices**
Knowledge and best practice in this field are constantly changing. As new research and experience broaden our understanding, changes in research methods, professional practices, or medical treatment may become necessary.

Practitioners and researchers must always rely on their own experience and knowledge in evaluating and using any information, methods, compounds, or experiments described herein. In using such information or methods they should be mindful of their own safety and the safety of others, including parties for whom they have a professional responsibility.

To the fullest extent of the law, neither the Publisher nor the authors, contributors, or editors, assume any liability for any injury and/or damage to persons or property as a matter of products liability, negligence or otherwise, or from any use or operation of any methods, products, instructions, or ideas contained in the material herein.

**Library of Congress Cataloging-in-Publication Data**
A catalog record for this book is available from the Library of Congress

**British Library Cataloguing-in-Publication Data**
A catalogue record for this book is available from the British Library

ISBN: 978-0-12-802024-1

For information on all Butterworth-Heinemann publications
visit our website at http://store.elsevier.com/

Working together
to grow libraries in
developing countries

www.elsevier.com • www.bookaid.org

*Publisher:* Joe Hayton
*Acquisition Editor:* Fiona Geraghty
*Editorial Project Manager:* Cari Owen
*Production Project Manager:* Lisa Jones
*Designer:* Maria Ines Cruz

Typeset by TNQ Books and Journals
www.tnq.co.in

Transferred to Digital Printing in 2015

# CONTENTS

# AUTHORS

**Alírio E. Rodrigues, Emeritus Professor**
Associate Laboratory LSRE – Laboratory of Separation and Reaction Engineering
Department of Chemical Engineering, Faculty of Engineering, University of Porto, Portugal

**Carla Pereira, Assistant Researcher**
Associate Laboratory LSRE – Laboratory of Separation and Reaction Engineering
Department of Chemical Engineering, Faculty of Engineering, University of Porto, Portugal

**Mirjana Minceva, Assistant Professor**
Assistant Professorship Biothermodynamics
Department of Life Science Engineering, TUM School of Life Sciences Weihenstephan
Technische Universität München

**Luís S. Pais, Professor**
Associate Laboratory LSRE – Laboratory of Separation and Reaction Engineering
School of Technology and Management, Polytechnic Institute of Bragança, Portugal

**Ana Mafalda Ribeiro, Assistant Researcher**
Associate Laboratory LSRE – Laboratory of Separation and Reaction Engineering
Department of Chemical Engineering, Faculty of Engineering, University of Porto, Portugal

**António Ribeiro, Assistant Professor**
Associate Laboratory LSRE – Laboratory of Separation and Reaction Engineering
School of Technology and Management, Polytechnic Institute of Bragança, Portugal

**Marta Silva, PhD**
Associate Laboratory LSRE – Laboratory of Separation and Reaction Engineering
Department of Chemical Engineering, Faculty of Engineering, University of Porto, Portugal

**Nuno Graça, PhD**
Associate Laboratory LSRE – Laboratory of Separation and Reaction Engineering
Department of Chemical Engineering, Faculty of Engineering, University of Porto, Portugal

**João Carlos Santos, Assistant Researcher**
Associate Laboratory LSRE – Laboratory of Separation and Reaction Engineering
Department of Chemical Engineering, Faculty of Engineering, University of Porto, Portugal

# PREFACE

This book is a result of more than 20 years research on Simulated Moving Bed (SMB) processes at the Laboratory of Separation and Reaction Engineering (LSRE) and teaching at undergraduate level at the Department of Chemical Engineering (ChE), Faculty of Engineering of University of Porto (FEUP), graduate courses at Technical University (TU) Eindhoven and TU Delft, and an in-house course for Companhia Petroquímica do Nordeste (COPENE) (now Brazchem) and Petrogal.

I graduated in ChE at University of Porto (U. Porto) in 1968, having never heard about SMB during those years. I heard about PAREX (and other Sorbex processes) in Nancy during my thesis work (1970–1973) with P. Le Goff and D. Tondeur. I found the idea of SMB—turning fixed-bed operation into continuous processes—a bright one. After my African endeavors (teaching at the University of Luanda in Angola and military service there), I landed again at FEUP in August 1976 as an Assistant Professor. An optional course on Petroleum Refining for Chemical Engineering (ChE) was offered to undergraduate students given by Lopes Vaz from Petrogal. He was working in Lisbon but coming to Porto every Saturday morning to teach that course. I asked permission to attend. Lopes Vaz was a very good lecturer. It was an opportunity to learn details of the PAREX unit existing in the aromatics plant in the Refinery of Petrogal in Matosinhos.

After the Revolution of April 1974, FEUP began offering evening courses allowing people with a "technical engineer" degree to get a diploma of Chemical Engineering (ChE) from U. Porto by following an additional two-year program. One of my students at that time was Soares Mota working for Petrogal and taking care of the PAREX unit.

In 1978, I organized my first NATO Advanced Study Institute (ASI) on "Percolation Processes: Theory and Applications." One of the lecturers I invited was D. Broughton from UOP (one of the inventors of SMB). He could not come, but instead A. De Rosset lectured in that ASI. I had the opportunity to travel to Des Plaines (Illinois) to visit Universal Oil Products (UOP) and meet D. Broughton at lunch. It was a business trip that I remember because I met some leaders in the Adsorption area (Vermeulen and Klein from University of California (UC) Berkeley, Wankat from Purdue, etc.).

In 1984, an opportunity arose for funding to work on PAREX and ISOMAR processes when Veiga Simão was Minister of Industry and Energy (MIE). He launched some Contracts for Industrial Development (CDIs) and I took the initiative of encouraging several engineers from Petrogal to join that initiative. The funding was supposed to be equivalent to 100,000 euro, but when the MIE came to Porto for the signing ceremony it seems he decided not to sign that CDI. I just found those documents while cleaning my office.

In the middle 1980s, we started a FEUP M.Sc. course on "Chemical Process Engineering." One of the participants was Engineer João Fugas (now with CUF) working for Petrogal at that time. We invited H. Hoffman from Erlangen, J. Villadsen (DTU), R. Mann (UMIST), and D. Cresswell (ICI) to teach in that program. Engineer Fugas was working with the PAREX process and was supposed to do his M.Sc. thesis on that topic. However, life moves very quickly. Before he could start his M.Sc. dissertation work, he moved to another company—and again the PAREX project was postponed.

In 1988, I organized another NATO ASI on "Adsorption: Science and Technology" at which Jim Johnson from UOP lectured about Sorbex processes. It was in that meeting that some university professors suggested that it would be nice if UOP could produce a video to teach the concept of SMB. They did! I have since then been using this video in my classes. However, the color was fading, so we found we could transform it into a digital version, so at least I can keep showing such a movie… it is better than words! One of the participants in that NATO ASI was Célio Cavalcante from COPENE with many years of experience in the operation of a PAREX unit in COPENE in Camaçari (near Salvador). He decided afterward to pursue a Ph.D. with Ruthven in Canadá, and he is now a Professor at UF Ceará (Brazil). Interestingly enough: when I gave a Honeywell/UOP Invitational Lecture in 2011 on "Cyclic adsorption/reaction processes: Old and new," I showed that movie at the end of the talk…Younger people present were unfamiliar with it…

Finally, I had the chance to do research on SMB processes in 1992. However, not in PAREX! I took part in an EU Brite project "ENANTIO-CHROM" dealing with chiral separations by SMB chromatography. The leader was Roger-Marc Nicoud (with Separex at the time). By the end of 1992, my pilot SMB arrived…and it is still working! It arrived in a Saturday morning; Nicoud was in Porto to install the equipment. It was too big to go into the lift. So I invited one student (J.A.C. Silva, who later did his Ph.D. with me) to help put the equipment into the lab on the first floor! When the project finished in 1995, Nicoud had started a new company, Novasep, which is today a leading company in SMB for pharmaceutical, fine chemicals, and bioseparations. The SMB we have is either the first or second unit made by the company (to check with J. Kinkel from Merck at that time and Nicoud). From LSRE, the Ph.D. student working on the project was Luís S. Pais (now at Instituto Politécnico Bragança [IPB]). A postdoc student, Dr. Lu Zuping, was also briefly involved in this project writing the first SMB simulation program. This project revealed a person (R.M. Nicoud) who wanted to run and develop a company; after some months, he resigned from the academic position he had at ENSIC and did a tremendous job bringing on people from pharmaceutical companies to teach them and demystify SMB. One of the persons who attended such workshops was Emile Cavoy from UCB Pharma, the first company to implement SMB at industrial scale in 1997.

So here at LSRE—FEUP we started from SMB "new" applications (Luís S. Pais, chiral separations), to "old" applications: sugars separation (Diana Azevedo) and xylene separations (Mirjana Minceva) and back to chiral separations (Michal Zabka, António Ribeiro from IPB) and back again to old applications with new adsorbents (MOFs)!

In the middle 1990s, I had the opportunity to go to COPENE (now Braskem) and teach a one-week course on SMB. Sergio Bello (working with PAREX in COPENE, now with TECHBIOS and UNIFACS) came with Diana Azevedo (UF Ceará, Brazil) for a two-month stay in LSRE to work on xylene separation. Later Diana came for Ph.D. work in Porto on sugar separation and Sérgio Bello defended his Ph.D. at UNICAMP.

With Mirjana Minceva (Macedonia), now at T. U. München, we could finally engage in long research on xylene separation involving experiments and simulation work. The last chapter of the Ph.D. on isomerization/separation of xylenes had to remain confidential during one year while Mirjana stopped her postdoc to raise her son Marco, and we did not apply for the patent. When I realized that "confidential" means nothing to some people who had signed a confidentiality agreement, I just published the work. Nevertheless, patents are coming out. This effort on PAREX was continued with Pedro Sá Gomes (now at BASF) in a study of aging effects in PAREX performance and revamping studies. During that time, we found people interested in our work like Dr. Jin Suk Lee from Samsung Chemicals (now Samsung TOTAL), who visited LSRE. Recently, I came across papers from our Korean colleague, who is very active in the field, on the PAREX process. I met him again in 2013 at WCCE9 in Seoul.

After Mirjana's work on xylenes, we went back again to chiral separations using new columns (monoliths). I was worried that our old SMB would breakdown and Pedro Sá Gomes built and tested, with various chiral separations and modes of operation, new equipment: the FlexSMB®. As the Nobel Prize winner Richard Ernst once said, "every student must leave his (her) fingerprint in the lab."

Some years ago after a visit of Luc Wolff from IFP (now with Axens), I had a project funded by IFP to look at propane/propylene separation by gas phase SMB (Ph.D. of Nabil Lamia). A patent was obtained and later with Pedro Sá Gomes a complete SMB design from scratch was published allowing him to receive in 2009 the Graduate Research Award in Separations from the Separations Division of the American Institute of Chemical Engineers (AIChE).

On March 16, 2010, a big day, I started my journey at 6:00 am driving to Lisbon to attend a meeting at GALP Energia headquarters. The morning session was busy with presentations by Ph.D. students who had started a new doctoral program in Portugal (EngIQ) involving five universities and the main chemical/petrochemical companies in the country (GALP, CUF, SONAE, Dow, Cires). The afternoon session was the usual show with the Minister of Science and Technology, presidents of GALP and CUF, and so forth, but with interesting views on the future by engineers from GALP (Soares Mota and Cavanna) and CUF (João Fugas). It was a long day, and I was home by 9:00 pm for the birthday of one of my granddaughters!

Marta Silva, one of my Ph.D. students, presented the scope of the project "Optimization of the PAREX unit" in 10 minutes and, at the end, a microphone was placed into my hands—I was supposed to say something. It was my opportunity to tell a long story in two minutes. I reported the above PAREX story and expanded it to SMB research at LSRE−FEUP. After three years, Marta Silva received her Ph.D. and now works for GALP in oil and gas exploration.

I became involved with the EngIQ doctoral program by coordinating and teaching a course on Separation Processes. It was an opportunity to have Mark Davis from UOP lecture on UOP Separations Technology and have Linda Cheng visit. In addition, Tim Golden from Air Products lectured in that doctoral program. Here we are, back again on the PAREX process sharing this project with Paulo Mota (UNL). You may guess: He is the son of Soares Mota from GALP and was also my undergraduate student at FEUP who later did a Ph.D. in collaboration with FEUP and LSGC−CNRS (Nancy). We shared the same co-adviser, Daniel Tondeur, in Nancy. Paulo Mota remembers a situation from the time he was an undergraduate at FEUP: he arrived late and asked permission to attend my class. I refused—in fact I was fed up with people arriving late that day and I had told the class that I would not allow more people coming in—then Paulo Mota arrived and knocked at the door—I had to say NO to his request. He is a guarantee that SMB research has a future in Portugal. Nevertheless, some years ago a young researcher, who did a postdoc in the SMB area in the USA, told me in one AIChE meeting "there is no future for SMB research…everything is done." I smiled—after that time, new operating modes (VARICOL, etc.) have arrived in practice in the pharmaceutical industry…

SMB remains a nice research area with many things to "discover" on the separation side, such as the gas-phase SMB. Process intensification by coupling reaction and separation was implemented as the Simulated Moving Bed Reactor (SMBR) during the Ph.D. of Viviana Silva (now with BASF) with acetalization reaction for the synthesis of diethyl acetal (DEE). This effort was followed by Ganesh Gandi (Godavari Biorefinery, Mumbai) for the synthesis of dimethyl acetal (DME), Nuno Graça working on dibutyl acetal (DBE), Carla Pereira (now with ExxonMobil) on ethyl lactate, Rui Faria on glycerol acetal, Mehamud Rahman on (1,1-Diethoxybutane [DEB]). Process reintensification by coupling SMBR and membrane pervaporation in the PermSMBR was developed by Viviana Silva and Carla Pereira, who won the PSE Model-based Innovation Prize in 2012.

The book is organized into 11 chapters. Chapter 1 deals with SMB principles: history, concept, applications, and modes of operation. Chapter 2 is about modeling and simulation of SMB separation processes, including modeling strategies, process performance indicators, and numerical tools. Chapter 3 addresses the design of SMB for binary or pseudo-binary separations; the concept of separation volume is illustrated with an example of sugar separation. The methodology for process development is addressed in Chapter 4 and illustrated with examples from chiral separations. The chapter also includes design, construction, modeling, and operation of the FlexSMB® as an example of Product Engineering. Chapter 5 concentrates on the PAREX process for $p$-xylene separation from a $C_8$ mixture. Multicomponent separations by SMB-based processes are handled in Chapter 6 and gas-phase SMB in Chapter 7. The Simulated Moving Bed Reactor (SMBR) is discussed in Chapter 8 as an example of process intensification, and in Chapter 9 process reintensification is illustrated with the concept of PermSMBR combining SMBR and permeable membranes. Chapter 10 deals with Sequential Centrifugal Partition Chromatography (sCPC), a concept of SMB without a solid phase, and Chapter 11 presents conclusions and perspectives.

LSRE—FEUP, November 25, 2014

# ACKNOWLEDGMENT

I would like to thank all Ph.D., pos-doc students and my colleague (J. M. Loureiro) who worked in this area with me at LSRE and in particular Viviana Silva and Diana Azevedo for starting SMBR work and implementing the concept of separation volume, and Pedro Sá Gomes for the conceptual design of Gas phase SMB and construction of the FlexSMB unit. Finally I acknowledge the collaboration of the co-authors in the writing effort of this book: Carla Pereira, Mirjana Minceva, Luís S. Pais, Marta Silva, Ana Mafalda Ribeiro, António Ribeiro, Nuno Graça, and João Carlos Santos (now at Jacobs Engineering). Last but not the least a special thank to my family for the support.

# CHAPTER 1

# Principles of Simulated Moving Bed

## 1.1 HISTORY: FROM BATCH CHROMATOGRAPHY TO CONTINUOUS COUNTERCURRENT CHROMATOGRAPHY

Adsorption and chromatographic processes are widely used in the chemical, pharmaceutical, and bioprocesses industries for separation, purification, or recovery purposes. Extensive reviews of adsorption techniques with applications in the field of preparative and industrial scale chromatography can be found in the literature (Giddings, 1965; Rodrigues and Tondeur, 1981; Ruthven, 1984; Rodrigues et al., 1989; Giddings, 1990; Suzuki, 1990; Dondi and Guiochon, 1992; Ganetsos and Barker, 1993; Guiochon et al., 1994).

The simulated moving bed (SMB) technology is one of the most powerful and promising techniques for preparative-scale chromatography. Briefly, the SMB technology allows for the continuous injection and separation of binary mixtures. The simulated countercurrent contact between the solid and the liquid phases maximizes the mass-transfer driving force, leading to a significant reduction in mobile and stationary phase consumption compared with elution chromatography.

Chromatography is a technique used to separate the components of a mixture carried by a mobile phase that flows through a stationary phase packed in a column. The separation is based on different partitioning of species between the mobile and the stationary phases. It is commonly divided into analytical, preparative, and industrial chromatography.

### 1.1.1 Who is the Father?

Chromatography was invented by the Russian botanist Mikhail Semenovich Tswett during his research on the physicochemical structure of plant chlorophylls. He produced a colorful separation of plant pigments (xanthophylls and chlorophylls) using a column of calcium carbonate as adsorbent and carbon disulfide as eluent. The method was described on 30 December 1901 at the XIth Congress of Naturalists and Physicians in St. Petersburg (Tswett, 1903). However, the term "chromatography" was not used at the time. This term, which literally means "to write with color," was introduced in print in 1906 by Tswett in two papers about chlorophyll in the German botanical journal *Berichte der Deutschen botanischen Gesellschaft* (Tswett, 1906a,b). He coined the term chromatography, inspired either by his experiment or, according to some authors, by his name (Tswett in Russian means color). The gradual steps in Tswett's work that led to the development of chromatography are conveniently described in *The Centenary of "Chromatography"* by Ettre (Ettre, 2006).

*Simulated Moving Bed Technology*
ISBN 978-0-12-802024-1

1

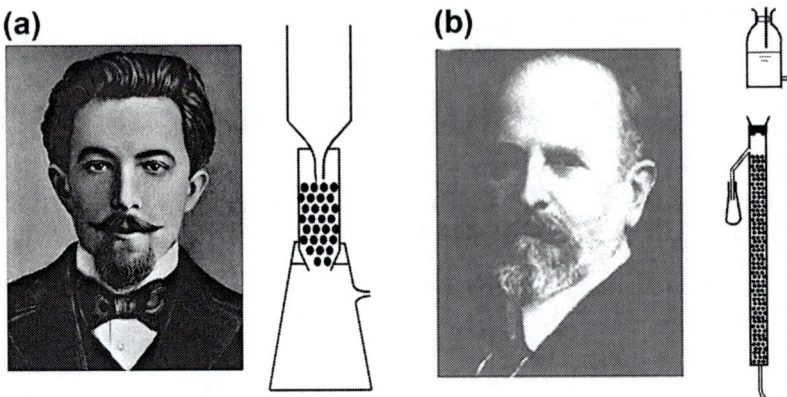

**Figure 1.1** The early days of chromatography: M. Tswett (a) and D.T. Day (b) experiment (Rodrigues et al., 1991).

Another view of the history of chromatography gives the first chromatography experiment as the one done by David Talbot Day (a geologist and engineer working for Mineral Resources of the U.S. Geological Survey). He presented a paper at the First International Petroleum Congress in Paris (1900) in which he claims that "crude oil forced upward through a column packed with limestone changed in colour and composition." This is the basis of PONA analysis (paraffinics, olefinics, naphtenics, and aromatics), established in 1914 and still used in the petroleum industry. The similarities in the equipment used by Tswett and Day are shown in Figure 1.1. David Talbot Day is the author of a historic book, *A Handbook of the Petroleum Industry,* vols. 1 and 2, published by Wiley (Day, 1922).

Over the years, the principles underlying Tswett's chromatography were applied in many different ways, giving rise to the different forms of chromatography and improving the technical performance of chromatographic processes for the separation of more and more similar molecules. The chromatographic methods can be classified based on the nature of the mobile phase (gas or liquid chromatography), the nature of the stationary phase (liquid, solid, or bonded liquid), the mechanism of separation (adsorption, ion exchange, exclusion, partition, modified partition), and the technique used (column or planar). Column chromatography in particular can be performed in various operation modes: elution, frontal, or displacement chromatography.

## 1.1.2 Column (Batch) Chromatography

Chromatographic separation processes can be operated in batch or continuous mode. In the elution mode, a column is filled with a suitable stationary phase (adsorbent) onto which is injected the mixture to be separated together with a mobile phase (eluent or desorbent). Additional desorbent is continuously fed to carry the mixture species through

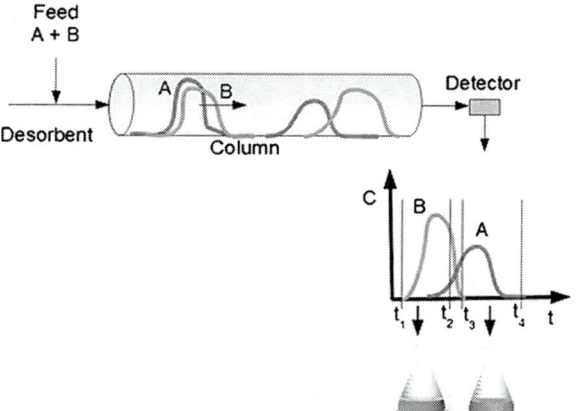

**Figure 1.2** Principle of elution chromatography.

the column. Owing to the different adsorption affinities of each component to the packing material (adsorbent), the components have different migration velocities and the mixture is gradually separated, while passing through the column. A scheme of elution chromatography for a binary separation is shown in Figure 1.2. At time $t_0$, a well-defined amount of a mixture comprising A and B is injected into the column inlet, and then with the help of the desorbent, the feed mixture passes through the column. Because A and B have different adsorption affinities to the stationary phase, they move along the column at different velocities, and therefore they become separated from each other. At the column outlet, the components B and A can be detected and separately collected between $t_1$ and $t_2$ and $t_3$ and $t_4$, respectively. The positions $t_2$ and $t_3$ are determined by the products' purity requirements, whereas $t_1$ and $t_4$ are determined by the minimum concentration threshold of the detector used. This procedure is repeated periodically.

This is not the most efficient mode of operation, especially for high quantities of mixture to be treated and difficult separations, because of the well-known drawbacks associated with batch processes, such as low yields and high product dilution.

### 1.1.2.1 Understanding Traveling of Concentration Peaks and Fronts: De Vault Equation

The basic equation for understanding chromatography is attributable to Don De Vault (De Vault, 1943). Based on a simple equilibrium model for a fixed-bed column, assuming diluted systems, plug flow, negligible pressure drop, and isothermal operation, the mass balance of species $i$ in a volume element of the column is

$$u_i \frac{\partial C_i}{\partial z} + \frac{\partial C_i}{\partial t} + \frac{1-\varepsilon}{\varepsilon} \frac{\partial q_i^*}{\partial t} = 0 \tag{1.1}$$

Combining with the adsorption equilibrium isotherm,

$$q_i^* = f(C_i) \tag{1.2}$$

it can be demonstrated, using the cyclic relation between partial derivatives, that the velocity of propagation of a concentration $C_i$ is

$$u_{C_i} = \left(\frac{\partial z}{\partial t}\right)_{C_i} = \frac{u_i}{1 + \frac{1-\varepsilon}{\varepsilon}f'(C_i)} \tag{1.3}$$

where $z$ is the axial position, $t$ is the time, $u_i$ is the interstitial velocity, $\varepsilon$ is the bed porosity, and $f'(C_i)$ is the slope of the isotherm relating the adsorbed phase concentration $q_i^*$ in equilibrium with $C_i$.

The equilibrium theory indicates that for favorable isotherms the concentration front in the adsorption step is *compressive* (higher concentrations travel at higher velocities), with a shock as the limit, and for unfavorable isotherms the concentration front is *dispersive* (higher concentrations travel at lower velocities); the reverse is true for the desorption step.

For linear isotherms,

$$q_i^* = K_i C_i \tag{1.4}$$

and assuming the equilibrium model is valid, a pulse injected onto the column will appear at the outlet with a retention time $t_{r,i}$:

$$t_{r,i} = \frac{L}{u_i}\left(1 + \frac{1-\varepsilon}{\varepsilon}K_i\right) \tag{1.5}$$

Therefore, the more retained components take longer time to exit the column.

## 1.1.3 Continuous Countercurrent Chromatography

The effectiveness of chromatographic processes in terms of stationary-phase usage and solvent consumption can be increased by using continuous countercurrent chromatography. This technique leads to higher productivity and less diluted product streams, compared with conventional batch chromatography.

To explain the concept of continuous countercurrent chromatography, let us first consider a single chromatographic column operated in a continuous way for the separation of a binary mixture (A and B), in which component A has higher affinity toward the stationary phase than component B (Figure 1.3(a)). The mixture, diluted in desorbent, is continuously fed to the packed column. Because the components have different affinities to the stationary phase, they will travel in the direction of the mobile phase at different velocities: A (the one with higher affinity to the stationary phase) will take longer time

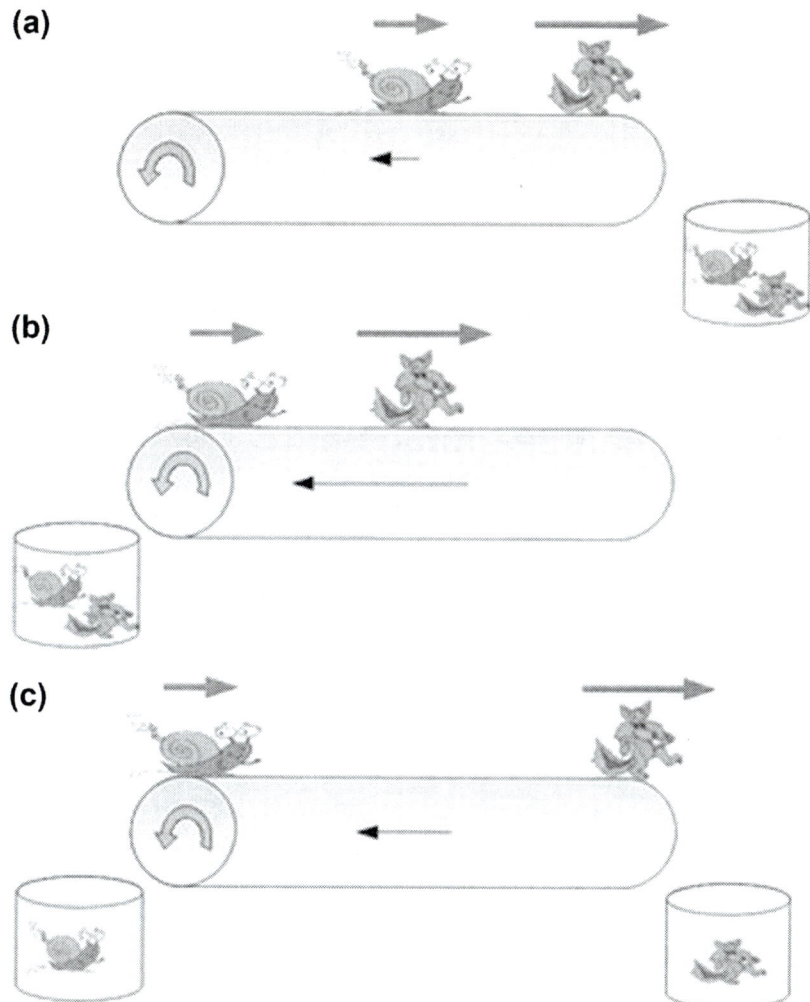

**Figure 1.3** Chromatographic column in which the more-retained species A is represented by a snail and less-retained species B by a fox: (a) low belt velocity, (b) high belt velocity, and (c) efficient belt velocity.

than B to reach the column outlet. If there is enough column length, all the B fed to the column will be collected at the end of the column before A starts coming out.

Now, let us consider that A is represented by a snail, B by a fox, and the column stationary phase by a moving belt. The fox and the snail start to run to the right, while the belt moves to the left (countercurrent movement). Depending on the moving belt's velocity, three scenarios are possible: (i) if the belt is moving slower than the snail, both snail and fox will be retarded but the result will be the same as in the previous case (both animals reach the right-hand side) (Figure 1.3(a)), (ii) if the belt is moving faster than the fox, both fox and snail reach the left-hand side (Figure 1.3(b)), and (iii) for a set of possible belt velocities, slower than the fox but faster than the snail, the animals are separated; the snail reaches the left-hand side and the fox reaches the right-hand side (Figure 1.3(c)). Bearing this in mind, we can conclude that snails and foxes can be continuously separated, if they are continuously fed to the center of the moving belt and the proper belt velocity is set. This is the principle of continuous countercurrent chromatography.

### 1.1.4 The True Moving Bed

A setup in which the solid actually moves in the direction opposite to that of the liquid is known as the true moving bed (TMB) and it is shown in Figure 1.4, in which the separation of a binary mixture is considered.

The TMB has two inlet streams (the desorbent and the feed, i.e., the mixture to be separated containing A and B species) and two outlet streams (the extract, where the more-adsorbed product is collected, and the raffinate, where the less-adsorbed product is withdrawn). These streams divide the unit into four different sections, of which each section may operate at a different flow rate and has a specific role:

Section I—regeneration of the solid, before being recycled to section IV, through the desorption of the more-adsorbed species (A);

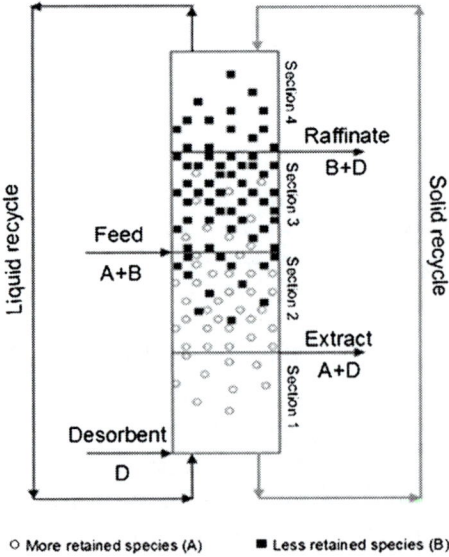

○ More retained species (A)    ■ Less retained species (B)

**Figure 1.4** Scheme of a true moving bed.

Section II—desorption of the less-adsorbed component (B) to avoid extract (X) contamination (B moves with the liquid);

Section III—adsorption of the more-adsorbed species (A) to avoid raffinate (R) contamination (A goes with the solid);

Section IV—regeneration of the desorbent by the adsorption of less-adsorbed component (B) from the fluid phase, before being recycled to section I.

The flow rates in each section of the TMB are

$$Q_I = Q_D + Q_{Rec} \tag{1.6a}$$

$$Q_{II} = Q_I - Q_X \tag{1.6b}$$

$$Q_{III} = Q_{II} + Q_F \tag{1.6c}$$

$$Q_{IV} = Q_{III} - Q_R \tag{1.6d}$$

Summing the previous relations and taking into account that $Q_{IV} = Q_{Rec}$ we obtain

$$Q_F + Q_D = Q_R + Q_X \tag{1.7}$$

The roles of each TMB section are guaranteed by setting the appropriate ratio in each section of liquid flow rate to solid flow rate as shown in Figure 1.5.

### 1.1.4.1 The Design of a True Moving Bed

The design problem of a TMB consists in setting the flow rates in each section to obtain the desired separation. Some constraints have to be met if one wants to recover the less-adsorbed component B in the raffinate and the more-retained component A in the extract. These constraints are expressed in terms of the net fluxes of components in each section (ratio of molar flow rates of a species transported by the liquid phase and by the solid phase). As was pointed out above, in section I both species must move upward, in sections II and III the light species must move upward, while the net flux of the

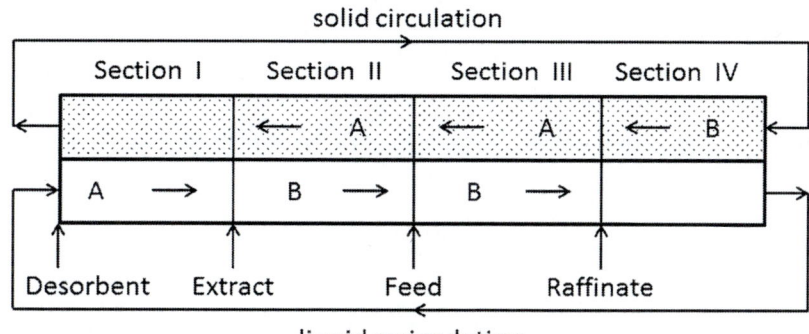

**Figure 1.5** Scheme of TMB indicating the desired net fluxes of A and B in each section.

more-retained component must be downward, and in section IV the net flux of both species has to be downward, that is,

$$\frac{Q_I C_{A,I}}{Q_S q_{A,I}} > 1 \tag{1.8a}$$

$$\frac{Q_{II} C_{A,II}}{Q_S q_{A,II}} < 1 \quad \text{and} \quad \frac{Q_{II} C_{B,II}}{Q_S q_{B,II}} > 1 \tag{1.8b}$$

$$\frac{Q_{III} C_{A,III}}{Q_S q_{A,III}} < 1 \quad \text{and} \quad \frac{Q_{III} C_{B,III}}{Q_S q_{B,III}} > 1 \tag{1.8c}$$

$$\frac{Q_{IV} C_{B,IV}}{Q_S q_{B,IV}} < 1 \tag{1.8d}$$

where $Q_I$, $Q_{II}$, $Q_{III}$, and $Q_{IV}$ are the volumetric liquid flow rates in the various sections of the TMB; $Q_S$ is the solid flow rate; and $C_{A,j}$ and $C_{B,j}$ are the concentrations of species A and B in the liquid phase and $q_{A,j}$ and $q_{B,j}$ are the average adsorbed concentrations of components A and B in section $j$.

For the case of a binary system with linear adsorption isotherms and assuming instantaneous equilibrium, very simple formulas can be derived to evaluate the better TMB flow rates (Ching et al., 1985; Ruthven and Ching, 1989). For the linear case, the net flux constraints presented in Eqn (1.8) are reduced to only four, which are

$$\frac{Q_I}{Q_S} > K_A \tag{1.9a}$$

$$\frac{Q_{II}}{Q_S} < K_A \quad \text{and} \quad \frac{Q_{II}}{Q_S} > K_B \tag{1.9b}$$

$$\frac{Q_{III}}{Q_S} < K_A \quad \text{and} \quad \frac{Q_{III}}{Q_S} > K_B \tag{1.9c}$$

$$\frac{Q_{IV}}{Q_S} < K_B \tag{1.9d}$$

where $K_B$ and $K_A$ are the coefficients of the linear isotherms for the less- and more-retained species, respectively.

If all these inequalities are satisfied by the same margin $\beta$ ($\beta > 1$), Eqn (1.9) can be rewritten as

$$\frac{Q_I}{Q_S K_A} = \beta \tag{1.10a}$$

$$\frac{Q_{II}}{Q_S K_B} = \beta \tag{1.10b}$$

$$\frac{Q_{III}}{Q_S K_A} = \frac{1}{\beta} \tag{1.10c}$$

$$\frac{Q_{IV}}{Q_S K_B} = \frac{1}{\beta} \tag{1.10d}$$

Noticing that

$$Q_D = Q_I - Q_{IV}; \quad Q_X = Q_I - Q_{II}; \quad Q_F = Q_{III} - Q_{II}; \quad Q_R = Q_{III} - Q_{IV} \tag{1.11}$$

where $Q_D$, $Q_X$, $Q_F$, and $Q_R$ are the desorbent or eluent, extract, feed, and raffinate volumetric flow rates, respectively, and considering that the volumetric flow rate in section IV represents the recycling flow rate, $Q_{Rec} = Q_{IV}$, the flow rates for the TMB operation can be evaluated by

$$Q_D = \left(\alpha\beta^2 - 1\right)Q_{Rec} \tag{1.12a}$$

$$Q_X = (\alpha - 1)\beta^2 Q_{Rec} \tag{1.12b}$$

$$Q_F = \left(\alpha - \beta^2\right)Q_{Rec} \tag{1.12c}$$

$$Q_R = (\alpha - 1)Q_{Rec} \tag{1.12d}$$

with

$$Q_{Rec} = \frac{K_B Q_S}{\beta} \tag{1.12e}$$

where $\alpha = K_A/K_B$ is the selectivity factor (equilibrium) of the binary linear system.

The total inlet or outlet volumetric flow rates are given by

$$Q_D + Q_F = Q_X + Q_R = (\alpha - 1)\left(1 + \beta^2\right)Q_{Rec} \tag{1.13}$$

Hence, the specification of the $\beta$ parameter and the solid flow rate (or, alternatively, one of the liquid flow rates) defines all the flow rates throughout the TMB system.

It should be pointed out that the $\beta$ parameter also has a higher limit. In fact, because the feed flow rate must be higher than zero, Eqn (1.12c) gives

$$\beta < \sqrt{\alpha} \tag{1.14}$$

and so

$$1 < \beta < \sqrt{\alpha} \tag{1.15}$$

is the interval of possible values for the $\beta$ parameter. The limiting case of $\beta = 1$ corresponds to the situation in which dilution of species is minimal, and the extract and raffinate product concentrations approach the feed concentrations. In fact, using

Eqn (1.12) and for $\beta = 1$, $Q_D = Q_X = Q_F = Q_R = (\alpha - 1)Q_{Rec} = (K_A - K_B)Q_S$. However, under these conditions, an infinite number of stages would be required in each section (Ruthven and Ching, 1989; Nicoud, 1992). To improve system stability, it is better to work with higher values, typically $1.02 < \beta < 1.05$ (Nicoud, 1993). For $\beta > 1$, from Eqn (1.10),

$$Q_I > Q_{III} > Q_{II} > Q_{IV} \tag{1.16}$$

or, from Eqn (1.11),

$$Q_D > Q_X > Q_R > Q_F \tag{1.17}$$

Considering a situation of complete separation, it follows that the concentrations of the less-retained component in the raffinate and of the more-retained component in the extract are, respectively,

$$C_B^R = \frac{Q_F}{Q_R}C_B^F = \frac{\alpha - \beta^2}{\alpha - 1}C_B^F \tag{1.18}$$

and

$$C_A^X = \frac{Q_F}{Q_X}C_A^F = \frac{\alpha - \beta^2}{(\alpha - 1)\beta^2}C_A^F \tag{1.19}$$

From Eqns (1.18) and (1.19) we conclude that the raffinate and extract concentrations in the TMB operation under linear conditions will never exceed the feed concentrations (Ruthven and Ching, 1989; Zhong and Guiochon, 1996).

### 1.1.4.2 Separation and Regeneration Regions

The separation region is obtained by representing the constraints of the ratio of the liquid to solid flow rate in section III (Eqn (1.9c)) versus that flow rate ratio in section II (Eqn (1.9b)) or, in terms of parameters $m_j = \frac{Q_j}{Q_S}$ (Storti et al., 1989), $m_3$ versus $m_2$ leading to

$$K_B < m_2, m_3 < K_A \tag{1.20}$$

which gives a rectangular triangle above the diagonal where the feed flow rate is zero (Figure 1.6).

Similarly the regeneration region is represented by the constraints in sections I and IV (Eqns (1.9a) and (1.9d)) in terms of $m_4$ versus $m_1$ leading to a rectangle (Figure 1.6):

$$m_1 > K_A \text{ and } m_4 < K_B \tag{1.21}$$

It should be noted that to guarantee a complete separation of a binary mixture with a TMB based on the assumption of instantaneous equilibrium, the operating point should be inside both the separation and the regeneration regions. Also, the vertex of

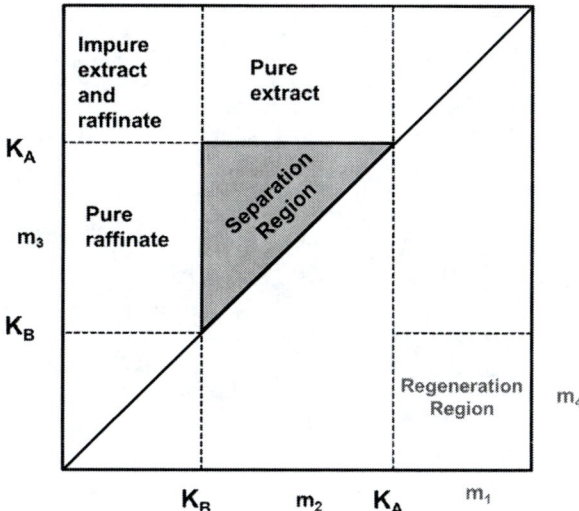

**Figure 1.6** Separation and regeneration regions.

the triangle is the point of maximum productivity, but not the more robust way of operating the TMB.

Some authors (Pais et al., 1998) represent the separation and regeneration regions in a diagram using the ratio of interstitial liquid and solid velocities in section $j$, that is, $\gamma_j = v_j/u_s$ instead of the ratio of liquid to solid flow rate; however, these parameters are related by $\gamma_j = \frac{1-\varepsilon}{\varepsilon} m_j$.

When steady-state operation is reached the concentration profiles in the TMB look like those shown in Figure 1.7.

The TMB allows for the attainment of high-purity products even if low-selectivity adsorbents are available, in contrast to batch chromatography, in which high selectivity

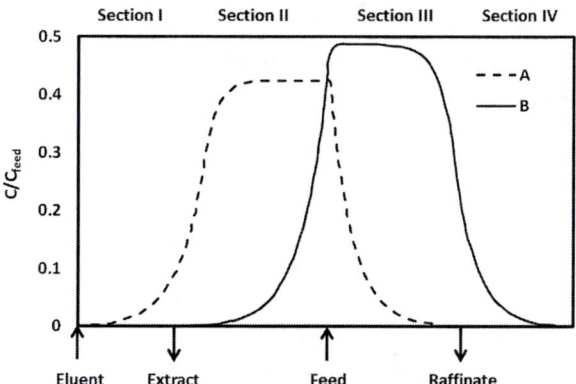

**Figure 1.7** Typical internal concentration profiles in a true moving bed. (A, more-retained component; B, less-retained component.)

**Figure 1.8** Donald B. Broughton, Clarence Gerhold, and Don Carson (James Johnson (UOP LLC, A Honeywell Company) is acknowledged for providing the photographs).

is critical to attain high-purity products. However, from an engineering point of view, the use of this type of equipment with the actual movement of the stationary phase is hard to implement, leading to some technical problems, namely, equipment abrasion, mechanical erosion of adsorbent, and difficulties in maintaining plug flow for the solid (especially in beds with large diameter).

These drawbacks motivated the development of the SMB technology, in 1961, by Broughton and Gerhold (Broughton and Gerhold, 1961) and Carson (rotary valve) (Carson and Purse, 1962) from Universal Oil Products (Figure 1.8).

## 1.2 THE SIMULATED MOVING BED CONCEPT

The SMB simulates the true countercurrent system, using a series of packed bed columns in which the inlet/outlet ports are periodically shifted in the direction of the fluid flow (Figure 1.9). In the conventional operating mode, at regular time intervals, called the switching time, $t^*$, the injection and withdrawal points all move one column ahead in the direction of the fluid flow. When the initial location of injection/collection of all the streams is reached, we have completed one cycle (in a four equally zoned SMB, it takes $4N_c t^*$ to complete one cycle, where $N_c$ is the number of columns in each of the four sections). In this way, during one cycle the same column is in different sections, assuming therefore different roles in the separation process.

### 1.2.1 Analogy with the True Moving Bed

#### 1.2.1.1 Equivalence between SMB and TMB

The SMB and TMB units are equivalent when the SMB sections consist of a large number of columns and the port switch occurs at high frequency, because the continuous countercurrent movement of the solid is better simulated under these conditions.

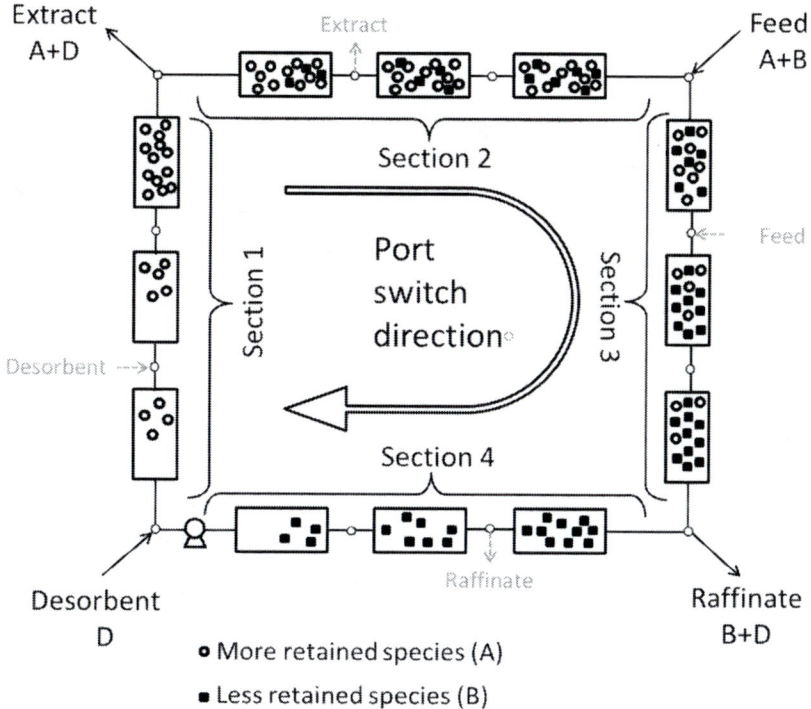

**Figure 1.9** Schematic diagram of a simulated moving bed.

Equivalence between the SMB and the TMB models can be made by keeping constant the liquid velocity relative to the solid velocity, that is, the liquid velocity in the SMB is

$$v_j^* = v_j + u_s \tag{1.22}$$

where $v_j^*$ and $v_j$ are the interstitial liquid velocities in the SMB and TMB, respectively, and $u_s$ is the interstitial solid velocity.

Also, the switching time $t^*$ in the SMB is related to the solid velocity $u_s$ in the TMB model by

$$t^* = \frac{L_c}{u_s} \tag{1.23}$$

where $L_c$ is the length of one SMB column. Alternatively, the equivalence can be made in terms of flow rates

$$Q_j^* = Q_j + \frac{\varepsilon}{1 - \varepsilon} Q_s \tag{1.24}$$

with

$$Q_s = \frac{(1 - \varepsilon) V_c}{t^*} \tag{1.25}$$

where $Q_j^*$ and $Q_j$ are the liquid flow rates in section $j$ of an SMB and TMB, respectively, and $V_c$ is the volume of one SMB column.

Hence, following the equivalence of internal flow rates results in the inlet and outlet flow rates being the same for TMB and SMB and

$$Q_{Rec}^* = Q_{Rec} + \frac{\varepsilon}{(1-\varepsilon)} Q_s = \left[\frac{K_A}{\beta} + \frac{\varepsilon}{(1-\varepsilon)}\right] Q_s = \left[1 + \frac{(1-\varepsilon)}{\varepsilon} \frac{K_A}{\beta}\right] \frac{\varepsilon V_c}{t^*} \quad (1.26)$$

where $Q_{Rec}^*$ is the recycling flow rate in the SMB operation.

The region for complete separation can be defined in terms of the flow-rate ratios in the four sections of the equivalent TMB unit:

$$m_j = \frac{Q_j^* t^* - \varepsilon V_c}{(1-\varepsilon) V_c} \quad (1.27)$$

### 1.2.1.2 Cyclic Steady State in Simulated Moving Bed versus Steady State in True Moving Bed

In the TMB operation, a steady state can be reached in which the concentrations of species at the extract and raffinate ports do not change with time. However, in the SMB operation a cyclic steady state (CSS) can be obtained in which the extract and raffinate concentrations change with time within a switching time interval but are repeated in subsequent switches. This is shown in Figure 1.10 for the case of an eight-column configuration (two columns per section).

## 1.2.2 Realizations of Simulated Moving Bed

In the conventional SMB system, the solid phase is fixed and the positions of the inlet and outlet streams move periodically in a synchronous way. This shift, carried out in the same direction of the liquid phase, simulates the movement of the solid phase in the opposite direction. However, it is impractical to move the liquid inlet and withdrawal positions continuously. Nevertheless, approximately the same effect can be obtained by dividing the adsorbent bed into a number of fixed bed columns and providing multiple access lines for the liquid streams between each two columns. Thereby, the four liquid access lines between each two columns can be used to perform a discrete movement of the inlet and outlet streams in the same direction of the liquid phase.

In the field of chemical engineering, the concept of the SMB has been known since 1961, when the first patent by UOP appeared (Broughton and Gerhold, 1961). This technology was originally developed in the areas of petroleum refining and petrochemicals and become known generally as the Sorbex process (Broughton, 1968; Broughton et al., 1970; de Rosset et al., 1981; Broughton, 1984; Johnson, 1989; Johnson and Kabza, 1993; Gembicki et al., 1997). Alternative processes for the production of p-xylene using

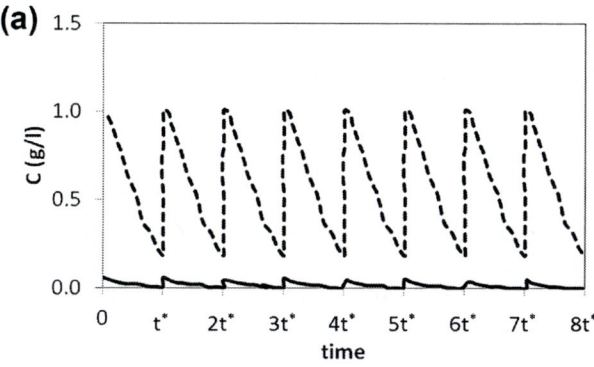

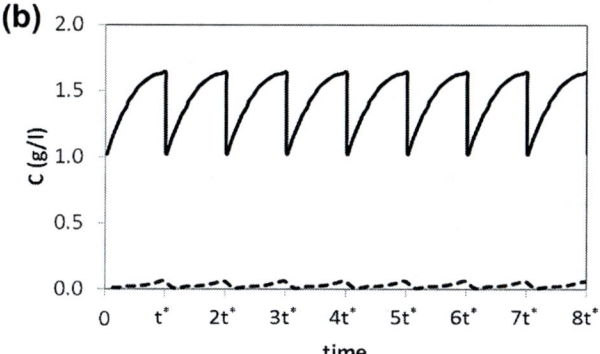

**Figure 1.10** Concentration profiles in (a) the extract and (b) the raffinate for SMB (2-2-2-2 configuration) at CSS during a full cycle. Dotted line, more-retained component; solid line, less-retained component.

the SMB technology were also developed in 1973 by the Toray Industries (Otani, 1973; Otani et al., 1973) and later by the Institut Français du Pétrole with the Eluxyl process (Hotier, 1997), now commercialized by Axens.

In the Sorbex SMB technology developed by UOP, a rotary valve is used to periodically change the position of the eluent, extract, feed, and raffinate lines along the adsorbent bed. At any particular moment, only four lines between the rotary valve and the adsorbent bed are active (this is a simplified description of the Sorbex process; in Chapter 5, dedicated to the Parex process, more details are given).

However, there are alternative techniques for performing the port switching. As a general rule, scaling down of the Sorbex flow sheet becomes less economical than using a set of individual on–off valves connecting the inlet and outlet streams to each node between columns (Keller II, 1995; Humphrey and Keller II, 1997; Nicoud, 1999a,b). This is the case not only in pharmaceutical applications of SMB technology but also in the Eluxyl process from Axens to recover *p*-xylene from a C8 mixture.

Other successful SMB processes in the carbohydrate industry are the production of high-fructose corn syrup (HFCS) and the recovery of sucrose from molasses. This technology was also originally developed by UOP in 1977, known as the Sarex process (de Rosset et al., 1981; Gembicki et al., 1997). With the expiration of the UOP patents, other companies developed alternative processes for fructose—glucose separation, such as Illinois Water Treatment (Rockford, IL, USA), Mitsubishi (Tokyo, Japan), Finnish Sugar (Kantvik, Finland), and Amalgamated Sugar Co. (Twin Falls, ID, USA) (Heikkilä, 1983; Nicoud, 1992; Rearick et al., 1997).

Developed for large-scale hydrocarbon and carbohydrate separations, the SMB technology has found new applications in the areas of biotechnology, pharmaceuticals, and fine chemistry (Gattuso et al., 1994; Gattuso et al., 1995, 1996; Keller II, 1995). Following the increasing interest in preparative chromatographic separations, other companies developed alternative SMB schemes. For example, Separex (now Novasep, Vandœuvre-lès-Nancy, France) and the Institut Français du Pétrole (Rueil-Malmaison, France) developed a commercial SMB plant in which the rotary valve was replaced by a combination of commonly used chromatography columns and commercial valves (Nicoud, 1992). Novasep now offers the Licosep SMB systems for a full range of production rates: from the Licosep Lab 12/26, which is capable of separating 10—1000 g per day, to the industrial systems, which can perform industrial enantiomer separation with a productivity of 5—50 tons per year. Other suppliers of SMB equipment are Mitsubishi (Tokyo, Japan), Advanced Separation Technologies, Inc. (Lakeland, FL, USA), Knauer (Berlin, Germany), SepTor Technologies (Leiden, the Netherlands), Tarpon Biosystems (Worcester, MA, USA and Leiden, the Netherlands), Semba Biosciences (Madison, WI, USA), and ChromaCon AG (Zurich, Switzerland).

Small-scale SMB units constitute a useful tool for the pharmaceutical industry. For preliminary biological tests, only a few grams of the chiral drug are needed. Furthermore, SMB can provide the two pure enantiomers that are required for comparative biological testing (Francotte, 1998). On the other hand, pharmaceutical companies work with short drug development times. SMB technology, combined with proper chromatographic chiral stationary phases now available, can be a quick system, easy to set up, at the same time enhancing a high throughput of drug material (Guest, 1997). However, the use of SMB technology in the pharmaceutical industry is not limited to laboratory tests. Its use at the production scale is an alternative to the up-until-now leading techniques such as enantioselective synthesis or diastereoisomeric crystallization. Large-scale chromatographic separations were in the past limited mainly because of the high cost of the adsorbent, the high dilution of products, and the large amounts of mobile phase needed. With the introduction of the SMB technology, large-scale separations can now be carried out under cost-effective conditions. Moreover, in large-scale chiral separation processes, new trends are being developed such as coupling SMB with other techniques such as racemization and enantioselective crystallization (Blehaut, 1997; Canvat et al., 1997).

SMB technology has found new preparative and industrial applications. Because it is basically a binary separation system, it is especially suitable for pharmaceutical process

**Figure 1.11** Another way of implementing SMB: columns moving in a circle and fixed inlet points.

development and bioprocessing. Furthermore, it offers many advantages over conventional preparative chromatography, leading to cleaner, smaller, safer, and faster processes (Nicoud, 1997).

There is also the possibility of keeping the inlet/outlet positions fixed and the columns moving around (SepTor technology, now Outotec, shown in Figure 1.11, CSEP from Advanced Separation Technologies).

## 1.3 SIMULATED MOVING BED APPLICATIONS: "OLD" AND "NEW"

The first SMB patent (Broughton and Gerhold, 1961) was licensed as the Sorbex process and issued for several large-scale separations that were difficult to perform using conventional techniques in the petrochemical and sugar industries. Examples from the petrochemical industries are the Parex units for the separation of $p$-xylene from its isomers on zeolites, the Molex for the separation of $n$-paraffins from branched and cyclic hydrocarbons, and the Olex process to separate olefins from paraffins. An example from the sugar industry is the Sarex for the separation of fructose from glucose in the production of HFCS using polystyrenedivinylbenzene resins in calcium form. Until the 1990s more than 100 units were licensed by UOP, mainly for petrochemical separations. In the early 1990s, the Institute Français du Pétrole developed the Eluxyl process for the separation of xylene isomers. The biggest SMB plant in the world (South Korea) is based on this process and comprises 24 columns with more than 9 m inner diameter and 1 m high.

The first applications of the SMB concept to fine chemistry separations appeared in the early 1990s; the SMB potential to perform chiral separations significantly increased interest in the development of this technology (Nicoud, 1999a,b). In 1997 (5 years after the first chiral separation demonstrations), UCB Pharma installed a multiton SMB unit for large-scale production (McCoy, 2000). Lundbeck's single enantiomer drug Lexapro was the first U.S. Food and Drug Administration-approved drug to be manufactured

**Table 1.1** Comparison of "Old" and "New" SMB Units (Sá Gomes et al., 2006)

|  | *p*-Xylene Separation | Chiral Separations |
|---|---|---|
| Number of columns $N_c$ | 24 | 6 |
| Column length $L_c$ (m) | 1.0 | 0.1 |
| Column internal diameter $D_c$ (m) | 9.5 | 1.0 |
| Particle radius $R_p$ ($\times 10^{-3}$ m) | 0.60 | 0.02 |
| Aspect ratio $D_c/L_c$ | 10 | 10 |
| Productivity (kg/(m$^3_{adsorbent}\cdot$h)) | 120 | 1–10 |
| Adsorbent capacity (kg/m$^3_{adsorbent}$) | 200 | 10 |

using SMB technology, in 2002 (Anon., 2003). SMB is now a well-established technology in the separation of fine chemicals, with several SMB units installed in pharmaceutical industries for the production of single-enantiomer drugs, for example, Bayer, Merck, Carbogen, GlaxoSmithKline, Novartis, Novasep, Pfizer, and UCB Pharma.

In the petrochemical field ("old" SMB applications), the SMB scale is much higher than that used in chiral separations ("new" SMB applications), as can be observed in Table 1.1. However, some design parameter relations are the same; for example, the aspect ratio between column diameter and column height, which is approximately 10. A summary of the SMB main industrial applications is shown in Figure 1.12. At Laboratory of Separation and Reaction Engineering (LSRE) there are two SMB units: the Licosep 12-26 and the home-made FlexSMB® shown in Figure 1.13.

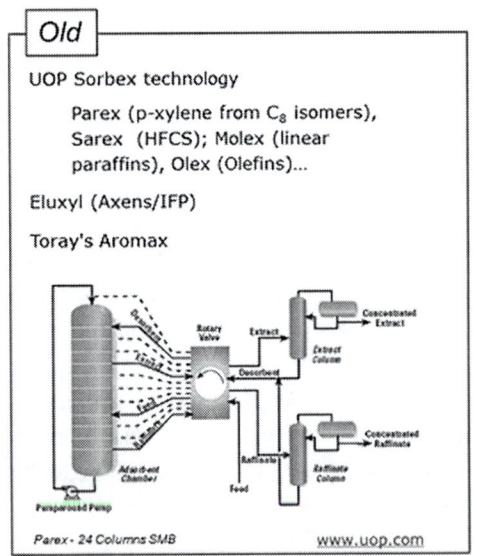

**Old**

UOP Sorbex technology

    Parex (p-xylene from C₈ isomers),
Sarex (HFCS); Molex (linear
paraffins), Olex (Olefins)...

Eluxyl (Axens/IFP)

Toray's Aromax

*Parex - 24 Columns SMB*       www.uop.com

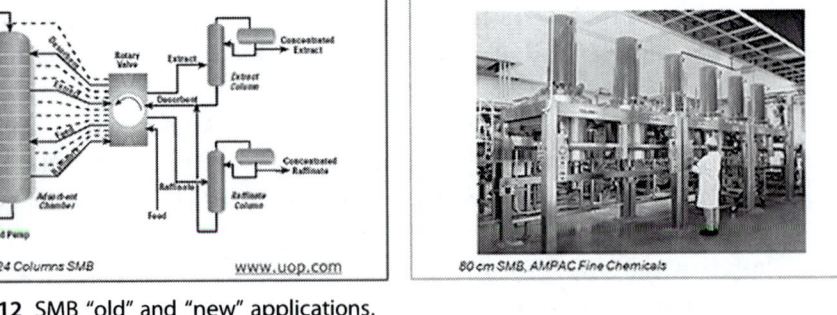

**New**

**Pharmaceutical and Fine Chemical:**

*Biltricide (Praziquantel), Cipralex/Lexapro (Escitalopram), Keppra (Levetiracetam), Modafinil/Provigil, Taxol (Paclitaxel), Xyzal (Levocetirizine), Zoloft (Sertraline), Zyrtec (Cetirizine), Celexa/Citrol/Cipram (Citalopram ), Prozac (Fluoxetine hydrochloride) and insulin among others...*

*80 cm SMB, AMPAC Fine Chemicals*

**Figure 1.12** SMB "old" and "new" applications.

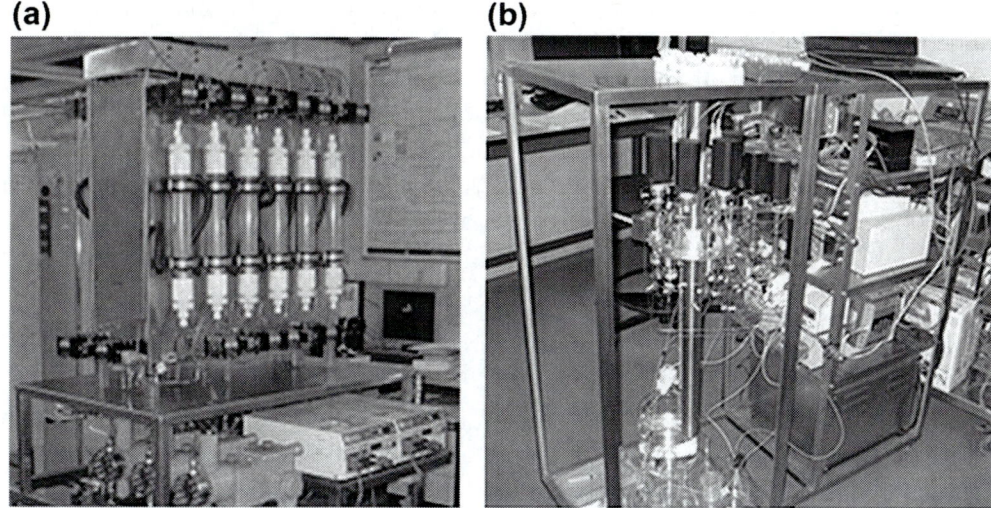

**Figure 1.13** SMB units at LSRE: (a) Licosep 12-16 (Novasep, France); (b) FlexSMB-LSRE®.

## 1.4 NONCONVENTIONAL MODES OF OPERATION

Since the 1990s, new SMB modes of operation have been developed, such as MultiFeed (Kim et al., 2005); ModiCon, in which the feed concentration is modulated (Schramm et al., 2002); Varicol (Adam et al., 2000; Ludemann-Hombourger et al., 2000); power feed, in which the feed flow rate is modulated (Kearney and Hieb, 1992; Zhang et al., 2003); outlet streams swing (Sá Gomes and Rodrigues, 2007); improved SMB (Yoritomi et al., 1981; Tanimura et al., 1991); partial-feed and partial-withdrawal/discard SMB (Zang and Wankat, 2002a,b); side-stream SMB (Beste and Arlt, 2002); two-zone, three-zone, and semicontinuous SMB (Grill, 1998; Abunasser et al., 2003; Hur and Wankat, 2006; Valery and Ludemann-Hombourger, 2007); gradient SMB (Clavier et al., 1996; Jensen et al., 2000; Antos and Seidel-Morgenstern, 2001); SMB—gas and supercritical phases (Storti et al., 1992; Mazzotti et al., 1996; Juza et al., 1998; Depta et al., 1999; Sá Gomes et al., 2009); and the JO process (Ando et al., 1990; Masuda et al., 1993; Mata and Rodrigues, 2001).

Mata and Rodrigues (2001) extended the range of SMB applications and improved even further this technology potential. In the following sections, some of the so-called nonconventional SMB modes of operation are highlighted.

### 1.4.1 Varicol

One of the most successful nonconventional SMB modes of operation is the Varicol process, also known as asynchronous shifts SMB (Adam et al., 2000; Ludemann-Hombourger et al., 2000), commercialized by Novasep S.A.S. (Pompey, France). The

Varicol process is characterized by the asynchronous shift of the inlet/outlet ports, providing a flexible use of the length of each SMB section; see Figure 1.14.

The implementation of the Varicol process can increase productivity up to 30%, compared with the conventional SMB mode of operation, for a given purity requirement, mainly for units with a reduced number of columns (Ludemann-Hombourger et al., 2002; Toumi et al., 2002; Zhang et al., 2002; Pais and Rodrigues, 2003; Subramani et al., 2003; Toumi et al., 2003; Rodrigues et al., 2007).

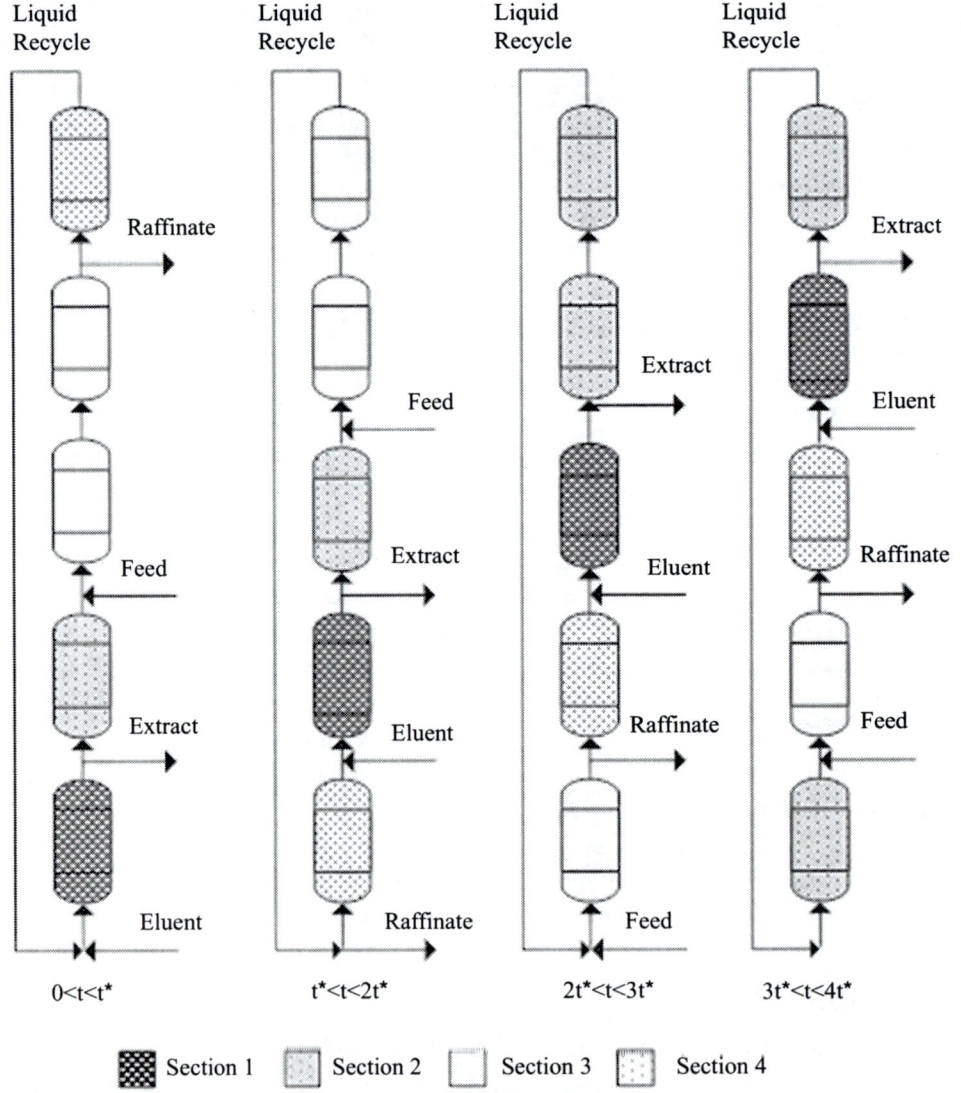

**Figure 1.14** Varicol process [1-1.5-1.5-1]; in terms of standard SMB units [1-1-2-1] during the first half of the switching time and [1-2-1-1] during the second half of the switching time.

## 1.4.2 MultiFeed

The MultiFeed process was first presented by Kim et al. (2005) and, as the name indicates, refers to the use of more than one feed stream in the SMB units, leading to five or more SMB sections. In Figure 1.15, a MultiFeed SMB unit scheme representation is presented as well as the corresponding conventional SMB.

The MultiFeed mode of operation was further investigated in the work of Sá Gomes and Rodrigues (2007). The main observations were: (i) when the two new feeds (Feeds 1 and 2; Figure 1.15(b)) have the same value as Feed 0 (conventional SMB; Figure 1.15(a)) and the unit is operated in a discontinuous mode by switching one of the feeds on when the other is switched off and vice versa; this will be equivalent to the Varicol scheme with the [1-1.5-1.5-1] configuration and (ii) when the MultiFeed configuration is set to continuous being Feed 1 = Feed 2 = 1/2 Feed 0, the TMB solution, compared with the equivalent TMB Varicol [1-1.5-1.5-1] configuration with Feed 0 = Feed 1 + Feed 2, gives the same results in terms of bulk concentration in the regions around the extract and raffinate ports (see Figure 1.16).

Indeed, Varicol, with a variable number of columns in sections II and III, can be considered as a particular case of the MultiFeed SMB operation strategy, as also observed

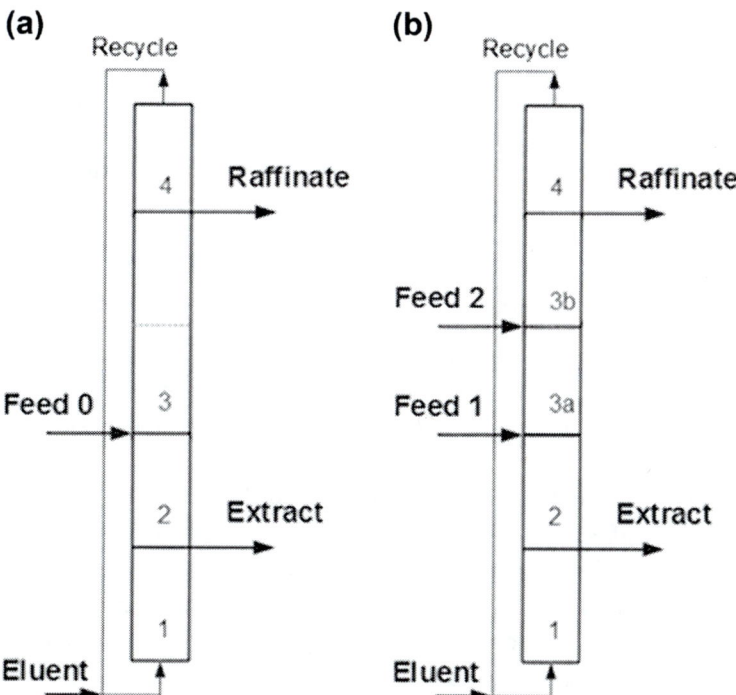

**Figure 1.15** (a) Classic [1-1-2-1] SMB unit scheme and (b) MultiFeed [1-1-1-1-1] SMB unit scheme.

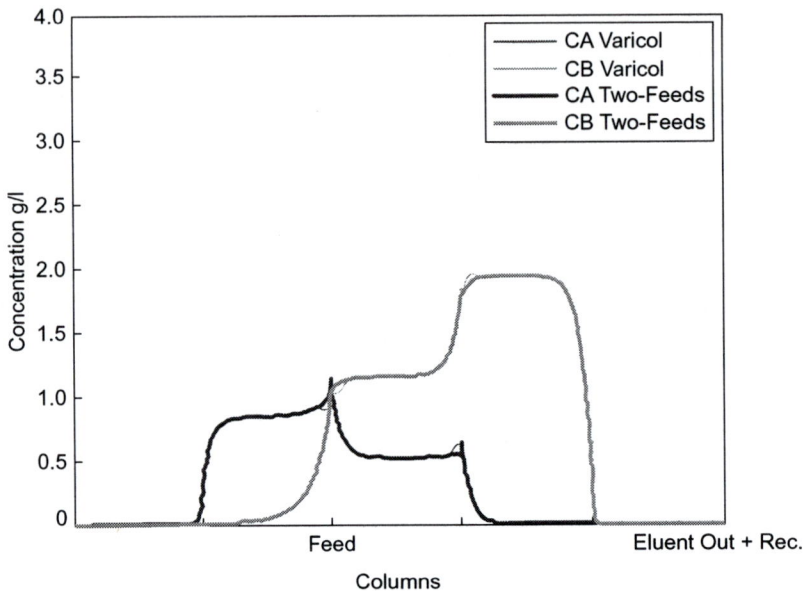

**Figure 1.16** MultiFeed (two feeds) and Varicol bulk concentration profiles, for the equivalent TMB model at cyclic steady state (Sá Gomes and Rodrigues, 2007). (Case study: separation of a racemic mixture of chiral epoxide enantiomers using microcrystalline cellulose triacetate as the CSP and pure methanol as eluent; Pais and Rodrigues, 2003.)

by Kim et al. (2005). Therefore, it will be expected that, under the same conditions, the separation region will be the same for the Varicol and MultiFeed operation modes.

The MultiFeed SMB technique combined with the distillation know-how for the optimum feeds location can provide a powerful methodology for the design/optimization of new SMB processes, in particular the Varicol process.

### 1.4.3 Outlet Streams Swing

Outlet streams swing (OSS) (Sá Gomes and Rodrigues, 2007) considers that the outlet stream (extract and raffinate) flow rates can be varied with time, but the overall operating conditions are maintained; the average flow rates in the sections over a switching time are equal to those in the conventional SMB. This is accomplished because a reduction in the extract or raffinate flow rate over an initial fraction of the switching time is compensated for by an increase in the same flow rate over the remaining part of the switching time period, or vice versa. The extract and raffinate flow rates can be varied at the same time, operating with high extract and low raffinate flow rates, during the first part of the switching time period, followed by low extract (or even none) and high raffinate flow rates (Figure 1.17), or vice versa (Figure 1.18). This technique allows the variation of the flow rates in sections I and IV, but maintains the flow rates in sections II and III by changing the flow rate of the eluent.

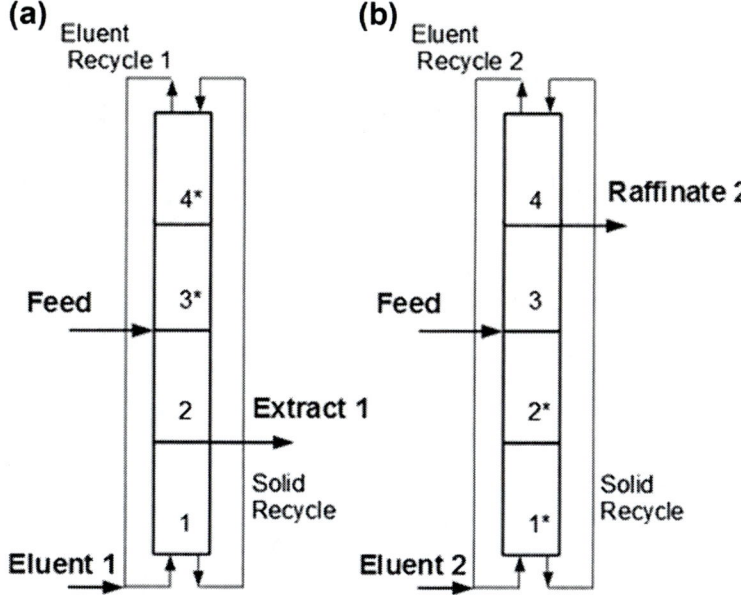

**Figure 1.17** OSS extract—raffinate (X—R) strategy, equivalent TMB scheme: (a) step 1, first fraction of the switching time, collection of extract; (b) step 2, remaining part of the switching time, collection of the raffinate.

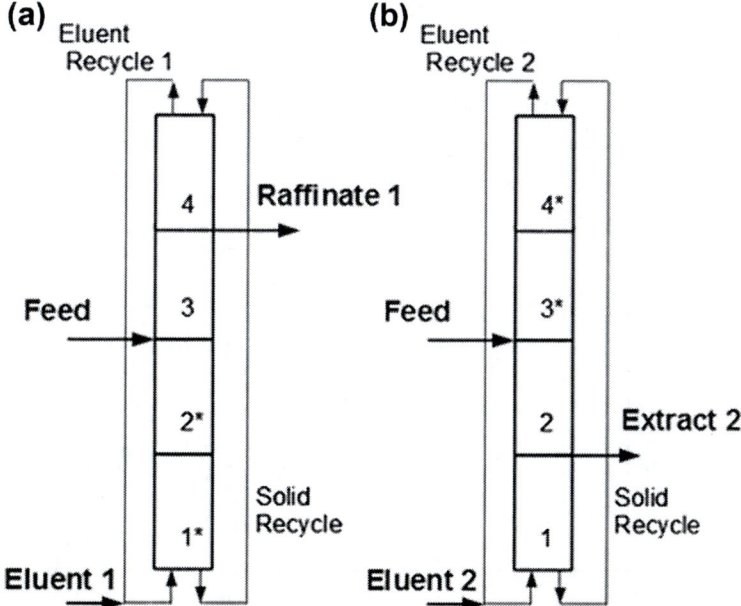

**Figure 1.18** OSS raffinate—extract (R—X) strategy, equivalent TMB scheme: (a) step 1, first fraction of the switching time, collection of raffinate; (b) step 2, remaining part of the switching time, collection of the extract.

Because it is possible to expand or contract the product fronts near to the outlet ports, the OSS technique allows increasing the extract/raffinate purity or decreasing the overall eluent consumption, improving the SMB process performance (see Figure 1.19).

The OSS experimental demonstration using the FlexSMB-LSRE® unit, its potentiality, and its main drawbacks (instability in the internal flow rates and considerable time required to reach the CSS as in most of the nonconventional SMB techniques based on flow rate variations) are reported in the literature (Sá Gomes and Rodrigues, 2010).

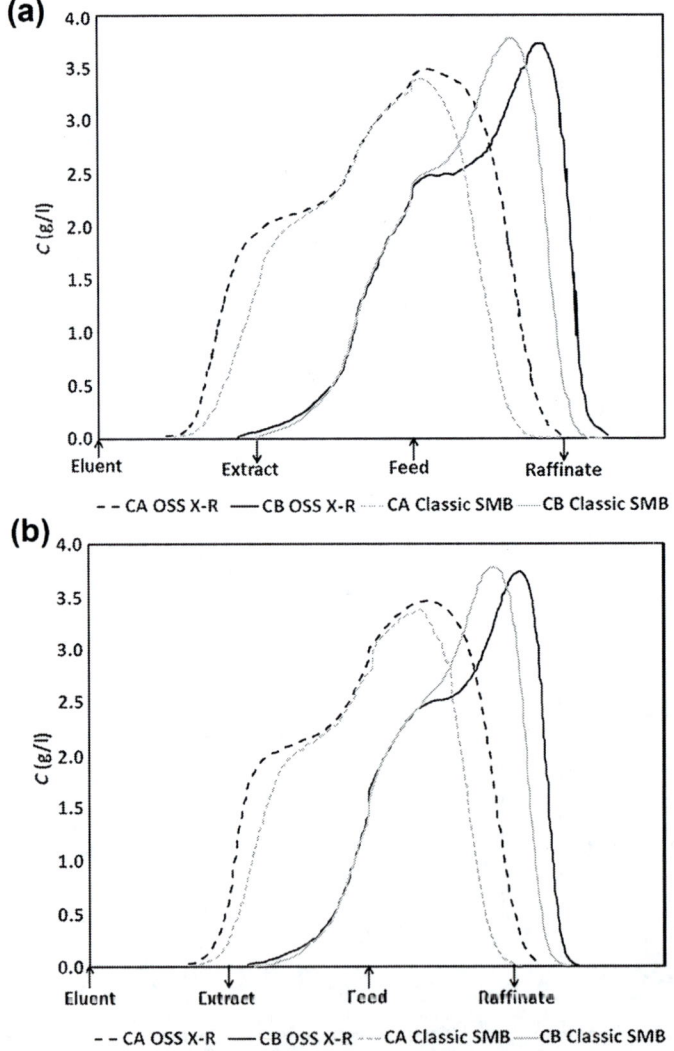

**Figure 1.19** Bulk concentration profiles for OSS raffinate—extract strategy (represented in Figure 1.18, in which each step is performed by half of the switching time), at the cyclic steady state: (a) at 23% of the switching time and (b) at 73% of the switching time (Sá Gomes and Rodrigues, 2007).

### 1.4.4 Backfill-Feed and Multicolumn M3C Processes

In the backfill-feed operation mode, a limited amount of products are refed to the SMB circuit as a backfill feed (Kim and Lee, 2013). The M3C process is a new multicolumn process integrating a concentration step in which, for example, a fraction of the purified extract is recycled to the system (Abdelmoumen et al., 2006). This is somewhat similar to the secondary flush in the Parex process with pure *p*-xylene.

### 1.4.5 JO Process, Improved SMB, Intermittent SMB, and Finnsugar Applexion Separation Technology

The JO process, also known as pseudo-SMB, is commercialized by the Japan Organo Company (www.organo.co.jp) and, in contrast to the classic SMB operation, allows the attainment of more than two high-purity products within the same unit. This process is divided into two steps: in the first step, the unit works as a series of fixed beds from the feed point to exactly the end of the column just before the feed point, where the product with intermediate affinity toward the adsorbent is withdrawn; in the second step, similar to the conventional SMB, the more-retained product is collected in the extract, and the less-retained product is collected in the raffinate, but the feed stream is closed (Figure 1.20).

Extensions of the JO process are multicolumn sequential processes such as the improved SMB (Tanimura et al., 1995), the intermittent SMB (Katsuo and Mazzotti,

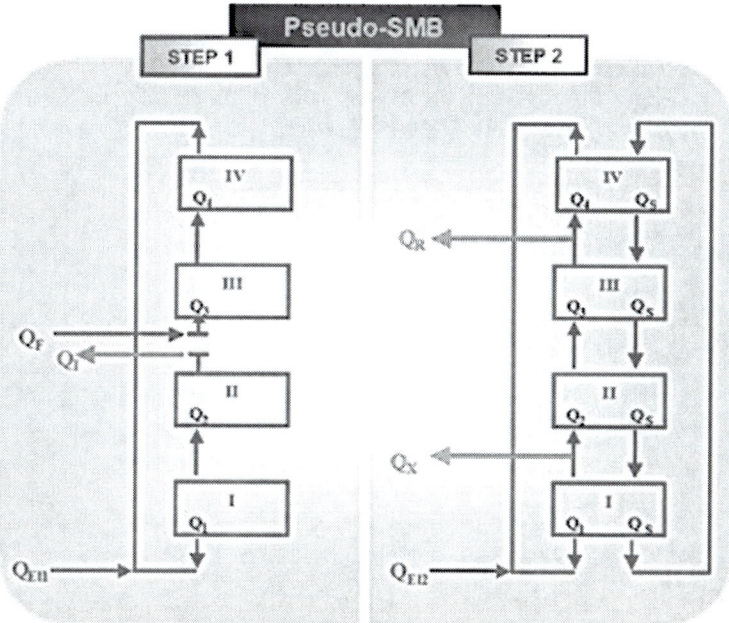

**Figure 1.20** JO process or pseudo-SMB (equivalent TMB scheme).

2010; Jermann and Mazzotti, 2014), and the Finnsugar Applexion separation technology (known as FAST) (Hyoky et al., 1999), in use for many years in the sugar industry.

## 1.5 CONCLUDING REMARKS

The concept of SMB is an excellent illustration of the Innovation Triangle, combining science and technology to develop sustainable processes. The scientific idea of simulating solid movement by shifting the inlet/outlet positions in the direction of fluid flow while keeping the adsorbent fixed in columns together with the technological development of adsorbent materials and the rotary valve allowed UOP to develop a series of Sorbex processes with application in the petrochemical and sugar industries and, much later, in the pharmaceutical industry and bioprocessing. The topic is not exhausted, as illustrated by the new modes of operation of SMB implemented in industry.

## NOMENCLATURE

| | |
|---|---|
| $C$ | Liquid-phase concentration (mol/m$^3$) |
| $K$ | Adsorption equilibrium parameter |
| $L$ | Column length (m) |
| $m$ | Ratio of liquid and solid flow rates |
| $N_c$ | Number of columns |
| $q_i^*$ | Adsorbed phase concentration (mol/m$^3$) |
| $Q$ | Volumetric flow rate (m$^3$/s) |
| $t$ | Time variable (s) |
| $t_{r,i}$ | Retention time (s) |
| $t^*$ | Switching time (s) |
| $u_{c_i}$ | Velocity of propagation of a concentration $C_i$ (m/s) |
| $u_i$ | Interstitial velocity (m/s) |
| $v$ | Fluid interstitial velocity in the TMB (m/s) |
| $v^*$ | Fluid interstitial velocity in the SMB (m/s) |
| $V_c$ | Column volume (m$^3$) |
| $z$ | Axial coordinate (m) |

### Greek Letters

| | |
|---|---|
| $\alpha$ | Selectivity factor |
| $\beta$ | Safety margin |
| $\gamma$ | Ratio of interstitial liquid and solid velocities |
| $\varepsilon$ | Bed porosity |

### Subscripts

| | |
|---|---|
| $c$ | Column |
| $D$ | Desorbent |
| $F$ | Feed |
| $i$ | Component $i$ ($i = A, B$) |
| $j$ | Section $j$ ($j = I, II, III, IV$ or 1, 2, 3, 4) |

| $R$ | Raffinate |
|-----|-----------|
| $Rec$ | Recycle |
| $s$ | Solid |
| $X$ | Extract |

## REFERENCES

Abdelmoumen, S., Muhr, L., Bailly, M., Ludemann-Hombourger, O., 2006. The M3C process: a new multicolumn chromatographic process integrating a concentration step. I - the equilibrium model. Sep. Sci. Technol. 41, 2639–2663.

Abunasser, N., Wankat, P.C., Kim, Y.S., Koo, Y.M., 2003. One-column chromatograph with recycle analogous to a four-zone simulated moving bed. Ind. Eng. Chem. Res. 42, 5268–5279.

Adam, P., Nicoud, R.N., Bailly, M., Ludemann-Hombourger, O., 2000. Process and Device for Separation with Variable-length. US Patent 6 136 198.

Ando, M., Tanimura, M., Tamura, M., 1990. Method of Chromatographic Separation. US Patent 4 970 002.

Anon, 2003. Chiral separations are enduring items in the toolbox. Chem. Eng. News Archive 83, 18.

Antos, D., Seidel-Morgenstern, A., 2001. Application of gradients in the simulated moving bed process. Chem. Eng. Sci. 56, 6667–6682.

Beste, Y.A., Arlt, W., 2002. Side-stream simulated moving-bed chromatography for multicomponent separation. Chem. Eng. Technol. 25, 956–962.

Blehaut, J., 1997. Large scale separation of optical isomers. In: Nicoud, R.-M. (Ed.), Recent Advances in Industrial Chromatographic Processes. Nancy, France, pp. 8–9.

Broughton, D.B., 1968. Molex - case history of a process. Chem. Eng. Prog. 64, 60–65.

Broughton, D.B., 1984. Production-scale adsorptive separations of liquid-mixtures by simulated moving-bed technology. Sep. Sci. Technol. 19, 723–736.

Broughton, D.B., Gerhold, C.G., 1961. Continuous Sorption Process Employing Fixed Bed of Sorbent and Moving Inlets and Outlets. US Patent 2 985 589.

Broughton, D.B., Neuzil, R.W., Pharis, J.M., Brearley, C.S., 1970. Parex process for recovering paraxylene. Chem. Eng. Prog. 66, 70–75.

Canvat, J.P., Deleers, M., Duchêne, G., Hamende, M., Cavoy, E., Zimmermann, V., 1997. Methodology and problems to be solved to implement an industrial GMP production using a SMB System. In: Nicoud, R.-M. (Ed.), Recent Advances in Industrial Chromatographic Processes. Nancy, France, pp. 14–17.

Carson, D.B., Purse, F.V., 1962. Rotary Valve. US Patent 3 040 777.

Ching, C.B., Ruthven, D.M., Hidajat, K., 1985. Experimental-study of a simulated counter-current adsorption system 3. Sorbex operation. Chem. Eng. Sci. 40, 1411–1417.

Clavier, J.Y., Nicoud, R.M., Perrut, M., 1996. A new efficient fractionation process: the simulated moving bed with supercritical eluent. In: von Rohr, P.R.V., Trepp, C. (Eds.), High Pressure Chemical Engineering. Elsevier Science, London.

Day, D.T., 1922. A Handbook of the Petroleum Industry. Wiley.

De Vault, D., 1943. The theory of chromatography. J. Am. Chem. Soc. 65, 532–540.

Depta, A., Giese, T., Johannsen, M., Brunner, G., 1999. Separation of stereoisomers in a simulated moving bed-supercritical fluid chromatography plant. J. Chromatogr. A 865, 175–186.

Dondi, F., Guiochon, G. (Eds.), 1992. Theoretical Advancement in Chromatography and Related Separation Techniques. NATO ASI Series, vol. 383. Kluwer Academic Publishers, The Netherlands.

Ettre, L.S., 2006. The centenary of "chromatography". LC-GC North Am. 24, 680–692.

Francotte, E., 1998. Enantioselective chromatography: a real alternative to enantioselective synthesis? In: Subramanian, G. (Ed.), Proceedings of the EUROTECH'98 Preparative and Process Scale Separations, Cambridge, England.

Ganetsos, G., Barker, P.E. (Eds.), 1993. Preparative and Production Scale Chromatography. Chromatographic Science Series, vol. 61. Marcel Dekker, New York, USA.

Gattuso, M.J., McCulloch, B., House, D.W., Baumann, W.M., 1995. UOP simulated moving bed technology - the preparation of single enantiomer drugs. In: Proceedings of Chiral USA'95, Boston, Massachusetts, USA.

Gattuso, M.J., McCulloch, B., House, D.W., Baumann, W.M., Gottschall, K., 1996. Simulated moving bed technology - the preparation of single enantiomer drugs. Pharm. Tech. Eur. 8, 20−25.

Gattuso, M.J., McCulloch, B., Priegnitz, J.W., 1994. UOP sorbex simulated moving bed technology. A cost effective route to chiral products. In: Proceedings of Chiral Europe'94, Nice, France.

Gembicki, S.A., Oroskar, A.R., Johnson, J.A., 1997. Adsorption, liquid chromatography. In: Ruthven, D.M. (Ed.), Encyclopedia of Separation Technology. John Wiley & Sons, New York, USA, pp. 172−199.

Giddings, J.C., 1965. Dynamics of Chromatography. Part I - Principles and Theory. Marcel Dekker, New York, USA.

Giddings, J.C., 1990. Unified Separation Science. John Wiley & Sons, New York, USA.

Grill, M.C., 1998. Single Column Closed-loop Recycling with Periodic Intra-profile Injection. WO Patent 9 851 391.

Guest, D.W., 1997. Evaluation of simulated moving bed chromatography pharmaceutical process development. J. Chromatogr. A 760, 159−162.

Guiochon, G., Shirazi, S.G., Katti, A.M., 1994. Fundamentals of Preparative and Nonlinear Chromatography. Academic Press, New York, USA.

Heikkilä, H., January 24, 1983. Separating sugars and amino acids with chromatography. Chem. Eng. 50−52.

Hotier, G., 1997. The eluxyl process: from the idea to the largest SMB unit in the world. In: Nicoud, R.-M. (Ed.), Recent Advances in Industrial Chromatographic Processes. Nancy, France.

Humphrey, J.L., Keller II, G.E., 1997. Separation Process Technology. McGraw-Hill, New York, USA.

Hur, J.S., Wankat, P.C., 2006. Two-zone SMB/chromatography for center-cut separation from ternary mixtures: linear isotherm systems. Ind. Eng. Chem. Res. 45, 1426−1433.

Hyoky, G., Paananen, H., Cotillon, M., Cornelius, G., 1999. FAST separation technology. In: 30th General Meeting of American Society of Sugar Beet Technologists, Orlando, USA.

Jensen, T.B., Reijns, T.G.P., Billiet, H.A.H., Van Der Wielen, L.A.M., 2000. Novel simulated moving-bed method for reduced solvent consumption. J. Chromatogr. A 873, 149−162.

Jermann, S., Mazzotti, M., 2014. Three column intermittent SMB chromatography. I - Process description and comparative assessment. J. Chromatogr. A 1361, 125−138.

Johnson, J.A., 1989. Sorbex: continuing innovation in liquid adsorption. In: Rodrigues, A.E., LeVan, M.D., Tondeur, D. (Eds.), Adsorption: Science and Technology. Kluwer Academic Publishers, The Netherlands, pp. 383−395.

Johnson, J.A., Kabza, R.G., 1993. Sorbex: industrial-scale adsorptive separation. In: Ganetsos, G., Barker, P.E. (Eds.), Preparative and Production Scale Chromatography. Marcel Dekker Inc, New York, pp. 257−271.

Juza, M., Di Giovanni, O., Biressi, G., Schurig, V., Mazzotti, M., Morbidelli, M., 1998. Continuous enantiomer separation of the volatile inhalation anesthetic enflurane with a gas chromatographic simulated moving bed unit. J. Chromatogr. A 813, 333−347.

Katsuo, S., Mazzotti, M., 2010. Intermittent simulated moving bed chromatography: 1. Design criteria and cyclic steady-state. J. Chromatogr. A 1217, 1354−1361.

Kearney, M.M., Hieb, K.L., 1992. Time Variable Simulated Moving Bed Process. US Patent 5 102 553.

Keller II, G.E., 1995. Adsorption - building upon a solid foundation. Chem. Eng. Prog. 91, 56−67.

Kim, J.K., Abunasser, N., Wankat, P.C., 2005. Use of two feeds in simulated moving beds for binary separations. Korean J. Chem. Eng. 22, 619−627.

Kim, K.M., Lee, C.H., 2013. Backfill-simulated moving bed operation for improving the separation performance of simulated moving bed chromatography. J. Chromatogr. A 1311, 79−89.

Ludemann-Hombourger, O., Nicoud, R.M., Bailly, M., 2000. The "VARICOL" process: a new multi-column continuous chromatographic process. Sep. Sci. Technol. 35, 1829−1862.

Ludemann-Hombourger, O., Pigorini, G., Nicoud, R.M., Ross, D.S., Terfloth, G., 2002. Application of the "VARICOL" process to the separation of the isomers of the SB-553261 racemate. J. Chromatogr. A 947, 59−68.

Masuda, T., Sonobe, T., Matsuda, F., Horie, M., 1993. Process for Fractional Separation of Multi-Component Fluid Mixture. US Patent 5 198 120.

Mata, V.G., Rodrigues, A.E., 2001. Separation of ternary mixtures by pseudo-simulated moving bed chromatography. J. Chromatogr. A 939, 23—40.

Mazzotti, M., Baciocchi, R., Storti, G., Morbidelli, M., 1996. Vapor-phase SMB adsorptive separation of linear/nonlinear paraffins. Ind. Eng. Chem. Res. 35, 2313—2321.

McCoy, M., 2000. SMB emerges as chiral technique. Chem. Eng. News Archive 78, 17—19.

Nicoud, R.-M., 1992. The simulated moving bed: a powerful chromatographic process. LC-GC Int. 5, 43—47.

Nicoud, R.-M., 1993. Simulated moving bed (SMB) in preparative chromatography: basics, limitations and use. In: Nicoud, R.-M. (Ed.), Simulated Moving Bed: Basics and Applications. Institut National Polytechnique de Lorraine. Nancy, France, pp. 54—64.

Nicoud, R.-M., 1997. Recent advances in industrial chromatographic processes. In: Nicoud, R.-M. (Ed.), Recent Advances in Industrial Chromatographic Processes. Nancy, France, pp. 4—5.

Nicoud, R.M., 1999a. The separation of optical isomers by simulated moving bed chromatography (Part I). Pharm. Technol. Eur. 11, 36—44.

Nicoud, R.M., 1999b. The separation of optical isomers by simulated moving bed chromatography (Part II). Pharm. Technol. Eur. 11, 28—34.

Otani, S., 1973. Adsorption separates xylenes. Chem. Eng. 80, 106—107.

Otani, S., Iwamura, T., Sando, K., Kanaoka, M., Matsumura, K., Akita, S., Yamamoto, T., Takeushi, I., Tsuchiya, T., Noguchi, Y., Mori, T., 1973. Separation Process of Components of Feed Mixture Utilizing Solid Sorbent. US Patent 3 761 533.

Pais, L.S., Loureiro, J.M., Rodrigues, A.E., 1998. Modeling strategies for enantiomers separation by SMB chromatography. AIChE J. 44, 561—569.

Pais, L.S., Rodrigues, A.E., 2003. Design of simulated moving bed and Varicol processes for preparative separations with a low number of columns. J. Chromatogr. A 1006, 33—44.

Rearick, D.E., Kearney, M., Costesso, D.D., 1997. Simulated moving-bed technology in the sweetener industry. Chemtech 27, 36—40.

Rodrigues, A.E., Dias, M.M., Lopes, J.C.B. (Eds.), 1991. Theory of Linear and Nonlinear Chromatography. NATO ASI Series, vol. 204. Springer, Netherlands. NATO ASI Series.

Rodrigues, A.E., LeVan, M.D., Tondeur, D. (Eds.), 1989. Adsorption: Science and Technology. NATO ASI Series, vol. 158. Kluwer Academic Publishers, The Netherlands.

Rodrigues, A.E., Tondeur, D. (Eds.), 1981. Percolation Processes: Theory and Applications. NATO ASI Series, vol. 33. Sijthoff & Noordhoff, The Netherlands.

Rodrigues, R.C.R., Araújo, J.M.M., Eusébio, M.F.J., Mota, J.P.B., 2007. Experimental assessment of simulated moving bed and Varicol processes using a single-column setup. J. Chromatogr. A 1142, 69—80.

de Rosset, A.J., Neuzil, R.W., Broughton, D.B., 1981. Industrial applications of preparative chromatography. In: Rodrigues, A.E., Tondeur, D. (Eds.), Percolation Processes: Theory and Applications. Sijthoff & Noordhoff, The Netherlands, pp. 249—281.

Ruthven, D.M., 1984. Principles of Adsorption and Adsorption Processes. John Wiley & Sons, New York, USA.

Ruthven, D.M., Ching, C.B., 1989. Countercurrent and simulated countercurrent adsorption separation processes. Chem. Eng. Sci. 44, 1011—1038.

Sá Gomes, P., Lamia, N., Rodrigues, A.E., 2009. Design of a gas phase simulated moving bed for propane/propylene separation. Chem. Eng. Sci. 64, 1336—1357.

Sá Gomes, P., Minceva, M., Rodrigues, A.E., 2006. Simulated moving bed technology: old and new. Adsorption 12, 375—392.

Sá Gomes, P., Rodrigues, A.E., 2007. Outlet Streams Swing (OSS) and Multifeed operation of simulated moving beds. Sep. Sci. Technol. 42, 223—252.

Sá Gomes, P., Rodrigues, A.E., 2010. Outlet stream swing simulated moving bed: separation and regeneration regions analysis. Sep. Sci. Technol. 45, 2259—2272.

Schramm, H., Kaspereit, M., Kienle, A., Seidel-Morgenstern, A., 2002. Improving simulated moving bed processes by cyclic modulation of the feed concentration. Chem. Eng. Technol. 25, 1151—1155.

Storti, G., Masi, M., Carra, S., Morbidelli, M., 1989. Optimal-design of multicomponent countercurrent adsorption separation processes involving nonlinear equilibria. Chem. Eng. Sci. 44, 1329—1345.

Storti, G., Mazzotti, M., Tadeu Furlan, L., Morbidelli, M., Carra, S., 1992. Performance of a six-port simulated moving-bed pilot plant for vapor-phase adsorption separations. Sep. Sci. Technol. 27, 1889—1916.

Subramani, H.J., Hidajat, K., Ray, A.K., 2003. Optimization of simulated moving bed and Varicol processes for glucose-fructose separation. Chem. Eng. Res. Des. 81, 549—567.

Suzuki, M., 1990. Adsorption Engineering, Chemical Engineering Monographs, vol. 25. Elsevier, Tokyo, Japan.

Tanimura, M., Tamura, M., Techima, M., 1991. Method of Chromatographic Separation. US Patent 5 064 539.

Tanimura, M., Tamura, M., Teshina, T., 1995. Japanese Patent JP-b-h07—046097.

Toumi, A., Engell, S., Ludemann-Hombourger, O., Nicoud, R.M., Bailly, M., 2003. Optimization of simulated moving bed and Varicol processes. J. Chromatogr. A 1006, 15—31.

Toumi, A., Hanisch, F., Engell, S., 2002. Optimal operation of continuous chromatographic processes: mathematical optimization of the Varicol process. Ind. Eng. Chem. Res. 41, 4328—4337.

Tswett, M., 1906a. Adsorption analysis and chromatographic method. Application to the chemistry of chlorophyll. Ber. Dtsch. Bot. Ges. 24, 384—392.

Tswett, M., 1906b. Physical-chemical studies of chlorophyll. Adsorption. Ber. Dtsch. Bot. Ges. 24, 316—326.

Tswett, M.S., 1903. Lecture at the March 8(21). In: On a New Category of Adsorption Phenomena and Their Application to Biochemical Analysis, Meeting of the Biological Section of the Warsaw Society of Natural Scientists, Warsaw.

Valery, E., Ludemann-Hombourger, O., 2007. Method and Device for Separating Fractions of a Mixture. WO Patent 2007 012 750 (A2).

Yoritomi, T., Kezuka, T., Moriya, M., 1981. Method for the Chromatographic Separation of Soluble Components in Feed Solution. US Patent 4 267 054.

Zang, Y., Wankat, P.C., 2002a. SMB operation strategy - partial feed. Ind. Eng. Chem. Res. 41, 2504—2511.

Zang, Y., Wankat, P.C., 2002b. Three-zone simulated moving bed with partial feed and selective withdrawal. Ind. Eng. Chem. Res. 41, 5283—5289.

Zhang, Z., Hidajat, K., Ray, A.K., Morbidelli, M., 2002. Multiobjective optimization of SMB and Varicol process for chiral separation. AIChE J. 48, 2800—2816.

Zhang, Z., Mazzotti, M., Morbidelli, M., 2003. PowerFeed operation of simulated moving bed units: changing flow-rates during the switching interval. J. Chromatogr. A 1006, 87—99.

Zhong, G.M., Guiochon, G., 1996. Analytical solution for the linear ideal model of simulated moving bed chromatography. Chem. Eng. Sci. 51, 4307—4319.

# CHAPTER 2

# Modeling and Simulation of Simulated Moving Bed Separation Processes

In this chapter we will discuss simulated moving bed (SMB) modeling and simulation using different strategies with the aim of design, optimization, and operation of SMB separation processes.

## 2.1 STRATEGIES OF MODELING

Modeling an SMB separation process can be analyzed by two different strategies: one by simulating the SMB system directly, taking into account its intermittent behavior; the other by representing its operation in terms of a true countercurrent system (true moving bed or TMB). The first model represents the real SMB and considers the periodic switch of the injection and collection points; the second model assumes equivalence with the TMB, in which solid and fluid phases flow in opposite directions.

The major difference between the two strategies is that, whereas the TMB strategy results in a continuous model with steady-state solutions, the real SMB strategy needs to be solved at the cyclic steady state (CSS), which can be achieved by solving the dynamic problem until the CSS is reached or by direct prediction techniques (Minceva et al., 2003). As a result, the first strategy usually leads to faster solutions, but it gives satisfactory results only when applied to SMB units having a significant number of columns in each section.

Several authors have developed models to predict the performance of an SMB separation process. These models can be classified according to the description of the fluid flow as continuous-flow models (plug or axial dispersed plug flow) or as mixing-cell models. Moreover, some authors considered mass transfer resistances by including an appropriate rate expression, usually by using the linear driving force (LDF) model (Glueckauf, 1955; Ruthven, 1984). Others used the equilibrium theory and neglected mass transfer resistances and axial mixing. Some references about the SMB modeling assuming the TMB equivalence or directly through the SMB intermittent model can be found in Ruthven and Ching (1989) and Pais et al. (1998a). Tables 2.1 and 2.2 present a literature survey of models for TMB and SMB, respectively.

### 2.1.1 Transient Models of SMB and TMB

The mathematical models of the real SMB and TMB systems are presented in the next section. They are based on the following assumptions: (1) An axially dispersed plug flow

*Simulated Moving Bed Technology*
ISBN 978-0-12-802024-1

**Table 2.1** Literature Survey on the Modeling Strategies for SMB Separation Processes: True Moving Bed Strategy

| Description of the Fluid Flow | Mass Transfer Resistance | References |
|---|---|---|
| Continuous-flow model | No<br>Equilibrium theory | Storti et al. (1989a,b, 1993a,b, 1995), Mazzotti et al. (1994, 1996a,b, 1997a,b), and Chiang (1998). |
| Continuous-flow model | Yes<br>Linear driving force model | Hashimoto et al. (1983, 1987, 1993), Ching and Ruthven (1985a), Ching et al. (1985, 1991, 1992), Kubota et al. (1989), Storti et al. (1989a,b, 1993a), Hidajat and Ching (1990), Ruthven and Ching (1993), Hassan et al. (1994), Rahman et al. (1994), Chu and Hashim (1995), Rodrigues et al. (1996), Dandekar et al. (1996), Hotier (1996), Zhong and Guiochon (1996), Schmidt-Traub and Strube (1996), Pais et al. (1997a,b, 1998a,b), Strube et al. (1997), Ma and Wang (1997), and Navarro et al. (1997). |
| Mixing-cell model | No<br>Equilibrium stage model | Ching and Ruthven (1985a,b), Ching et al. (1985), Ernst and Hsu (1989, 1992), Charton and Nicoud (1995), and Navarro et al. (1997). |

**Table 2.2** Literature Survey on the Modeling Strategies for SMB Separation Processes: Simulated Moving Bed Strategy

| Description of the Fluid Flow | Mass Transfer Resistance | References |
|---|---|---|
| Continuous-flow model | No<br>Ideal and equilibrium-dispersive model | Zhong and Guiochon (1996, 1998), Zhong et al. (1997), and Yun et al. (1997a,b,c). |
| Continuous-flow model | Yes<br>Linear driving force model | Hashimoto et al. (1983, 1987, 1993), Carta and Pigford (1986), Storti et al. (1988, 1989b), Lameloise and Viard (1993), Chu and Hashim (1995), Hassan et al. (1995), Lim and Ching (1995), Hotier (1996), Schmidt-Traub and Strube (1996), Strube et al. (1997), Zhong and Guiochon (1997a,b), Ma and Wang (1997), Pais et al. (1998a), and Azevêdo et al. (1998). |
| Mixing-cell model | No<br>Equilibrium stage model | Barker et al. (1983), Ching (1983), Hidajat et al. (1986a,b), Ching et al. (1987, 1988, 1993), Ruthven and Ching (1993), and Charton and Nicoud (1995). |

model is used to describe the fluid flow. (2) A plug flow model is used to represent the countercurrent solid flow in the TMB approach. (3) The adsorbent particles are considered homogeneous, and mass transfer between fluid and solid is described by the LDF model. (4) The model can handle any kind of adsorption equilibrium isotherm. (5) Isothermal operation. (6) No bed radial concentration gradients. (7) Constant fluid velocity. (8) Uniform particle size and constant bed porosity.

The model equations for the real SMB result from the mass balances over a volume element of the bed and at a particle level. The transient SMB model equations are summarized below, with initial and boundary conditions, as well as with the necessary mass balances at the nodes between each two columns.

### 2.1.1.1 Model Equations for the Transient SMB

*Mass balance in a volume element of the bed k:*

$$\frac{\partial C_{ik}}{\partial t} = D_{L_k}\frac{\partial^2 C_{ik}}{\partial z^2} - v_k^*\frac{\partial C_{ik}}{\partial z} - \frac{(1-\varepsilon)}{\varepsilon}k_h\left(q_{ik}^* - q_{ik}\right) \tag{2.1}$$

where the subscript $i$ ($i = A, B$) refers to the species in the mixture, and subscript $k$ ($k = 1, 2, ..., N_c$) is the column number; $C_{ik}$ and $q_{ik}$ are the fluid-phase and average adsorbed-phase concentrations of species $i$ in column $k$ of the SMB unit, respectively; $z$ is the axial coordinate; $t$ is the time variable; $\varepsilon$ is the bed porosity; $v_k^*$ is the interstitial fluid velocity in the $k$th SMB column; $D_{Lk}$ is the axial dispersion coefficient; and $k_h$ is the intraparticle mass transfer coefficient.

*Mass balance in the particle:*

$$\frac{\partial q_{ik}}{\partial t} = k_h\left(q_{ik}^* - q_{ik}\right) \tag{2.2}$$

where $q_{ik}^*$ is the adsorbed phase concentration in equilibrium with $C_{ik}$.

*Initial conditions:*

$$t = 0 \qquad C_{ik} = q_{ik} = 0 \tag{2.3}$$

*Boundary conditions for column k:*

$$z = 0 \qquad C_{ik} - \frac{D_{L_k}}{v_k^*}\frac{\partial C_{ik}}{\partial z} = C_{ik,0} \tag{2.4}$$

where $C_{ik,0}$ is the inlet concentration of species $i$ in column $k$.

$z = L_k$

for a column inside a section, and for extract and raffinate nodes,

$$C_{ik} = C_{i(k+1),0} \tag{2.5a}$$

for the eluent node

$$C_{ik} = \frac{v_I^*}{v_{IV}^*} C_{i(k+1),0} \tag{2.5b}$$

for the feed node

$$C_{ik} = \frac{v_{III}^*}{v_{II}^*} C_{i(k+1),0} - \frac{v_F}{v_{II}^*} C_i^F \tag{2.5c}$$

*Global balances:*

Eluent node

$$v_I^* = v_{IV}^* + v_E \tag{2.6a}$$

Extract node

$$v_{II}^* = v_I^* - v_X \tag{2.6b}$$

Feed node

$$v_{III}^* = v_{II}^* + v_F \tag{2.6c}$$

Raffinate node

$$v_{IV}^* = v_{III}^* - v_R \tag{2.6d}$$

*Multicomponent adsorption equilibrium isotherm:*

$$q_{Ak}^* = f_A(C_{Ak}, C_{Bk}) \quad \text{and} \quad q_{Bk}^* = f_A(C_{Ak}, C_{Bk}) \tag{2.7}$$

Introducing the dimensionless variables $x = z/L_k$ and $\theta = t/t^*$, where $t^*$ is the switching time and $L_k$ is the length of one SMB column, the model Eqns (2.1) and (2.2) become

$$\frac{\partial C_{ik}}{\partial \theta} = \gamma_k^* \left\{ \frac{1}{Pe_k} \frac{\partial^2 C_{ik}}{\partial x^2} - \frac{\partial C_{ik}}{\partial x} \right\} - \frac{(1-\varepsilon)}{\varepsilon} \alpha_k (q_{ik}^* - q_{ik}) \tag{2.8}$$

and

$$\frac{\partial q_{ik}}{\partial \theta} = \alpha_k (q_{ik}^* - q_{ik}) \tag{2.9}$$

The initial and boundary conditions are the same presented before (Eqns (2.3)−(2.5(a−c))) and, for $x = 0$ ($z = 0$), Eqn (2.4) becomes

$$C_{ik} - \frac{1}{Pe_k} \frac{\partial C_{ik}}{\partial x} = C_{ik,0} \tag{2.10}$$

The resulting model parameters are

Ratio between solid and fluid volumes, $\qquad \dfrac{(1-\varepsilon)}{\varepsilon} \tag{2.11}$

Ratio between fluid and solid interstitial velocities, $\quad \gamma_k^* = \dfrac{v_k^*}{u_s} = \dfrac{v_k^*}{L_k/t^*} \tag{2.12}$

Peclet number, $\qquad\qquad\qquad\qquad\qquad . \qquad Pe_k = \dfrac{v_k^* L_k}{D_{L_k}} \tag{2.13}$

Number of mass transfer units, $$\alpha_k = \frac{k_h L_k}{u_s} = k_h t^* \qquad (2.14)$$

where $u_s$ is the interstitial solid velocity in the equivalent TMB model. To complete the list of model parameters, the adsorption equilibrium parameters have to be added.

Owing to the switch of inlet and outlet lines, each column has different functions during a whole cycle, depending on its location (section). As a consequence, the boundary conditions for each column change after the end of each switch time interval. This time dependence of the boundary conditions leads to a CSS for this system, instead of a real steady state present in the TMB model. This means that, after CSS is reached, the internal concentration profiles vary during a given cycle, but they are identical at the same time within the cycle for two successive cycles.

### 2.1.1.2 The Transient True Moving Bed Model

The TMB concept can be used as an alternative strategy to predict the SMB operation. In the TMB model, the solid phase is assumed to move in plug flow in the opposite direction of the fluid phase, while the inlet and outlet lines remain fixed. As a consequence, each column has the same function, depending on its location. As seen in Chapter 1, an equivalence between the TMB and the SMB models can be made by keeping constant the liquid velocity relative to the solid velocity (Chapter 1, Eqn (1.22)) and the solid velocity in the TMB model must be evaluated from the value of the switch time interval of the SMB model according to Eqn (1.23) in Chapter 1. In terms of volumetric liquid and solid flow rates in TMB, they are

$$Q_j = Q_j^* - \frac{\varepsilon}{1 - \varepsilon} Q_s \qquad (2.15)$$

and

$$Q_s = \frac{(1 - \varepsilon) V_c}{t^*} \qquad (2.16)$$

where $Q_j$ and $Q_j^*$ are the volumetric liquid flow rates in section $j$ of an SMB and TMB, respectively, $Q_s$ is the solid flow rate in the equivalent TMB model, and $V_c$ is the volume of one SMB column.

The resulting model equations for the transient TMB are shown below.

*Mass balance in a volume element of section $j$:*

$$\frac{\partial C_{ij}}{\partial t} = D_{L_j} \frac{\partial^2 C_{ij}}{\partial z^2} - v_j \frac{\partial C_{ij}}{\partial z} - \frac{(1 - \varepsilon)}{\varepsilon} k_h (q_{ij}^* - q_{ij}) \qquad (2.17)$$

*Mass balance in the particle:*

$$\frac{\partial q_{ij}}{\partial t} = u_s \frac{\partial q_{ij}}{\partial z} + k_h (q_{ij}^* - q_{ij}) \qquad (2.18)$$

*Initial conditions:*

$$t = 0 \qquad C_{ij} = q_{ij} = 0 \tag{2.19}$$

*Boundary conditions for section j:*

$$z = 0 \quad C_{ij} - \frac{D_{L_j}}{v_j} \frac{\partial C_{ij}}{\partial z} = C_{ij,0} \tag{2.20}$$

where $C_{ij,0}$ is the inlet concentration of species $i$ in section $j$.

$z = L_j$

for the eluent node $\qquad\qquad C_{iIV} = \dfrac{v_I}{v_{IV}} C_{iI,0} \tag{2.21a}$

for the extract node $\qquad\qquad C_{iI} = C_{iII,0} \tag{2.21b}$

for the feed node $\qquad\qquad C_{iII} = \dfrac{v_{III}}{v_{II}} C_{iIII,0} - \dfrac{v_F}{v_{II}} C_i^F \tag{2.21c}$

for the raffinate node $\qquad\qquad C_{iIII} = C_{iIV,0} \tag{2.21d}$

and

$$q_{iIV} = q_{iI,0}, \quad q_{iI} = q_{iII,0}, \quad q_{iII} = q_{iIII,0}, \quad q_{iIII} = q_{iIV,0} \tag{2.22}$$

*Global balances:*

Eluent node $\qquad\qquad\qquad v_I = v_{IV} + v_E \tag{2.23a}$

Extract node $\qquad\qquad\qquad v_{II} = v_I - v_X \tag{2.23b}$

Feed node $\qquad\qquad\qquad v_{III} = v_{II} + v_F \tag{2.23c}$

Raffinate node $\qquad\qquad\qquad v_{IV} = v_{III} - v_R \tag{2.23d}$

*Multicomponent adsorption equilibrium isotherm:*

$$q_{Aj}^* = f_A(C_{Aj}, C_{Bj}) \quad \text{and} \quad q_{Bj}^* = f_A(C_{Aj}, C_{Bj}) \tag{2.24}$$

Introducing the dimensionless variables $x = z/L_j$ and $\theta = t/\tau_s$, with $\tau_s = L_j/u_s = N_s t^*$, where $\tau_s$ is the solid space time in a section of a TMB unit, $L_j$ is the length of a TMB section, and $N_s$ is the number of columns per section in an SMB unit, the model Eqns (2.17) and (2.18) become

$$\frac{\partial C_{ij}}{\partial \theta} = \gamma_j \left\{ \frac{1}{Pe_j} \frac{\partial^2 C_{ij}}{\partial x^2} - \frac{\partial C_{ij}}{\partial x} \right\} - \frac{(1 - \varepsilon)}{\varepsilon} \alpha_j (q_{ij}^* - q_{ij}) \tag{2.25}$$

and

$$\frac{\partial q_{ij}}{\partial \theta} = \frac{\partial q_{ij}}{\partial x} + \alpha_j (q_{ij}^* - q_{ij}) \tag{2.26}$$

The initial and boundary conditions are the same presented before (Eqns (2.19)–(2.22)) and, for $x = 0$ ($z = 0$), Eqn (2.20) becomes

$$C_{ij} - \frac{1}{Pe_j} \frac{\partial C_{ij}}{\partial x} = C_{ij,0} \tag{2.27}$$

The resulting model parameters are similar to those presented for the SMB model, except that they are expressed in terms of the length of the TMB sections:

Ratio between solid and fluid volumes,     $\dfrac{(1 - \varepsilon)}{\varepsilon}$     (2.28)

Ratio between fluid and solid interstitial velocities,     $\gamma_j = \dfrac{v_j}{u_s}$     (2.29)

Peclet number,     $Pe_j = \dfrac{v_j L_j}{D_{L_j}}$     (2.30)

Number of mass transfer units,     $\alpha_j = \dfrac{k_h L_j}{u_s}$     (2.31)

Again, adsorption equilibrium parameters have to be added to the list above.

### 2.1.1.3 TMB and SMB Models with Porous Adsorbent Particles

Alternatively, if we assume adsorbent particles with porous structure in the model equations for TMB and SMB systems, a new variable must be considered: the average concentration of species $i$ in the particle pores $\overline{C}_{p_i}$. Model equations for TMB and SMB are summarized respectively in Tables 2.3 and 2.4. These models assume that the total concentration in the fluid phase is constant.

### 2.1.1.4 Comparing Transient SMB and TMB Models

The chromatographic resolution of bi-naphthol enantiomers (Figure 2.1) was considered for simulation purposes. The chiral stationary phase used in this system was 3,5-dinitrobenzoyl phenylglycine bonded to silica gel and a mixture of 72/28 (v/v) heptane/isopropanol was used as eluent. The limit of solubility in this eluent is 3 g/l of each enantiomer.

The adsorption equilibrium isotherms, measured at 25 °C, are of bi-Langmuir type and were proposed by the Separex group (personal communication):

$$q_A^* = \frac{2.69 C_A}{1 + 0.0336 C_A + 0.0466 C_B} + \frac{0.10 C_A}{1 + C_A + 3 C_B} \tag{2.32a}$$

$$q_B^* = \frac{3.73 C_B}{1 + 0.0336 C_A + 0.0466 C_B} + \frac{0.30 C_B}{1 + C_A + 3 C_B} \tag{2.32b}$$

**Table 2.3** Model Equations for the Transient TMB Model with Porous Adsorbent Particles

Mass balance in a volume element of the section $j$

$$\varepsilon \frac{\partial C_{ij}}{\partial t} = \varepsilon D_{axj} \frac{\partial^2 C_{ij}}{\partial z^2} - \varepsilon v_j \frac{\partial C_{ij}}{\partial z} - (1-\varepsilon) \frac{3}{R_p} k_{ov,ij}(C_{ij} - \overline{C}_{p_{ij}})$$

Particle mass balance

$$\varepsilon_p \frac{\partial \overline{C}_{pij}}{\partial t} + \frac{\partial q_{ij}}{\partial t} = u_s \left[ \varepsilon_p \frac{\partial \overline{C}_{pij}}{\partial z} + \frac{\partial q_{pij}}{\partial z} \right] + \frac{3}{R_p} k_{ov,ij}(C_{ij} - \overline{C}_{p_{ij}})$$

Initial conditions

$$t = 0; \quad C_{ij} = \overline{C}_{p_{ij}} = 0; \quad q_{ij} = 0$$

Boundary conditions

$$z = 0: \ v_j C_{ij} - D_{ax,j} \frac{\partial C_{ij}}{\partial z}\bigg|_{z=0} = v_j C_{ij}^{in}$$

$$z = L_j: \frac{\partial C_{ij}}{\partial z}\bigg|_{z=L_j} = 0; \quad \overline{C}_{pij,L_j} = \overline{C}_{pij+1,0}(j = 1:3);$$

$$q_{ij,L_j} = q_{ij+1,0}(j = 1:3); \quad \overline{C}_{pi4,L_j} = \overline{C}_{pi1}; \quad q_{i4,L_j} = q_{i1}$$

Adsorption equilibrium isotherm

$$q_{ij} = f(\overline{C}_{pij})$$

Global mass balances in sections
   Eluent node:
$$v_4 + v_E = v_1$$
$$C_{i,4}^{out} v_4 = C_{i,1}^{in} v_1$$

   Feed node:
$$v_2 + v_F = v_3$$
$$C_{i,2}^{out} v_2 + C_{i,F} v_F = C_{i,3}^{in} v_3$$

   Extract node:
$$v_1 - v_X = v_2$$
$$C_{i,1}^{out} = C_{i,2}^{in} = C_{i,X}$$

   Raffinate node:
$$v_3 - v_R = v_4$$
$$C_{i,3}^{out} = C_{i,4}^{in} = C_{i,R}$$

In these equations $i$ relates to the component in the mixture and $j$ relates to the TMB section number ($j = 1, 2, 3, 4$).

**Table 2.4** Model Equations for the Transient SMB Model with Porous Adsorbent Particles

Mass balance in a volume element of column $k$

$$\varepsilon \frac{\partial C_{ik}}{\partial t} = \varepsilon D_{axk} \frac{\partial^2 C_{ik}}{\partial z^2} - \varepsilon v_k^* \frac{\partial C_{ik}}{\partial z} - (1 - \varepsilon) \frac{3}{R_p} k_{ov,ij} (C_{ik} - \overline{C}_{p_{ik}})$$

Particle mass balance

$$\varepsilon_p \frac{\partial \overline{C}_{pik}}{\partial t} + \frac{\partial q_{ik}}{\partial t} = \frac{3}{R_p} k_{ov,ij} (C_{ik} - \overline{C}_{p_{ik}})$$

Initial conditions

$$t = 0; \quad C_{ik} = \overline{C}_{p_{ik}} = 0; \quad q_{ik} = 0$$

Boundary conditions

$$z = 0: \quad v_k^* C_{ik} - D_{ax,k} \frac{\partial C_{ik}}{\partial z}\bigg|_{z=0} = v_k^* C_{ik}^{in}$$

$$z = L_j: \quad \frac{\partial C_{ik}}{\partial z}\bigg|_{z=L_k} = 0$$

Adsorption equilibrium isotherm

$$q_{ik} = f(\overline{C}_{pik})$$

Global mass balances in sections

Eluent node:

$$v_4^* + v_E = v_1^*$$
$$C_{i,4}^{out} v_4^* = C_{i,1}^{in} v_1^*$$

Feed node:

$$v_2^* + v_F = v_3^*$$
$$C_{i,2}^{out} v_2^* + C_{i,F} v_F = C_{i,3}^{in} v_3^*$$

Extract node:

$$v_1^* - v_X = v_2^*$$
$$C_{i,1}^{out} = C_{i,2}^{in} = C_{i,X}$$

Raffinate node:

$$v_3^* - v_R = v_4^*$$
$$C_{i,3}^{out} = C_{i,4}^{in} = C_{i,R}$$

In these equations $i$ relates to the component in the mixture and $k$ relates to the SMB column number ($k = 1, 2, \ldots$).

The operating conditions and model parameters used in the simulations for the TMB approach are presented in Table 2.5. Tables 2.6 and 2.7 present the equivalencies in terms of flow rates and model parameters that have to be made for the SMB systems with different subdivisions of the bed.

Note that for SMB systems, the internal flow rates in the four different sections are independent of the degree of subdivision of the bed and they are related to the TMB flow rates by the equivalencies presented before (Eqns (2.15) and (2.16)). Three cases

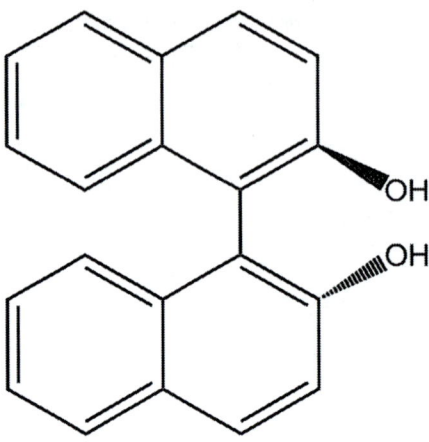

**Figure 2.1** Bi-naphthol enantiomers.

**Table 2.5** Operating Conditions and Model Parameters for the TMB Approach

| TMB Operating Conditions | Model Parameters |
|---|---|
| Feed concentration: 2.9 g/l each | Solid/fluid volumes, $(1 - \varepsilon)/\varepsilon = 1.5$ |
| Solid flow rate: 11.15 ml/min | Ratio between fluid and solid velocities: |
| Recycling flow rate: 27.95 ml/min | $\gamma_I = 6.65$; $\gamma_{II} = 4.23$ |
| Eluent flow rate: 21.45 ml/min | $\gamma_{III} = 4.72$; $\gamma_{IV} = 3.76$ |
| Extract flow rate: 17.98 ml/min | Number of mass transfer units |
| Feed flow rate: 3.64 ml/min | $\alpha = 36.0$ ($k_h = 0.1$ s$^{-1}$) |
| Raffinate flow rate: 7.11 ml/min | Peclet number, $Pe = 2000$ |
| Columns: | |
| Diameter: 2.6 cm | Section length: 21.0 cm |

**Table 2.6** Equivalence between TMB and SMB Flow Rates

| Section | TMB | | SMB | |
|---|---|---|---|---|
| | $Q_j$ | $\gamma_j$ | $Q_j^*$ | $\gamma_j^*$ |
| I | 49.40 | 6.65 | 56.83 | 7.65 |
| II | 31.42 | 4.23 | 38.85 | 5.23 |
| III | 35.06 | 4.72 | 42.49 | 5.72 |
| IV | 27.95 | 3.76 | 35.38 | 4.76 |

are analyzed for the SMB system: SMB4, constituted by four columns, one in each section; SMB8 with two columns per section; and SMB12, with three columns per section. The length of each fixed-bed column in these cases was chosen by keeping constant the total length of each section. The value for the switch time interval was then evaluated keeping constant the ratio $L_c/t^*$, the simulated solid velocity. Also, the number of mass transfer units per section is the same for the TMB and SMB cases and is evaluated

**Table 2.7** Equivalence between TMB and SMB with Different Subdivision of the Bed

| Case | $N_s$ | $L_c$ (cm) | $L_j$ (cm) | $t^*$ (min) | $u_s = L_c/t^*$ (cm/min) | $\alpha$ | $N_s \alpha$ | Pe |
|------|-------|-----------|-----------|-------------|--------------------------|----------|--------------|-----|
| TMB  | –     | –         | 21        | –           | 3.5                      | 36       | 36           | 2000 |
| SMB4 | 1     | 21        | 21        | 6           | 3.5                      | 36       | 36           | 2000 |
| SMB8 | 2     | 10.5      | 21        | 3           | 3.5                      | 18       | 36           | 1000 |
| SMB12| 3     | 7         | 21        | 2           | 3.5                      | 12       | 36           | 667 |

for $k_h = 0.1$ s$^{-1}$ (Table 2.5). Summarizing, all the SMB cases present the same operating conditions and model parameters at a section scale (equivalent to the TMB case), except for the degree of subdivision of the bed.

The CSS behavior, characteristic of an SMB operation, is shown in Figure 2.2 and Figure 1.10, Chapter 1, for the case of an eight–column configuration (two columns per section). Figure 1.10 in Chapter 1 presents the concentration of the two enantiomers

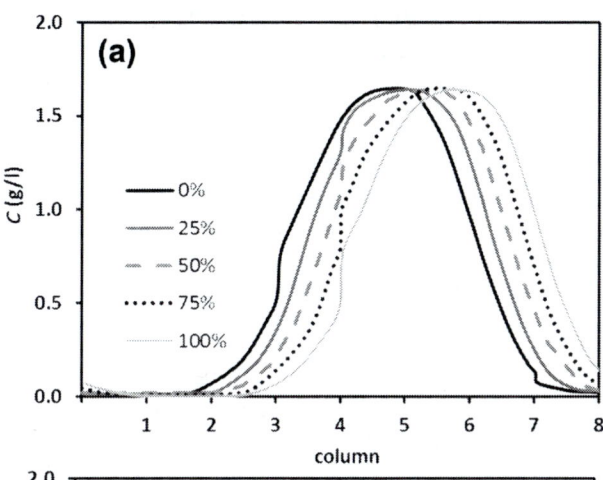

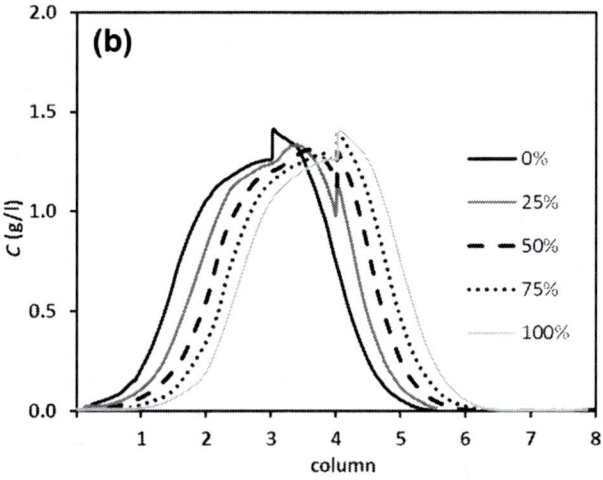

**Figure 2.2** Cyclic steady-state internal concentration profiles during a switching time interval (fraction) for SMB8 configuration 2-2-2-2. (a) Less-retained component and (b) more-retained component. *(Reprinted from AIChE Journal 44, Pais, L.S., Loureiro, J.M., Rodrigues, A.E. Modeling strategies for enantiomers separation by SMB chromatography, pp. 561–569, Copyright (1998), with permission from John Wiley & Sons Inc.)*

in extract and raffinate after CSS is reached (in this case, after 10 full cycles). Extract and raffinate exhibit concentration profiles that are reproduced in the same way fraction after fraction.

For the same case (SMB8), Figure 2.2 shows the evolution of the internal concentration profiles after CSS is reached, during a switch time interval. Note that, because steady state is achieved, the concentration profiles at the end of a switch time interval are the same as the ones at the beginning of this interval, but that they are advanced one column. Again, these profiles will be reproduced in the same way fraction after fraction and column after column.

The influence of the degree of subdivision of the bed in the transient concentration of extract and raffinate is shown in Figures 2.3 and 2.4, which also show the comparison with the TMB approach. The behavior of the SMB is predicted in three ways: the exact transient evolution of concentration profiles, the average concentration evaluated at each switch time interval, and the instantaneous concentration evaluated at half-time between two successive switchings. These figures show the transient evolution during the first five cycles. Although the switching time depends on the degree of subdivision of the bed, the duration of a full cycle will be 24 min for all SMB cases. It is clear that there are differences between SMB and TMB predictions and that they are attenuated with the increase in the number of subdivisions. In fact, the similarity would be perfect if the adsorbent bed were divided into an infinite number of fixed-bed columns and using an infinitesimal switch time interval. Figure 2.5 compares the steady-state internal concentration profiles, evaluated at half-time between switchings, for TMB and SMB cases. The major difference appears for the SMB4 case, whereas small deviations occur between SMB8 and SMB12.

For practical aims, the more important question may be what is the difference between TMB and SMB performances and what is the influence of the degree of subdivision of the adsorbent bed. Table 2.8 shows the predictions obtained for extract and raffinate purities after steady state is reached. The raffinate and extract purities in SMB units with 4, 8, and 12 columns are increasing toward the one obtained in the equivalent TMB unit.

It is obvious that a system with more columns per section will be more expensive. On the other hand, a system constituted by a higher number of columns will be able to perform separations with higher purities and/or higher productivities. The optimum degree of subdivision of the SMB unit will depend on the difficulty of the separation and the product purity requirement (Hidajat et al., 1986b; Tondeur and Bailly, 1993; Bauer et al., 1996).

## 2.1.2 The TMB Steady-State Model

It is clear from this study that the prediction of the performance of an SMB operation, and so the flow-rate optimization, can be done using the TMB approach, although small differences will appear between these two strategies of modeling. These differences will be more significant if a low degree of subdivision of the adsorbent bed is used, especially in the case of an SMB with only one column per section. Nevertheless, the SMB model will always be useful to characterize the dynamic cyclic behavior of the concentration profiles.

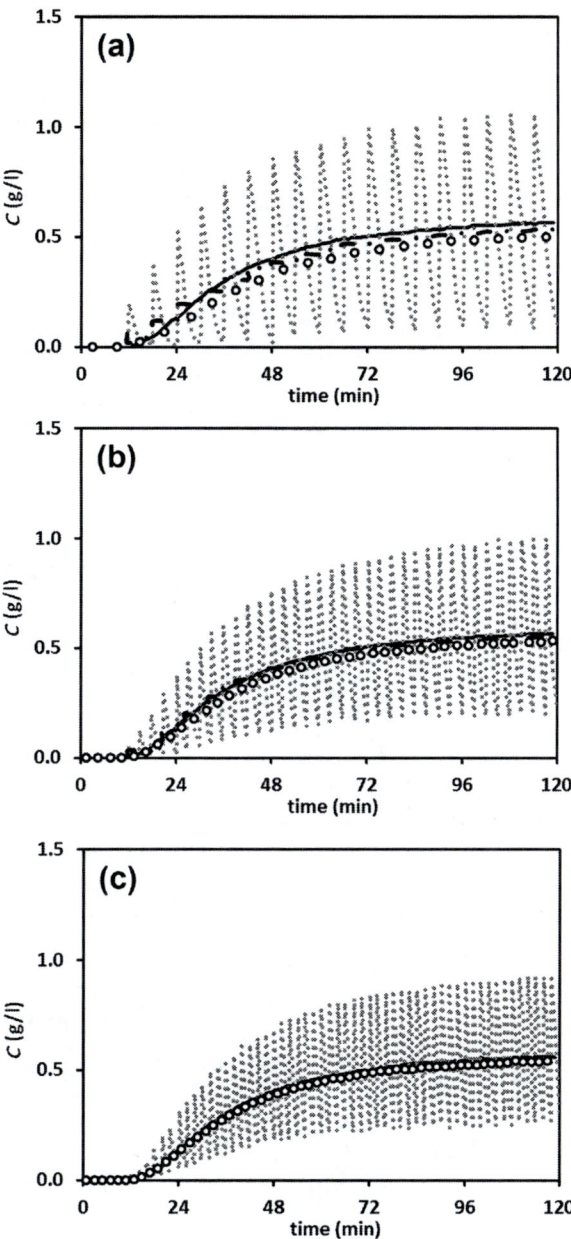

**Figure 2.3** Transient evolution (first five cycles) of the concentration of the more-retained component in the extract for (a) SMB4, (b) SMB8, and (c) SMB12 (solid line, TMB approach; dotted line, SMB approach; step-dotted line, SMB approach with average concentration over a switch time interval; open circles, the SMB instantaneous concentration evaluated between switchings). *(Reprinted from AIChE Journal 44, Pais, L.S., Loureiro, J.M., Rodrigues, A.E. Modeling strategies for enantiomers separation by SMB chromatography, pp. 561–569, Copyright (1998), with permission from John Wiley & Sons Inc.)*

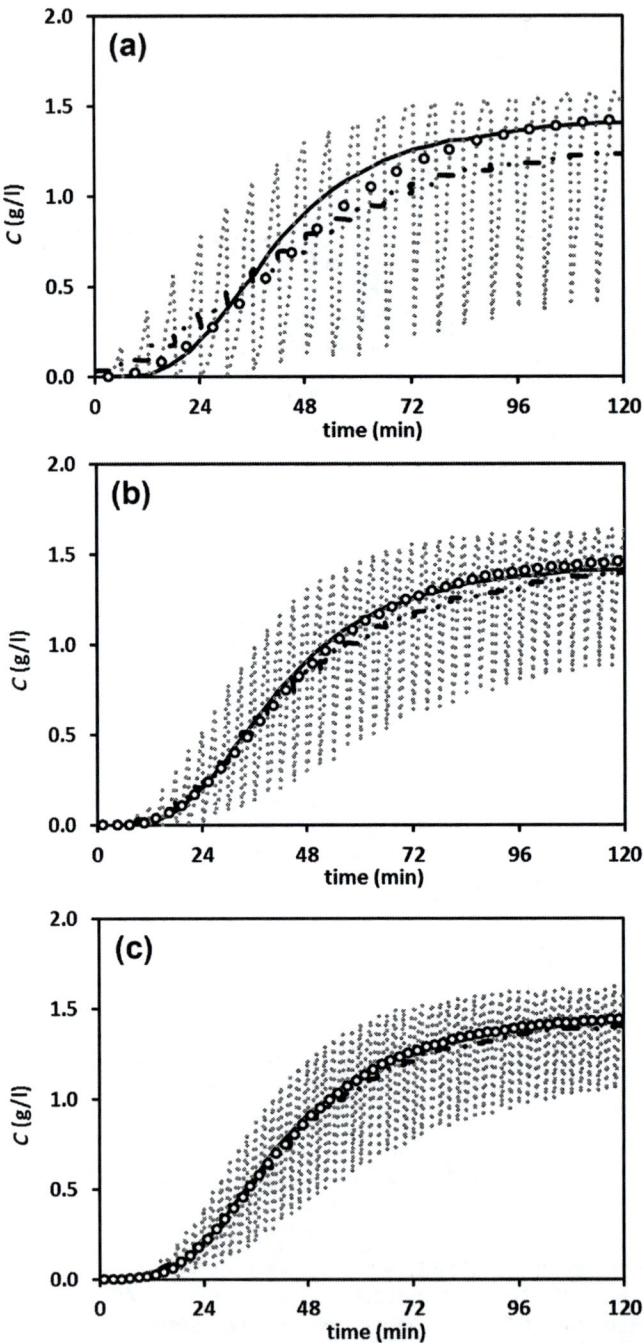

**Figure 2.4** Transient evolution (first five cycles) of the concentration of the less-retained component in the raffinate for (a) SMB4, (b) SMB8, and (c) SMB12 (solid line, TMB approach; dotted line, SMB approach; step-dotted line, SMB approach with average concentration over a switch time interval; open circles, the SMB instantaneous concentration evaluated between switchings). *(Reprinted from AIChE Journal 44, Pais, L.S., Loureiro, J.M., Rodrigues, A.E. Modeling strategies for enantiomers separation by SMB chromatography, pp. 561–569, Copyright (1998), with permission from John Wiley & Sons Inc.)*

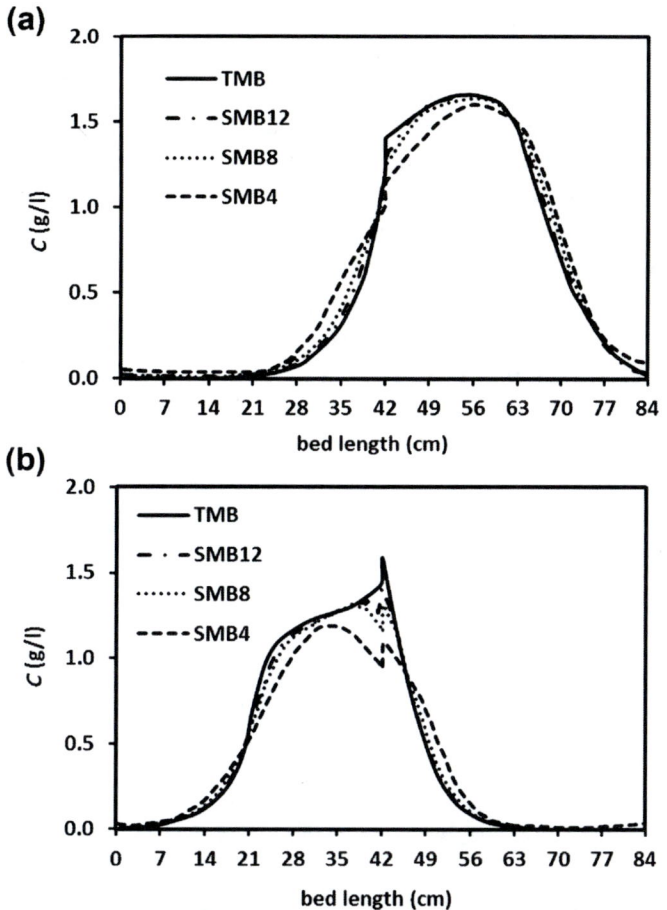

**Figure 2.5** Steady-state internal concentration profiles for TMB and SMB cases at half-time of switching interval in CSS. (a) Less-retained species; (b) more-retained species. *(Reprinted from AIChE Journal 44, Pais, L.S., Loureiro, J.M., Rodrigues, A.E. Modeling strategies for enantiomers separation by SMB chromatography, pp. 561–569, Copyright (1998), with permission from John Wiley & Sons Inc.)*

**Table 2.8** Comparison between Extract and Raffinate Purities in TMB and SMB Cases

| Case | Extract Purity (%) | Raffinate Purity (%) |
| --- | --- | --- |
| SMB4 | 89.5 | 95.2 |
| SMB8 | 95.9 | 98.7 |
| SMB12 | 96.8 | 99.1 |
| TMB | 97.7 | 99.3 |

To predict the steady-state performance of a SMB separation process, one can use the TMB model with obvious advantages in computing time savings. Moreover, if we are interested only in the steady-state operation, we can develop a steady-state TMB model, which is simpler to implement. In fact, the original problem represented by a set of partial differential equations (PDEs) will be simplified to a set of ordinary differential equations (ODEs). This strategy will be followed to study the influence of the various operating variables and model parameters on the SMB performance.

The equations for the steady-state TMB model are just those of the transient TMB model with time derivatives equal to zero. The steady-state TMB model equations are summarized below, with boundary conditions, as well as with the necessary mass balances at the nodes between each two sections.

*Mass balance in a volume element of the section j:*

$$D_{L_j} \frac{d^2 C_{ik}}{dz^2} - v_j \frac{dC_{ij}}{dz} - \frac{(1 - \varepsilon)}{\varepsilon} k_h \left( q_{ij}^* - q_{ij} \right) = 0 \qquad (2.33)$$

*Mass balance in the particle:*

$$u_s \frac{dq_{ij}}{dz} + k_h \left( q_{ij}^* - q_{ij} \right) = 0 \qquad (2.34)$$

*Boundary conditions for section j:*

$$z = 0 \qquad C_{ij} - \frac{D_{L_j}}{v_j} \frac{dC_{ij}}{dz} = C_{ij,0} \qquad (2.35)$$

where $C_{ij,0}$ is the inlet concentration of species $i$ in section $j$.

$z = L_j$

for the eluent node $\qquad\qquad C_{iIV} = \frac{v_I}{v_{IV}} C_{iI,0} \qquad (2.36a)$

for the extract node $\qquad\qquad C_{iI} = C_{iII,0} \qquad (2.36b)$

for the feed node $\qquad\qquad C_{iII} = \frac{v_{III}}{v_{II}} C_{iIII,0} - \frac{v_F}{v_{II}} C_i^F \qquad (2.36c)$

for the raffinate node $\qquad\qquad C_{iIII} = C_{iIV,0} \qquad (2.36d)$

and

$$q_{iIV} = q_{iI,0}, \quad q_{iI} = q_{iII,0}, \quad q_{iII} = q_{iIII,0}, \quad q_{iIII} = q_{iIV,0} \qquad (2.37)$$

*Global balances:*

Eluent node $\qquad\qquad v_I = v_{IV} + v_F \qquad (2.38a)$

Extract node $\qquad\qquad v_{II} = v_I - v_X \qquad (2.38b)$

Feed node $\qquad v_{III} = v_{II} + v_F \qquad$ (2.38c)

Raffinate node $\qquad v_{IV} = v_{III} - v_R \qquad$ (2.38d)

Multicomponent adsorption equilibrium isotherm:

$$q_{Aj}^* = f_A(C_{Aj}, C_{Bj}) \quad \text{and} \quad q_{Bj}^* = f_A(C_{Aj}, C_{Bj}) \qquad (2.39)$$

By introducing the dimensionless variable $x = z/L_j$, the model Eqns (2.33) and (2.34) become

$$\gamma_j \left\{ \frac{1}{Pe_j} \frac{d^2 C_{ij}}{dx^2} - \frac{dC_{ij}}{dx} \right\} - \frac{(1-\varepsilon)}{\varepsilon} \alpha_j (q_{ij}^* - q_{ij}) = 0 \qquad (2.40)$$

and

$$\frac{dq_{ij}}{dx} + \alpha_j \left( q_{ij}^* - q_{ij} \right) = 0 \qquad (2.41)$$

The boundary conditions for $x = 1$ ($z = L_j$) are the same as presented before (Eqns (2.36a−d) and (2.37)) and, for $x = 0$ ($z = 0$), Eqn (2.35) becomes

$$C_{ij} - \frac{1}{Pe_j} \frac{dC_{ij}}{dx} = C_{ij,0} \qquad (2.42)$$

The resulting model parameters are the same as for the transient TMB model presented in Eqns (2.28)−(2.31). Again, adsorption equilibrium parameters have to be added to the list.

## 2.2 PROCESS PERFORMANCE PARAMETERS: PURITY, RECOVERY, PRODUCTIVITY, AND DESORBENT CONSUMPTION

The SMB performance parameters are important indicators of the process feasibility under different operating conditions.

Because the SMB is a cyclic separation technology, in the calculation of the performance parameters average concentrations of the species collected in the extract or raffinate streams over a cycle are used, which are determined using the integral of the concentration over a complete cycle; for instance, the average concentration of species $B$ in the raffinate stream is $\langle C_B^R \rangle = \int_t^{t+N_c t^*} C_B^R dt / N_c t^*$.

Figure 2.6(a) and (b) show, as an example, the transient evolution profiles of species $A$ (more-retained compound) and species $B$ (less-retained compound) in the extract and raffinate streams, respectively, as well as the average concentrations calculated over one cycle, obtained in the SMB separation of a racemic mixture of chiral epoxide enantiomers

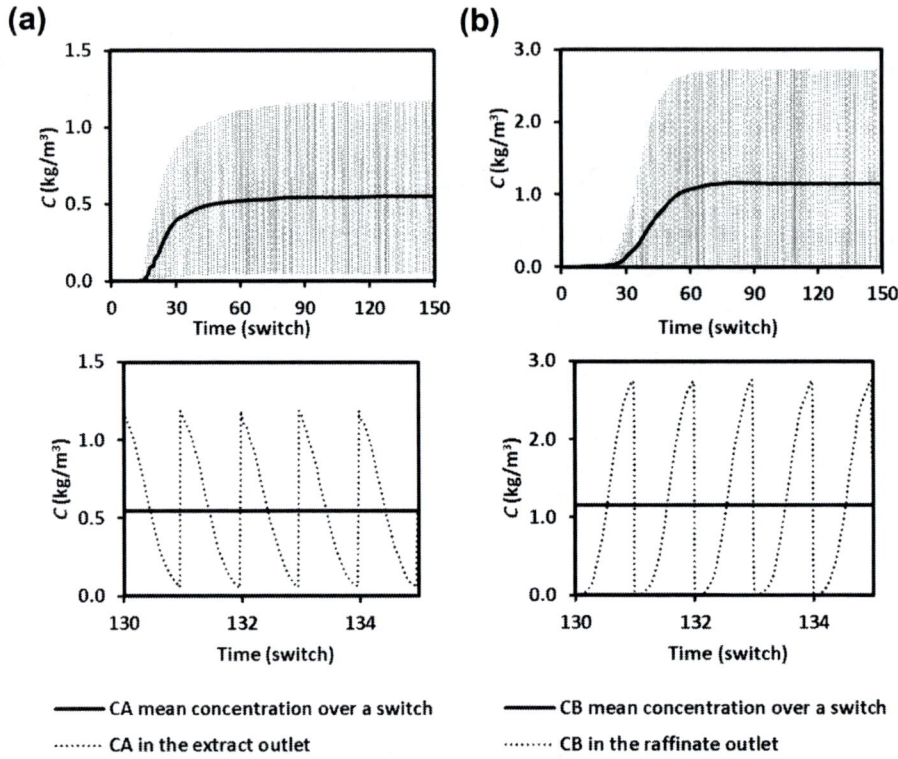

**Figure 2.6** Concentration history of (a) more-retained compound (A) in the extract and (b) less-retained compound (B) in the raffinate. *(Reprinted from Adsorption 12, Sá Gomes, P., Minceva, M., Rodrigues, A.E. Simulated moving bed technology: old and new, pp. 375–392, Copyright (2006), with permission from Springer.)*

with microcrystalline cellulose triacetate particles as chiral stationary phase and pure methanol as eluent (Sá Gomes et al., 2006).

### 2.2.1 Purity

The purity of the raffinate and extract streams at CSS over a complete cycle is defined as

$$\text{Raffinate Purity: } PUR = \frac{\langle C_B^R \rangle}{\langle C_A^R \rangle + \langle C_B^R \rangle} \tag{2.43a}$$

$$\text{Extract Purity: } PUR = \frac{\langle C_A^X \rangle}{\langle C_A^X \rangle + \langle C_B^X \rangle} \tag{2.43b}$$

The numerator of the above equations indicates the concentration of the target component collected in the outlet streams (extract, X, or raffinate, R) within a complete cycle, whereas the denominator indicates the sum of the concentrations of the components collected (these indicators are calculated on a desorbent-free basis).

### 2.2.2 Recovery

The recovery of the less-retained species ($B$) in the raffinate stream, $RE_R$, and of the more-retained species ($A$) in the extract stream, $RE_X$, at CSS over a complete cycle is defined as

$$\text{Raffinate Recovery: } RE_R = Q_R \langle C_B^R \rangle / (Q_F C_B^F) \tag{2.44a}$$

$$\text{Extract Recovery: } RE_X = Q_X \langle C_A^X \rangle / (Q_F C_A^F) \tag{2.44b}$$

where the values in brackets are the average concentrations collected during a cycle time at CSS operation.

### 2.2.3 Productivity

The productivity, $PR$, is the amount of species $B$ (in the raffinate) or $A$ (in the extract) produced per unit time by mass of adsorbent, calculated by:

$$PR_R \left[ \frac{\text{kg}}{\text{kg}_{ads}\text{day}} \right] = RE_R Q_F C_B^F / ((1 - \varepsilon) V_c \rho_{ads} N_c) \tag{2.45a}$$

$$PR_X \left[ \frac{\text{kg}}{\text{kg}_{ads}\text{day}} \right] = RE_X Q_F C_A^F / ((1 - \varepsilon) V_c \rho_{ads} N_c) \tag{2.45b}$$

This is an important indicator from the economic point of view, because it tells the mass flow of species produced per amount of adsorbent used in the unit.

### 2.2.4 Desorbent Consumption

Another important performance parameter is the desorbent consumption, $DC$, which reflects on the costs involved in the separation of the solvent from the products. If the desired product is $A$, recovered in the extract, the desorption consumption is

$$DC \left[ \frac{L_{desorbent}}{\text{kg}_{product}} \right] = \frac{Q_D}{Q_X C_A^X} \tag{2.46}$$

In chiral separations or sugar separations the desorbent or eluent is the mobile phase also present in the feed. In this case the desorbent consumption for dilute feeds can be calculated by

$$DC \left[ \frac{L_{desorbent}}{\text{kg}_{product}} \right] = \frac{Q_D + Q_F}{Q_F (C_A^F + C_B^F)} \tag{2.47}$$

These parameters are valid for both SMB and TMB model approaches; however, in the TMB model approach there is no need for the integrals calculation, because at the steady state the concentrations of products are constant and thus the performance parameters are constant along time. In the SMB, when the unit is symmetrical, that is, there are no differences between the switching time periods (owing to the implementation of nonconventional modes of operation or to the use of more detailed models accounting for dead volumes or switching time asymmetries), the performance equations can be simplified for a switching time period (from $t$ to $t + t^*$).

## 2.3 EFFECT OF MODEL PARAMETERS AND OPERATING VARIABLES ON PROCESS PERFORMANCE

In this section, the effects of the operating conditions on the SMB performance are analyzed by simulation, using the steady-state TMB model presented in Section 2.1.2. The resolution of the bi–naphthol enantiomers was used for simulation purposes. The adsorption equilibrium isotherms are given in Eqn (2.32a) and (2.32b). A reference case relative to an eight-column configuration (2-2-2-2) of the SMB, based on the values of operating variables and model parameters shown in Table 2.9 was chosen. Table 2.5 presents the equivalent TMB operating conditions and model parameters for the reference case.

Figure 2.7 presents the corresponding steady-state internal concentration profiles obtained with the simulation package. The performance parameters obtained for this reference case are the following: extract purity 97.6%, raffinate purity 99.3%, recovery of $A$ in the extract 99.3%, productivity 68.2 g/(day. l of bed), and desorbent consumption 1.19 l/g.

### 2.3.1 Effect of Switching Time on Process Performance

The influence of the time interval on the system performance is shown in Figure 2.8. A change in the switch time interval will lead to a change in the equivalent solid flow

**Table 2.9** SMB Operating Conditions and Model Parameters for the Reference Case

| SMB Operation Conditions | Model Parameters |
|---|---|
| Feed concentration: 2.9 g/l each | Solid/fluid volumes, $(1 - \varepsilon)/\varepsilon = 1.5$ |
| Switch time interval: 3 min | Ratio between fluid and solid velocities: |
| Recycling flow rate: 35.38 ml/min | $\gamma_I^* = 7.65$; $\gamma_{II}^* = 5.23$ |
| Eluent flow rate: 21.45 ml/min | $\gamma_{III}^* = 5.72$; $\gamma_{IV}^* = 4.76$ |
| Extract flow rate: 17.98 ml/min | Number of mass transfer units: |
| Feed flow rate: 3.64 ml/min | $\alpha_k = 18.0$ ($k = 0.1$ s$^{-1}$) |
| Raffinate flow rate: 7.11 ml/min | Peclet number, $Pe_k = 1000$ |

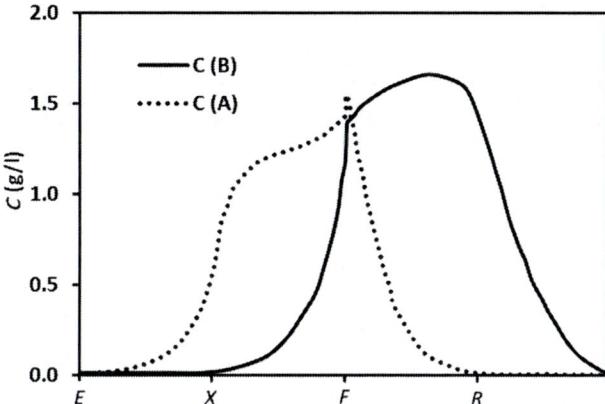

**Figure 2.7** Steady-state internal concentration profiles for the reference case. *(Reprinted from Chemical Engineering Science 52, Pais, L.S., Loureiro, J.M., Rodrigues, A.E. Separation of 1,1'-bi-2-naphthol enantiomers by continuous chromatography in simulated moving bed, pp. 245–257, Copyright (1996), with permission from Elsevier.)*

rate throughout the system. In all runs the inlet and outlet flow rates, as well as the internal liquid flow rates in all the four sections of the SMB unit are kept constant.

Increasing the switch time interval is equivalent to decreasing the solid flow rate, and the net fluxes of components in all four sections of the TMB unit will be pushed in the same direction of the liquid phase. This implies that, first, the more-retained component will move upward in section III and will contaminate the raffinate stream and second, the less-retained species will move upward in section IV, will be recycled to section I, and will contaminate the extract stream.

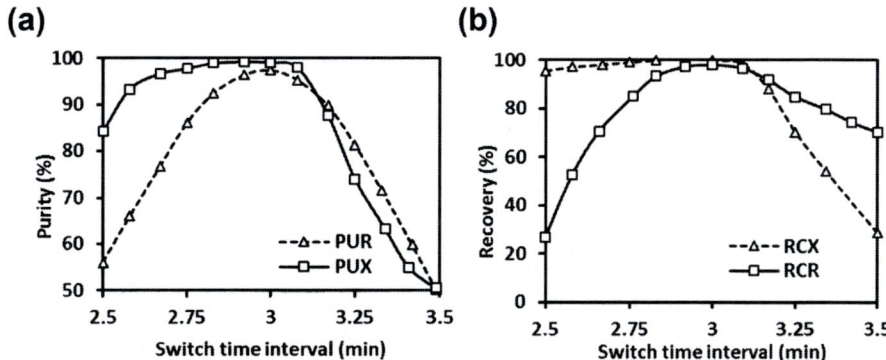

**Figure 2.8** Effect of switching time on (a) purity and (b) recovery. *(Reprinted from Chemical Engineering Science 52, Pais, L.S., Loureiro, J.M., Rodrigues, A.E. Separation of 1,1'-bi-2-naphthol enantiomers by continuous chromatography in simulated moving bed, pp. 245–257, Copyright (1996), with permission from Elsevier.)*

A decrease in the switch time interval will have similar consequences. The equivalent solid flow rate will increase and the net fluxes of components in all four sections of the TMB unit will be pushed in the opposite direction of the liquid phase. This implies that, first, the less-retained species will move downward in section II and will contaminate the extract stream and second, the more-retained component will also move downward in section I, will be recycled with the solid phase to section IV, and will contaminate the raffinate stream.

We can conclude that high purities and recoveries can be obtained only in a narrow window of switch time interval values. Figure 2.9 emphasizes this idea in a purity versus recovery plot. It is clear from this figure that it is possible to obtain simultaneously high purities and recoveries in a SMB system, but the tuning has to be carefully done.

## 2.3.2 Effect of the Feed Flow Rate on Process Performance

The effect of the feed flow rate on the SMB performance is shown in Figure 2.10; in this case, the extract and raffinate flow rates remain constant, and a change in the feed flow rate is followed by a change of opposite signal in the eluent flow rate. For all runs, the total inlet or outlet flow rates remain constant and equal to 25.09 ml/min. Both the switching time and the liquid recycling flow rate are kept constant. Because the feed flow rate is changing, both the desorbent consumption and the adsorbent productivity will also change.

In terms of the internal liquid flow rates, a change in the feed and eluent, keeping constant the extract and raffinate as well as the recycling flow rates, will lead to a change in the internal liquid flow rates in sections I and II, whereas the internal liquid flow rates in the two last sections remain constant.

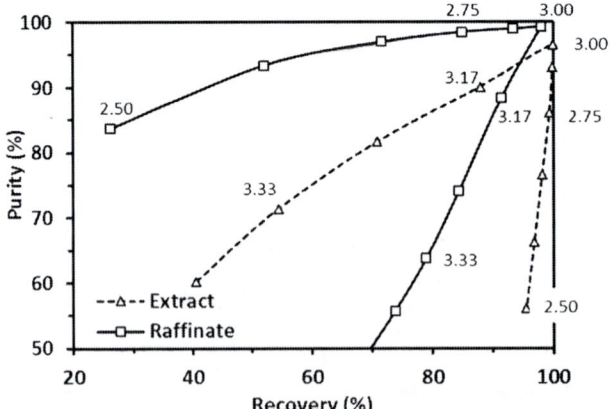

**Figure 2.9** Purity versus recovery: window of switching time allowing high purity and recovery simultaneously. *(Reprinted from Chemical Engineering Science 52, Pais, L.S., Loureiro, J.M., Rodrigues, A.E. Separation of 1,1'-bi-2-naphthol enantiomers by continuous chromatography in simulated moving bed, pp. 245–257, Copyright (1996), with permission from Elsevier.)*

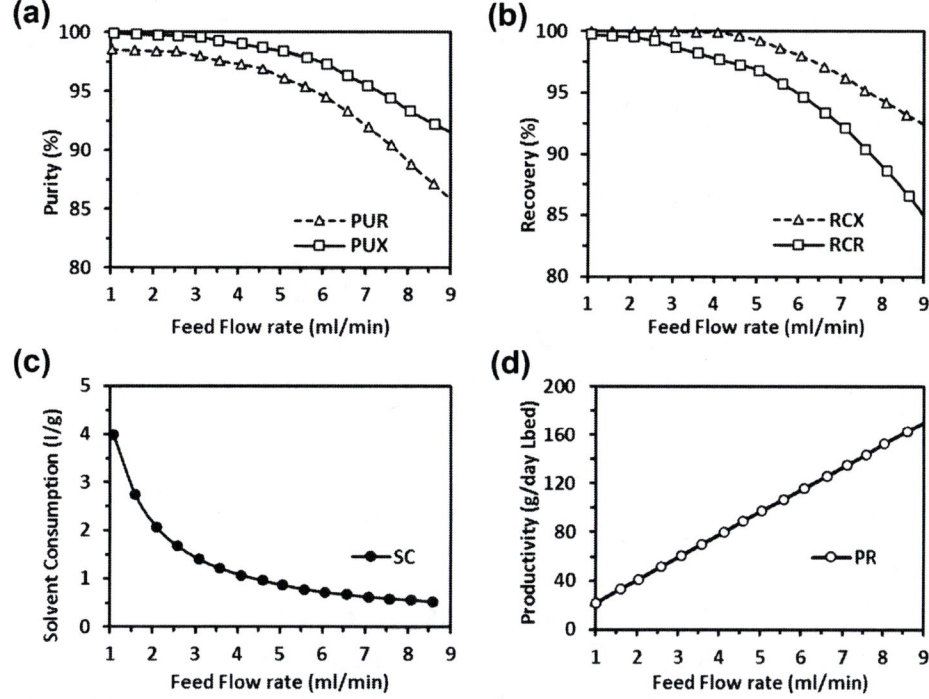

**Figure 2.10** Effect of the feed flow rate on the performance parameters: (a) purity; (b) recovery; (c) solvent consumption; (d) productivity. *(Reprinted from Chemical Engineering Science 52, Pais, L.S., Loureiro, J.M., Rodrigues, A.E. Separation of 1,1'-bi-2-naphthol enantiomers by continuous chromatography in simulated moving bed, pp. 245–257, Copyright (1996), with permission from Elsevier.)*

As a general conclusion we can say that increasing the feed flow rate improves productivity and solvent consumption but reduces both purity and recovery. In fact, increasing the feed, keeping constant the recycling flow rate, means that the internal liquid flow rate in section II decreases and the less-retained species $B$ will move downward and contaminate the extract stream. Also, the internal flow rate in the first section decreases and species $A$ eventually moves downward, is recycled with the solid phase to section IV, and will contaminate the raffinate stream.

A decrease in the feed, keeping constant the recycling flow rate, will lead to an increase in the internal liquid flow rates in the two first sections, whereas in the two last sections they remain constant. These changes will improve the separation because the net fluxes of components $B$ and $A$ upward in sections II and I, respectively, will be intensified, and the net fluxes in the two last sections will not be significantly affected.

## 2.3.3 Effect of the Axial Dispersion on the SMB Performance

The influence of the axial dispersion on the performance of an SMB adsorber in terms of purity is shown in Figure 2.11. The Peclet number refers to a whole section of the

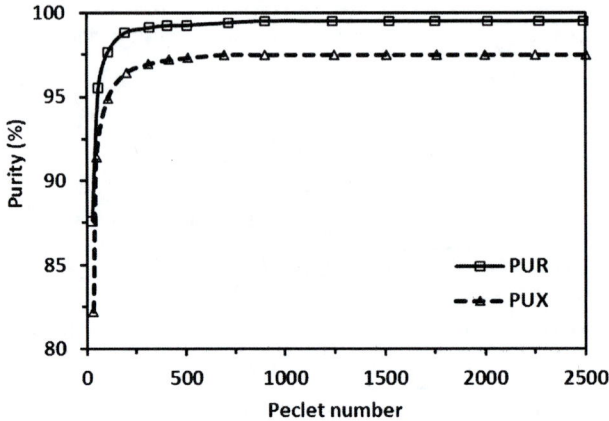

**Figure 2.11** Effect of the axial dispersion on purity.

equivalent TMB system. It is obvious that the influence of the axial dispersion on the SMB performance is negligible for Peclet numbers higher than 500.

### 2.3.4 Effect of Mass Transfer Resistance on the SMB Performance

A higher value for the mass transfer coefficient corresponds to a situation in which mass transfer resistance is less important. Obviously, the performance of the SMB will be better for higher values of the mass transfer coefficient.

As was proposed by Glueckauf (1955), the mass transfer coefficient used in the LDF approximation depends only on the intraparticle diffusivity of species and particle size. Therefore, increasing $k_h$ (or the corresponding dimensionless number, $\alpha$) by decreasing the particle size improves the performance of the SMB, provided the constraint of acceptable pressure drop is met. Some applications, namely in the area of protein processing, will eventually use large-pore permeable particles in which intraparticle mass transport by convective flow is important, leading to an enhancement of the mass transfer rate. The LDF models presented can still be used in that case if the mass transfer coefficient $k_h$ is replaced by an augmented mass transfer coefficient (Leitão and Rodrigues, 1995, 1996).

In Figure 2.12 we can observe the influence of the mass transfer coefficient on the steady-state internal concentration profiles. Following the enhancement of the SMB performance, a higher mass transfer coefficient will lead to sharper internal concentration profiles.

### 2.4 NUMERICAL SOLUTION OF SMB AND TMB MODELS

There are packages commercially available to solve SMB models, such as ASPEN Chromatography (http://www.aspentech.com/publication_files/chromatography.pdf).

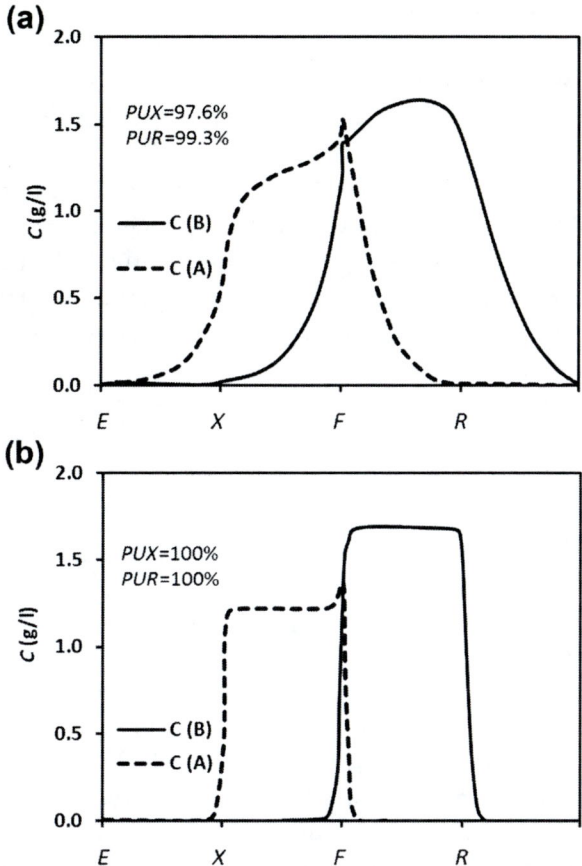

**Figure 2.12** Effect of the mass transfer resistance on the internal concentration profiles: (a) $k = 0.1$ s$^{-1}$; (b) $k = 1$ s$^{-1}$. *(Reprinted from Chemical Engineering Science 52, Pais, L.S., Loureiro, J.M., Rodrigues, A.E. Separation of 1,1'-bi-2-naphthol enantiomers by continuous chromatography in simulated moving bed, pp. 245—257, Copyright (1996), with permission from Elsevier.)*

ProSim (2001) launched a simulator for SMB and still provides ProSim DAC (http://www.prosim.net/en/software-prosim-dac-12.php) for dynamics of adsorption columns.

In this section we present homemade simulators using public domain software or simulators implemented in gPROMS (Process System Enterprise, 2014) or MATLAB (Mathworks, 2014).

## 2.4.1 Homemade Simulators for Solving SMB and TMB Transient Models

The SMB and TMB models, defined by a set of PDEs, were numerically solved by using the *PDECOL* software (Madsen and Sincovec, 1979) based on the method of orthogonal collocation in finite elements.

For the TMB model, there are four PDEs for each section: for each component there is a PDE resulting from the mass balance in a volume element of the section and others resulting from the mass balance in the particle. Because the TMB unit is composed of four sections, and considering a binary separation, the TMB system is defined by a set of 16 PDEs.

In the SMB model, four equivalent PDEs must be written, but now for each column. The four-section SMB system for a binary separation is then defined by a set of four $N_c$ PDEs, where $N_c$ is the total number of columns used in the SMB unit.

In this treatment we are assuming that the algebraic equations describing adsorption equilibrium isotherms are already substituted in the interface condition in the mass balance for the adsorbent particle.

The *PDECOL* software is based on the method of lines and uses a finite element collocation procedure for the discretization of the spatial variable. The collocation procedure reduces the PDE system to an initial-value ODE system, which then depends only on the time variable. The time integration is then accomplished by using the ODE solver *STIFIB*, which is a modified version of the *GEARIB* ODE package developed by Hindmarsh (1976).

The user must specify the piecewise polynomial to be used to compute the approximate solution. In selecting this space, the order, *KORD*, of the polynomials to be used must first be specified (*KORD* = polynomial degree + 1). Next, the number of intervals, *NINT*, into which the spatial domain is to be divided, is chosen. The approximate solution at any time will be a polynomial of order (*KORD* − 1) in each $i$ subinterval ($i = 1, 2, ..., NINT$). The common value *KORD* = 4 was used and an equal number of intervals, *NINT* = 20, was used to divide the spatial domain of each section or column in the TMB and SMB models, respectively. The number of continuity conditions, *NCC*, to be imposed on the polynomial pieces across all of the interior breakpoints is then chosen to complete the definition of the piecewise polynomial space. The common value *NCC* = 2 was used, which requires that the approximate solution and its first spatial derivative are continuous at the breakpoints and hence on the entire spatial domain. The dimension of this linear space is known and finite and is $NCPTS = KORD^* NINT - NCC^* (NINT - 1)$, where *NCPTS* means the number of collocation points used for each equation. Summarizing, *PDECOL* software reduces a set of *NPDE* PDEs to a set of $N = NPDE^* NCPTS$ time-dependent ODEs.

The model equations for the transient SMB or TMB operation are solved, from an initial condition under which no feed components are present (columns are filled with the eluent) until steady state is reached. In the TMB and SMB packages developed, the steady state is considered to be achieved when (1) the relative errors between the average concentrations (evaluated during a full cycle, for the SMB cases, and during a solid space time interval, for the TMB model) of each component in the extract and raffinate streams for two successive iterations are less than a maximum error defined by the user and (2) the relative errors between the total amounts of each component that enters (in the feed stream) and leaves (in the extract and raffinate streams) the system (evaluated during a full cycle, for the SMB cases, and during a solid space time interval, for the TMB model) are less than a maximum error defined by the user.

In fact, a unique global error is defined as

$$e = e_X + e_R + e_A + e_B \tag{2.48}$$

where

$e_X$ is the relative error of the average concentrations of the two components in the extract for two successive iterations,

$$e_X = \frac{\left|\overline{C}_{A[i]}^{X} - \overline{C}_{A[i-1]}^{X}\right|}{\overline{C}_{A[i]}^{X}} + \frac{\left|\overline{C}_{B[i]}^{X} - \overline{C}_{B[i-1]}^{X}\right|}{\overline{C}_{B[i]}^{X}} \tag{2.49a}$$

$e_R$ is the relative error of the average concentrations of the two components in the raffinate in two successive iterations,

$$e_R = \frac{\left|\overline{C}_{A[i]}^{R} - \overline{C}_{A[i-1]}^{R}\right|}{\overline{C}_{A[i]}^{R}} + \frac{\left|\overline{C}_{B[i]}^{R} - \overline{C}_{B[i-1]}^{R}\right|}{\overline{C}_{B[i]}^{R}} \tag{2.49b}$$

$e_A$ is the relative error between the amounts of the more-retained component that enters (in the feed stream) and leaves (in the extract and raffinate streams) the system,

$$e_A = \frac{\left|Q_F C_A^F - \left(Q_X \overline{C}_A^X + Q_R \overline{C}_A^R\right)\right|}{Q_F C_A^F} \tag{2.49c}$$

$e_B$ is the relative error between the amounts of the less-retained component that enters (in the feed stream) and leaves (in the extract and raffinate streams) the system

$$e_B = \frac{\left|Q_F C_B^F - \left(Q_X \overline{C}_B^X + Q_R \overline{C}_B^R\right)\right|}{Q_F C_B^F} \tag{2.49d}$$

The steady state is considered to be achieved when the global error $e$, defined by Eqn (2.48), is less than a maximum error defined by the user, $\delta$. For both modeling strategies, TMB and SMB, a value of $\delta = 0.01$ was used.

As was pointed out before, the countercurrent motion of fluid and solid phases in the SMB operation is achieved with a discrete jump of the injection (feed and eluent) and collection (extract and raffinate) points. Owing to this switch of the inlet and outlet points, the boundary conditions for each column vary with time, changing with the end of each switch time interval. Hence, the SMB model must take into account this time dependence of the boundary conditions.

If the algebraic equations describing the adsorption equilibrium isotherms are kept as such we should choose an adequate software to handle a system of partial differential algebraic equations such as DASSL solver (http://www.netlib.org/toms/690).

## 2.4.2 Homemade Simulators for Solving the Steady-State TMB Model

The steady-state TMB model was numerically solved by using the COLNEW software (Bader and Ascher, 1987). This package solves a general class of mixed-order systems of boundary value ODEs and is a modification of the COLSYS package developed by Ascher et al. (1979, 1981).

Each section of the TMB unit is defined by four ODEs: for each component there is an ODE resulting from the mass balance in a volume element of the section and others resulting from the mass balance in the particle. Because the TMB unit is composed of four sections, and considering a binary separation, the steady-state TMB model is defined by a set of 16 ODEs. The COLNEW software (http://www.netlib.org/ode/colnew.f) incorporates a new basis representation replacing B-splines and improvements for the linear and nonlinear algebraic equation solvers. The numerical approximation of mixed-order systems of multipoint value ODEs by collocation requires appropriate representation of the piecewise polynomial solutions. B-splines were originally implemented in the general purpose code COLSYS, whereas a Runge—Kutta monomial basis is used for the COL-NEW software. Turner and Mills (1990) applied these packages to chemical engineering systems and concluded that the COLSYS software requires more CPU time and is less stable than the newer version of COLNEW. The user must specify a list of parameters to implement the COLNEW software. These parameters are IPAR(1) for the type of problem (linear 0, nonlinear 1), IPAR(2) for the number of collocation points per interval, IPAR(3) for the number of subintervals in the mesh, and TOL for the tolerance (e.g., $10^{-5}$). No initial guess to the solution is provided and we let the COLNEW software generate the initial mesh and control it during the numerical resolution. A unique global error, similar to the one presented before in Section 2.4.1, was used. The convergence of the method is considered to be achieved when the global error, as defined by Eqn (2.48), is less than 0.01.

The COLNEW software proved to be very efficient at the numerical resolution of the steady-state TMB model. The computing time for the solution of the steady-state model is much faster than the running times for the transient models; we easily conclude there are obvious advantages to using the COLNEW software if we want to characterize only the steady-state TMB operation.

Again, if the algebraic equations describing the adsorption equilibrium isotherms are not substituted in the mass balance for the particle, we get a system of differential algebraic equation, which should be handled with an appropriate solver such as COLDAE (http://www.netlib.org/ode/coldae.f). More details can be found in Leão and Rodrigues (2004).

### 2.4.3 Implementation in gPROMS and MATLAB

The SMB and TMB models can also be implemented in commercial platforms, such as gPROMS (Process System Enterprise, 2014) or MATLAB (Mathworks, 2014).

The modeling in gPROMS involves the definition of several entities within a *project*, namely *variable types*, *models*, and *processes*. The equations, including the boundary conditions, are written in the *model* along with the declaration of the quantities (such as parameters and variables) that appear in these equations. All variables declared in the *model* must be of a *variable type*. In the entity *variable types* each type is given a name, a default value (used as an initial guess for any iterative calculation), and upper and lower bounds. In the *process*, the *model's* parameters, input variables (degrees of freedom), and initial conditions are specified. The operating procedure is also set in the *process* in the form of a schedule. A schedule may simply specify the execution of an undisturbed simulation for a period of time (as for example the simulation of a breakthrough curve or TMB process) or represent a more complex operation (such as that of an SMB: in this case the operation mode is changed from time to time, to reproduce the switching of the columns and to complete the schedule of a cycle, and the cycles are repeated one after the other). The numerical method for the axial discretization is also defined in the *process*. The user can choose finite difference methods (centered, backward, or forward) or an orthogonal collocation on finite elements method. The latter is the most commonly used for these applications. The number of finite elements in which the axial coordinate is to be divided and the order of the polynomials to be used should be set. After the axial discretization step, the time integration is performed by the ODE solver SRADAU, a fully implicit Runge–Kutta method that implements a variable time step. The result is then solved by gPROMS BDNSOL (Block Decomposition Nonlinear Solver).

In MATLAB, the system of PDEs was solved by the *pdepe* function from the MAT-LAB library. This function solves initial-boundary value problems for systems of parabolic and elliptic PDEs in the one space variable $x$ and time. The syntax is the following: sol − = *pdepe* (*m, pdefun, icfun, bcfun, xmesh, tspan*), where *m* is a parameter corresponding to the symmetry of the problem (slab = 0, cylindrical = 1, or spherical = 2); *pdefun, icfun,* and *bcfun* are functions where the PDEs and initial and boundary conditions are defined respectively; *xmesh* is the vector that defines the points at which the discretization of the axial coordinate is made; and *tspan* is the vector that defines the times at which the solution is requested. After axial discretization, the ODEs are solved by stiff ODE solvers, such as *ode115s*, also available in the MATLAB library.

## 2.5 CONCLUDING REMARKS

In this chapter a strategy of modeling for SMB and TMB was presented, starting with a base model for transient situations in which LDF for particle mass transfer was assumed as well as axial dispersed flow and constant total concentration of species. Steady-state TMB and CSS of SMB were discussed and the influence of model parameters in process performance (purity, recovery, desorbent consumption, and productivity) was analyzed in detail. Numerical solution of the various model equations was discussed with reference to packages commercially available and homemade simulators using public domain software.

## NOMENCLATURE

| | |
|---|---|
| $C$ | Liquid-phase concentration (mol/m$^3$) |
| $\langle C \rangle$ | Average concentration (mol/m$^3$) |
| $\overline{C}_p$ | Average concentration in the particle pores (mol/m$^3$) |
| $D_L, D_{ax}$ | Axial dispersion coefficient (m$^2$/s) |
| $k_h$ | Intraparticle mass transfer coefficient (s$^{-1}$) |
| $k_{ov}$ | Global mass transfer coefficient (m/s) |
| $L$ | Column length (m) |
| $N_c$ | Number of columns |
| $N_s$ | Number of columns per section |
| $Pe$ | Peclet number |
| $PR$ | Productivity |
| $PUR$ | Purity |
| $q$ | Average adsorbed-phase concentration (mol/m$^3$) |
| $q^*$ | Adsorbed-phase concentration in equilibrium with $C$ (mol/m$^3$) |
| $Q$ | Volumetric flow rate (m$^3$/s) |
| $RE$ | Recovery |
| $R_p$ | Particle radius (m) |
| $t$ | Time variable (s) |
| $t^*$ | Switching time (s) |
| $u_s$ | Interstitial solid velocity (m/s) |
| $v$ | Fluid interstitial velocity in the TMB (m/s) |
| $v^*$ | Fluid interstitial velocity in the SMB (m/s) |
| $V_c$ | Column volume (m$^3$) |
| $x$ | Dimensionless axial coordinate |
| $z$ | Axial coordinate (m) |

## Greek Letters

| | |
|---|---|
| $\alpha$ | Number of mass transfer units |
| $\varepsilon$ | Bed porosity |
| $\varepsilon_p$ | Particle porosity |
| $\gamma$ | Ratio between fluid and solid interstitial velocities |
| $\theta$ | Dimensionless time |
| $\rho_{ads}$ | Particle density (kg/m$^3$) |
| $\tau$ | Space time |

## Subscripts

| | |
|---|---|
| **0** | Inlet |
| **c** | Column |
| **E** | Eluent |
| **F** | Feed |
| **i** | Component $i$ ($i = A, B$) |
| **j** | Section $j$ ($j = I, II, III, IV$ or 1, 2, 3, 4) |
| **k** | Column ($k = 1, 2, ..., N_c$) |
| **R** | Raffinate |
| **s** | Solid |
| **X** | Extract |

## REFERENCES

Ascher, U., Christiansen, J., Russell, R.D., 1979. Collocation solver for mixed order systems of boundary-value problems. Math. Comput. 33, 659–679.

Ascher, U., Christiansen, J., Russell, R.D., 1981. Collocation software for boundary-value odes. ACM Trans. Math. Softw. 7, 209–222.

Azevêdo, D.C.S., Neves, S.B., Ravagnani, S.P., Cavalcante Jr., C.L., Rodrigues, A.E., 1998. The influence of dead zones on simulated moving bed units. In: Proceedings of the 6th International Conference on Fundamentals of Adsorption, Presqu'île de Giens, France.

Bader, G., Ascher, U., 1987. A new basis implementation for a mixed order boundary-value ode solver. Siam J. Sci. Stat. Comput. 8, 483–500.

Barker, P.E., England, K., Vlachogiannis, G., 1983. Mathematical-model for the fractionation of dextran on a semi-continuous countercurrent simulated moving bed chromatograph. Chem. Eng. Res. Des. 61, 241–247.

Bauer, J., Priegnitz, A., Chandhok, A., Wilcher, S., 1996. UOP sorbex simulated moving bed technology: the effect of bed number on simulated moving bed performance. In: Poster Presented at PREP'96 International Symposium On Preparative and Industrial Chromatography and Related Techniques. Basel, Switzerland.

Carta, G., Pigford, R.L., 1986. Periodic countercurrent operation of sorption processes applied to water desalination with thermally regenerable ion-exchange resins. Ind. Eng. Chem. Fundam. 25, 677–685.

Charton, F., Nicoud, R.M., 1995. Complete design of a simulated moving-bed. J. Chromatogr. A 702, 97–112.

Chiang, A.S.T., 1998. Complete separation conditions for a local equilibrium TCC adsorption unit. AIChE J. 44, 332–340.

Ching, C.B., 1983. A theoretical-model for the simulation of the operation of the semi-continuous chromatographic refiner for separating glucose and fructose. J. Chem. Eng. Jpn. 16, 49–53.

Ching, C.B., Chu, K.H., Hidajat, K., Ruthven, D.M., 1993. Experimental-study of a simulated countercurrent adsorption system. 7. Effects of nonlinear and interacting isotherms. Chem. Eng. Sci. 48, 1343–1351.

Ching, C.B., Chu, K.H., Hidajat, K., Uddin, M.S., 1991. Experimental and modeling studies on the transient-behavior of a simulated countercurrent adsorber. J. Chem. Eng. Jpn. 24, 614–621.

Ching, C.B., Chu, K.H., Hidajat, K., Uddin, M.S., 1992. Comparative-study of flow schemes for a simulated countercurrent adsorption separation process. AIChE J. 38, 1744–1750.

Ching, C.B., Ho, C., Hidajat, K., Ruthven, D.M., 1987. Experimental-study of a simulated countercurrent adsorption system. 5. Comparison of resin and zeolite absorbents for fructose glucose separation at high-concentration. Chem. Eng. Sci. 42, 2547–2555.

Ching, C.B., Ho, C., Ruthven, D.M., 1988. Experimental-study of a simulated countercurrent adsorption system. 6. Non-linear systems. Chem. Eng. Sci. 43, 703–711.

Ching, C.B., Ruthven, D.M., 1985a. An experimental-study of a simulated countercurrent adsorption system. 1. Isothermal steady-state operation. Chem. Eng. Sci. 40, 877–885.

Ching, C.B., Ruthven, D.M., 1985b. An experimental-study of a simulated countercurrent adsorption system. 2. Transient-response. Chem. Eng. Sci. 40, 887—891.

Ching, C.B., Ruthven, D.M., Hidajat, K., 1985. Experimental-study of a simulated countercurrent adsorption system. 3. Sorbex operation. Chem. Eng. Sci. 40, 1411—1417.

Chu, K.H., Hashim, M.A., 1995. Simulated countercurrent adsorption processes — a comparison of modeling strategies. Chem. Eng. J. Biochem. Eng. J. 56, 59—65.

Dandekar, H.W., Chandhok, A.K., Priegnitz, J.W., 1996. Modeling and simulation of SMB technology for pharmaceutical and fine chemical applications. In: LeVan, M.D. (Ed.), Proceedings of the 5th International Conference on Fundamentals of Adsorption. Kluwer Academic Publishers, Boston, Massachusetts.

Ernst, U.P., Hsu, J.T., 1989. Study of simulated moving-bed separation processes using a staged model. Ind. Eng. Chem. Res. 28, 1211—1221.

Ernst, U.P., Hsu, J.T., 1992. Theoretic study of backmixing in simulated moving-bed adsorption process with multiple equilibrium stages between ports. Sep. Technol. 2, 197—207.

Glueckauf, E., 1955. Theory of chromatography. 10. Formulae for diffusion into spheres and their application to chromatography. Trans. Faraday Soc. 51, 1540—1551.

Hashimoto, K., Adachi, S., Noujima, H., Maruyama, H., 1983. Models for the separation of glucose fructose mixture using a simulated moving-bed adsorber. J. Chem. Eng. Jpn. 16, 400—406.

Hashimoto, K., Adachi, S., Shirai, Y., Morishita, M., 1993. Operation and design of simulated moving-bed adsorbers. In: Ganetsos, G., Barker, P.E. (Eds.), Preparative and Production Scale Chromatography. Marcel Dekker Inc., New York, pp. 273—300.

Hashimoto, K., Yamada, M., Shirai, Y., Adachi, S., 1987. Continuous separation of glucose-salts mixture with nonlinear and linear adsorption-isotherms by using a simulated moving-bed adsorber. J. Chem. Eng. Jpn. 20, 405—410.

Hassan, M.M., Rahman, A.K.M.S., Loughlin, K.F., 1994. Numerical-simulation of unsteady continuous countercurrent adsorption system with nonlinear adsorption-isotherm. Sep. Technol. 4, 15—26.

Hassan, M.M., Rahman, A.K.M.S., Loughlin, K.F., 1995. Modeling of simulated moving-bed adsorption system — a more precise approach. Sep. Technol. 5, 77—89.

Hidajat, K., Ching, C.B., 1990. Simulation of the performance of a continuous counter-current adsorption system by the method of orthogonal collocation with nonlinear and interacting adsorption-isotherms. Chem. Eng. Res. Des. 68, 104—108.

Hidajat, K., Ching, C.B., Ruthven, D.M., 1986a. Numerical-simulation of a semicontinuous countercurrent adsorption unit for fructose glucose separation. Chem. Eng. J. Biochem. Eng. J. 33, B55—B61.

Hidajat, K., Ching, C.B., Ruthven, D.M., 1986b. Simulated countercurrent adsorption processes — a theoretical-analysis of the effect of subdividing the adsorbent bed. Chem. Eng. Sci. 41, 2953—2956.

Hindmarsh, A.C., 1976. Preliminary Documentation of GEARIB. Solution of Implicit Systems of Ordinary Differential Equations with Banded Jacobians. Report UCID — 30130, Lawrence Livermore Laboratory, Livermore.

Hotier, G., 1996. Physically meaningful modeling of the 3-zone and 4-zone simulated moving bed processes. AIChE J. 42, 154—160.

Kubota, K., Hata, C., Hayashi, S., 1989. A study of a simulated moving bed adsorber based on the axial-dispersion model. Can. J. Chem. Eng. 67, 1025—1029.

Lameloise, M.L., Viard, V., 1993. Modelling and simulation of a glucose-fructose simulated moving bed adsorber. Trans. Int. Chem. Eng 71, 27—32.

Leão, C.P., Rodrigues, A.E., 2004. Transient and steady-state models for simulated moving bed processes: numerical solutions. Comput. Chem. Eng. 28, 1725—1741.

Leitão, A., Rodrigues, A., 1995. Adsorptive processes using "large-pore" materials: analysis of a criterion for equivalence of diffusion-convection "apparent" diffusion and "extended" linear driving force models. Chem. Eng. J. Biochem. Eng. J. 60, 81—87.

Leitão, A., Rodrigues, A., 1996. The performance of the extended linear driving force model to simulate adsorptive processes using large pore materials. In: LeVan, M.D. (Ed.), Proceedings of the 5th International Conference on Fundamentals of Adsorption. Kluwer Academic Publishers, Boston, Massachusetts.

Lim, B.G., Ching, C.B., Tan, R.B.H., 1995. Determination of competitive adsorption isotherms of enantiomers on a dual-site adsorbent. Sep. Technol. 5, 213–228.

Ma, Z., Wang, N.H.L., 1997. Standing wave analysis of SMB chromatography: linear systems. AIChE J. 43, 2488–2508.

Madsen, N.K., Sincovec, R.F., 1979. PDECOL: general collocation software for partial differential equations. ACM Trans. Math. Softw. 5, 326–351.

Mathworks, 2014. MATLAB. http://www.mathworks.com/products/matlab/.

Mazzotti, M., Storti, G., Morbidelli, M., 1994. Robust design of countercurrent adsorption separation processes. 2. Multicomponent systems. AIChE J. 40, 1825–1842.

Mazzotti, M., Pedeferri, M., Morbidelli, M., 1996a. Design of optimal and robust operating conditions for chiral separations using simulated moving beds. In: Proceedings of the Chiral Europe'96 Symposium. Spring Innovations Limited, Stockport, UK.

Mazzotti, M., Storti, G., Morbidelli, M., 1996b. Robust design of countercurrent adsorption separation. 3. Nonstoichiometric systems. AIChE J. 42, 2784–2796.

Mazzotti, M., Storti, G., Morbidelli, M., 1997a. Optimal operation of simulated moving bed units for nonlinear chromatographic separations. J. Chromatogr. A 769, 3–24.

Mazzotti, M., Storti, G., Morbidelli, M., 1997b. Robust design of countercurrent adsorption separation processes. 4. Desorbent in the feed. AIChE J. 43, 64–72.

Minceva, M., Pais, L.S., Rodrigues, A.E., 2003. Cyclic steady state of simulated moving bed processes for enantiomers separation. Chem. Eng. Process. 42, 93–104.

Navarro, A., Caruel, H., Rigal, L., Phemius, P., 1997. Continuous chromatographic separation process: simulated moving bed allowing simultaneous withdrawal of three fractions. J. Chromatogr. A 770, 39–50.

Pais, L.S., Loureiro, J.M., Rodrigues, A.E., 1997a. Modeling, simulation and operation of a simulated moving bed for continuous chromatographic separation of 1,1'-bi-2-naphthol enantiomers. J. Chromatogr. A 769, 25–35.

Pais, L.S., Loureiro, J.M., Rodrigues, A.E., 1997b. Separation of 1,1'-bi-2-naphthol enantiomers by continuous chromatography in simulated moving bed. Chem. Eng. Sci. 52, 245–257.

Pais, L.S., Loureiro, J.M., Rodrigues, A.E., 1998a. Modeling strategies for enantiomers separation by SMB chromatography. AIChE J. 44, 561–569.

Pais, L.S., Loureiro, J.M., Rodrigues, A.E., 1998b. Separation of enantiomers of a chiral epoxide by simulated moving bed chromatography. J. Chromatogr. A 827, 215–233.

Process System Enterprise, 2014. gPROMS. www.psenterprise.com/gproms.

Rahman, A.K.M.S., Hassan, M.M., Loughlin, K.F., 1994. Unsteady-state simulation of sorbex system with nonlinear adsorption-isotherms. Sep. Technol. 4, 27–37.

Rodrigues, A.E., Loureiro, J.M., Lu, Z.P., Pais, L.S., 1996. Modeling and operation of a simulated moving bed for the separation of optical isomers. In: LeVan, M.D. (Ed.), Proceedings of the 5th International Conference on Fundamentals of Adsorption. Kluwer Academic Publishers, Boston, Massachusetts.

Ruthven, D.M., 1984. Principles of Adsorption and Adsorption Processes. John Wiley & Sons, New York.

Ruthven, D.M., Ching, C.B., 1989. Countercurrent and simulated countercurrent adsorption separation processes. Chem. Eng. Sci. 44, 1011–1038.

Ruthven, D.M., Ching, C.B., 1993. Modeling of chromatographic processes. In: Ganetsos, G., Barker, P.E. (Eds.), Preparative and Production Scale Chromatography. Marcel Dekker Inc., New York, pp. 629–672.

Sá Gomes, P., Minceva, M., Rodrigues, A.E., 2006. Simulated moving bed technology: old and new. Adsorption 12, 375–392.

Schmidt-Traub, H., Strube, J., 1996. Dynamic simulation of simulated-moving-bed chromatographic processes. Comput. Chem. Eng. 20, S641–S646.

Storti, G., Baciocchi, R., Mazzotti, M., Morbidelli, M., 1995. Design of optimal operating-conditions of simulated moving-bed adsorptive separation units. Ind. Eng. Chem. Res. 34, 288–301.

Storti, G., Masi, M., Carra, S., Morbidelli, M., 1989a. Optimal-design of multicomponent countercurrent adsorption separation processes involving nonlinear equilibria. Chem. Eng. Sci. 44, 1329–1345.

Storti, G., Masi, M., Morbidelli, M., 1989b. On countercurrent adsorption separation processes. In: Rodrigues, A.E., LeVan, M.D., Tondeur, D. (Eds.), Adsorption: Science and Technology. Kluwer Academic Publishers, The Netherlands, pp. 357–381.

Storti, G., Masi, M., Morbidelli, M., 1993a. Modeling of countercurrent adsorption processes. In: Ganetsos, G., Barker, P.E. (Eds.), Preparative and Production Scale Chromatography. Marcel Dekker, New York, pp. 673–700.

Storti, G., Mazzotti, M., Morbidelli, M., Carra, S., 1993b. Robust design of binary countercurrent adsorption separation processes. AIChE J. 39, 471–492.

Storti, G., Masi, M., Paludetto, R., Morbidelli, M., Carra, S., 1988. Adsorption separation processes – countercurrent and simulated countercurrent operations. Comput. Chem. Eng. 12, 475–482.

Strube, J., Altenhoner, U., Meurer, M., SchmidtTraub, H., Schulte, M., 1997. Dynamic simulation of simulated moving-bed chromatographic processes for the optimization of chiral separations. J. Chromatogr. A 769, 81–92.

Tondeur, D., Bailly, M., 1993. Simulated countercurrent, fixed-bed and column switching schemes. In: Nicoud, R.-M. (Ed.), Simulated Moving Bed: Basics and Applications. Institut National Polytechnique de Lorraine, Nancy, France, pp. 97–117.

Turner, J.R., Mills, P.L., 1990. Evaluation of finite-element collocation techniques for boundary-value ODEs and their application to chemical engineering systems. Presented at the AIChE Annual Meeting, Chicago, USA.

Yun, T., Bensetiti, Z., Zhong, G.M., Guiochon, G., 1997a. Effect of column efficiency on the internal concentration profiles and the performance of a simulated moving-bed unit in the case of a linear isotherm. J. Chromatogr. A 758, 175–190.

Yun, T., Zhong, G.M., Guiochon, G., 1997b. Experimental study of the influence of the flow rates in SMB chromatography. AIChE J. 43, 2970–2983.

Yun, T., Zhong, G.M., Guiochon, G., 1997c. Simulated moving bed under linear conditions: experimental vs. calculated results. AIChE J. 43, 935–945.

Zhong, G.M., Guiochon, G., 1996. Analytical solution for the linear ideal model of simulated moving bed chromatography. Chem. Eng. Sci. 51, 4307–4319.

Zhong, G.M., Guiochon, G., 1997a. Simulated moving bed chromatography. Comparison between the behaviors under linear and nonlinear conditions. Chem. Eng. Sci. 52, 4403–4418.

Zhong, G.M., Guiochon, G., 1997b. Simulated moving bed chromatography. Effects of axial dispersion and mass transfer under linear conditions. Chem. Eng. Sci. 52, 3117–3132.

Zhong, G.M., Guiochon, G., 1998. Steady-state analysis of simulated moving-bed chromatography using the linear, ideal model. Chem. Eng. Sci. 53, 1121–1130.

Zhong, G.M., Smith, M.S., Guiochon, G., 1997. Effect of the flow rates in linear, ideal, simulated moving-bed chromatography. AIChE J. 43, 2960–2969.

# CHAPTER 3

# Design of Simulated Moving Bed for Binary or Pseudo-Binary Separations

The design of a simulated moving bed (SMB) separation to obtain the desired performance involves the choice of various parameters, from the configuration of the unit (number of columns per section, column length, and diameter, particle size) to the operating conditions (switching time, flow rates in each section, and feed concentration). This can be defined by performing extensive simulation work, but this is time-consuming, and so it is useful to have a design methodology that provides an estimation of the optimum operating point, followed by simulation and/or optimization.

In Chapter 1, we presented the design fundamentals of SMB units for binary separations based on the equilibrium theory in which the separation region has a "triangle" shape and therefore is known as the "triangle" theory (Storti et al., 1989, 1993; Mazzotti et al., 1994, 1996, 1997b; Chiang, 1998; Migliorini et al., 2000; Mazzotti, 2006).

Another design method uses the "standing wave" analysis of Lind Wang's group (Ma and Wang, 1997; Xie et al., 2000). The idea is that if adsorption/desorption waves migrate with a velocity equal to the average port velocity, they remain "standing" over a switching time, enabling high purity and yield. The separation occurs when the flow rates in the four zones of the SMB are such that the average port velocity is higher than the migration speed of the more-retained species and lower than the migration speed of the less-retained component.

Considering an already existing SMB unit (with defined geometry and number of columns), the first step to obtain an efficient separation is to define the basic operating conditions. This can be made considering first the true moving bed (TMB) unit and then applying the equivalence between TMB and SMB (Eqns (1.22)−(1.26) in Chapter 1).

A quantitative analysis of the TMB unit was presented in detail in Chapter 1, Section 1.1.4.1.

Taking into account the role of each section of the TMB unit for the separation of A and B (A being the more-retained species and B the less-retained one, Figure 1.5, Chapter 1), a set of constraints can be written in terms of net fluxes (ratio of flux of species transported by the liquid phase and flux of species transported by the solid phase), which will limit the feasible operating region to allow a complete separation; the fluxes of each species must be set so that in section I, the more-retained component (A) moves with the liquid (upward direction), to regenerate the solid; in sections II and III, the more-retained component (A) moves with the solid and the less-retained one (B) moves with the liquid,

*Simulated Moving Bed Technology*
ISBN 978-0-12-802024-1

and in section IV, the less-retained component (B) moves with the solid (downward), to regenerate the desorbent (or eluent) before being recycled.

$$\text{Section I:} \quad \frac{Q_1 C_{A1}}{Q_s q_{A1}} > 1 \tag{3.1}$$

$$\text{Section II:} \quad \frac{Q_2 C_{B2}}{Q_s q_{B2}} > 1 \cap \frac{Q_2 C_{A2}}{Q_s q_{A2}} < 1 \tag{3.2}$$

$$\text{Section III:} \quad \frac{Q_3 C_{B3}}{Q_s q_{B3}} > 1 \cap \frac{Q_3 C_{A3}}{Q_s q_{A3}} < 1 \tag{3.3}$$

$$\text{Section IV:} \quad \frac{Q_4 C_{B4}}{Q_s q_{B4}} < 1 \tag{3.4}$$

where $C_{i,j}$ is the bulk fluid concentration of species $i$ in section $j$ and $q_{i,j}$ is the average solid concentration of species $i$ in section $j$.

Based on these constraints (Eqns (3.1)–(3.4)), several design methodologies have been developed, which are typically approximated and/or graphical, as for example, the plate theory and the McCabe–Thiele diagrams (Ruthven and Ching, 1989) or the analytical solutions for a linear adsorption isotherm system in the presence of mass transfer resistances (Silva et al., 2004) or for reactive systems (Minceva et al., 2005).

## 3.1 EQUILIBRIUM THEORY FOR LINEAR ISOTHERMS (TRIANGLE THEORY)

In the triangle theory approach, the term $\frac{q_{i,j}}{C_{i,j}}$ is simplified by the assumption that the adsorption equilibrium is reached everywhere at every time ($q_{i,j} = q_{i,j}^*$) (Helfferich, 1967; Klein et al., 1967; Tondeur and Klein, 1967). So, when adsorption is described by linear isotherms,

$$q_A^* = K_A C_A \tag{3.5}$$

$$q_B^* = K_B C_B \tag{3.6}$$

where $K$ is Henry's adsorption constant, it is possible to write constraints 3.1–3.4 as follows:

$$\text{Section I:} \quad \frac{Q_1}{Q_s} > K_A \tag{3.7}$$

$$\text{Section II:} \quad \frac{Q_2}{Q_s} > K_B \cap \frac{Q_2}{Q_s} < K_A \tag{3.8}$$

$$\text{Section III:} \quad \frac{Q_3}{Q_s} > K_B \cap \frac{Q_3}{Q_s} < K_A \tag{3.9}$$

$$\text{Section IV: } \frac{Q_4}{Q_s} < K_B \tag{3.10}$$

Usually, the fluid and solid velocities in each section are combined into one dimensionless parameter, such as $m_j = \frac{Q_j}{Q_s}$ from Morbidelli and his co-workers (Mazzotti et al., 1997a) or $\gamma_j = \frac{1-\varepsilon}{\varepsilon}\left(\frac{Q_j}{Q_s}\right)$ as used by Ruthven and Ching, (1989). In Figure 3.1, separation and regeneration regions are represented in the $\gamma_2 \times \gamma_3$ plane and the $\gamma_1 \times \gamma_4$ plane, respectively, for the case of linear isotherms. Depending on the values of $\gamma_2$ and $\gamma_3$, four regions can be identified: one region in which none of the extract and raffinate streams are pure, two in which only one of the two streams is pure, and one region in which both extract and raffinate streams are pure (the separation region). The SMB must be operated with the conditions inside this last region, because obviously the natural goal is to obtain simultaneously pure extract and raffinate streams. In addition, the system should be operated using operating conditions near the vertex point of this separation region. The vertex point is the point at the boundary of the separation region most distant from the diagonal $\gamma_3 = \gamma_2$, representing the best operating conditions in terms of system productivity.

Combining Eqns (3.7)–(3.10) with the global balances and using a safety factor, $\beta, 1 < \beta < \sqrt{\alpha}$, where $\alpha = \frac{K_A}{K_B}$, it is possible to determine the basic operating parameters for a successful separation:

$$Q_S = \frac{Q_F}{\frac{K_A}{\beta} - K_B\beta} \tag{3.11}$$

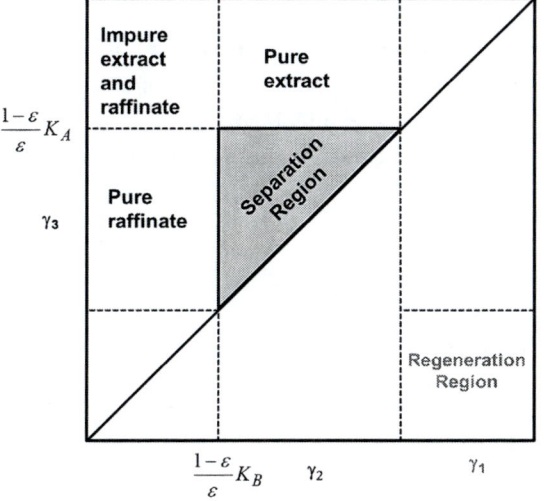

**Figure 3.1** Triangle theory: separation and regeneration regions for linear isotherms, where A is the more-retained species and B is the less-retained one.

$$Q_4 = Q_{Rec} = Q_S \frac{K_B}{\beta} \tag{3.12}$$

$$Q_X = Q_S(K_A - K_B)\beta \tag{3.13}$$

$$Q_R = Q_S(K_A - K_B)\frac{1}{\beta} \tag{3.14}$$

The safety factor is considered, because the triangle theory states that the critical flow-rate ratios required to achieve separation depend exclusively on the equilibrium data, neglecting the effects of mass transfer resistance. However, the presence of mass transfer resistance can affect significantly the performance of the SMB operation, reducing the size of the separation region and modifying the optimum SMB operating conditions.

The triangle theory is a very useful design methodology that allows easy estimation of the boundaries of separation/regeneration regions, which then can be refined using more precise models that take into account the resistances to mass transfer and axial dispersion such as those presented in Chapter 2, Section 2.1.1. Because of its importance, the equilibrium theory was extended for the design of SMB units under reduced purity requirements, in which the separation triangle boundaries are stretched to account for different extract and/or raffinate purities (Kaspereit et al., 2007; Rajendran, 2008).

## 3.2 EXTENSION TO NONLINEAR ISOTHERMS

In the case of nonlinear isotherms described by the binary Langmuir model,

$$q_i^* = \frac{q_m b_i C_i}{1 + \sum\limits_i b_i C_i} \tag{3.15}$$

where $i = A, B$ it is possible to construct the separation region for these systems based on the equilibrium theory. In the case of a desorbent very weakly retained, the "triangle" shown in Figure 3.2 is constructed as follows (Mazzotti et al., 1996).

If $C_A^F$ and $C_B^F$ are the feed concentrations of species A and B, respectively, and $\lambda_i = q_m b_i$,

$$\lambda_A = m_{1,\min} < m_1 < \infty \tag{3.16}$$

$$m_{2,\min}(m_2, m_3) < m_2 < m_3 < m_{3,\max}(m_2, m_3) \tag{3.17}$$

$$0 < m_4 < m_{4,\max}(m_2, m_3)$$

$$= \frac{1}{2}\left\{ \lambda_B + m_3 + b_B C_B^F(m_3 - m_2) - \sqrt{\left[\lambda_B + m_3 + b_B C_B^F(m_3 - m_2)\right]^2 - 4\lambda_B m_3} \right\} \tag{3.18}$$

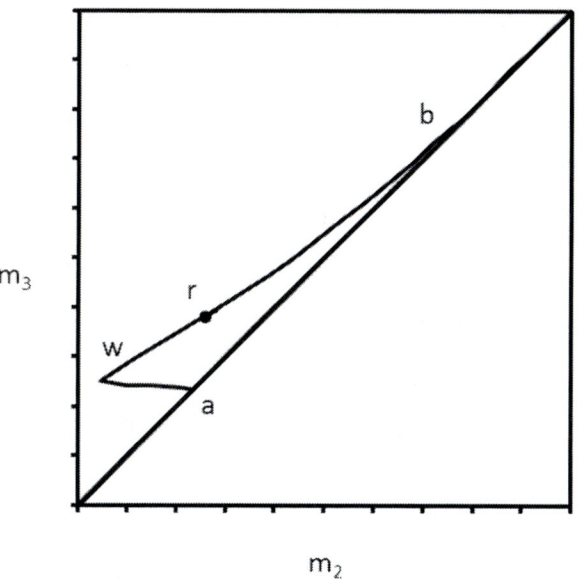

**Figure 3.2** Region of complete separation in the ($m_2$, $m_3$) plane for a nonadsorbable desorbent. *(Reprinted from AIChE Journal 42, Mazzotti, M., Storti, G., Morbidelli, M. Robust Design of Countercurrent Adsorption Separation: 3. Nonstoichiometric Systems, pp. 2784–2796, Copyright (1996), with permission from John Wiley & Sons Inc.)*

Boundaries of the complete separation region in the ($m_2$,$m_3$) plane are as follows: Straight line wr:

$$\left[\lambda_A - \omega_G\left(1 + b_A C_A^F\right)\right]m_2 + b_A C_A^F \omega_G m_3 = \omega_G(\lambda_A - \omega_G) \qquad (3.19)$$

Straight line wa:

$$\left[\lambda_A - \lambda_B\left(1 + b_A C_A^F\right)\right]m_2 + b_A C_A^F \lambda_B m_3 = \lambda_B(\lambda_A - \lambda_B) \qquad (3.20)$$

Curve rb:

$$m_3 = m_2 + \frac{\left(\sqrt{\lambda_A} - \sqrt{m_2}\right)^2}{b_A C_A^F} \qquad (3.21)$$

Straight line ab:

$$m_3 = m_2 \qquad (3.22)$$

The coordinates of the intersection points are given by:

$$\text{Point a: } (\lambda_B, \lambda_B) \qquad (3.23)$$

$$\text{Point b: } (\lambda_A, \lambda_A) \qquad (3.24)$$

$$\text{Point r: } \left(\frac{\omega_G^2}{\lambda_A}, \frac{\omega_G[\omega_F(\lambda_A - \omega_G)(\lambda_A - \lambda_B) + \lambda_B \omega_G(\lambda_A - \omega_G)]}{\lambda_A \lambda_B(\lambda_A - \omega_F)}\right) \qquad (3.25)$$

$$\text{Point w:} \left( \frac{\lambda_B \omega_G}{\lambda_A}, \frac{\omega_G[\omega_F(\lambda_A - \lambda_B) + \lambda_B(\lambda_B - \omega_F)]}{\lambda_B(\lambda_A - \omega_F)} \right) \tag{3.26}$$

with $\omega_G > \omega_F > 0$, which are given by the roots of the following quadratic equation:

$$\left(1 + b_A C_A^F + b_B C_B^F\right)\omega^2 - \left[\lambda_B\left(1 + b_A C_A^F\right) + \lambda_A\left(1 + b_B C_B^F\right)\right]\omega + \lambda_A\lambda_B = 0 \tag{3.27}$$

These equations are also valid when a modified Langmuir isotherm, Linear + Langmuir (L + L), (Eqn (3.28)), is considered:

$$q_i^* = HC_i + \frac{q_m b_i C_i}{1 + \sum_i b_i C_i} \tag{3.28}$$

For this type of isotherm, the complete separation region, given by Eqns (3.16)−(3.27), is shifted using the relation

$$m_j^{L+L} = m_j^L + H \tag{3.29}$$

where $m_j^L$ is the value obtained when only the Langmuir term is considered and $H$ is the linear coefficient in the Linear + Langmuir isotherm.

## 3.3 MASS TRANSFER EFFECTS

The SMB design can be affected by the presence of mass transfer resistances and so, in such cases, the triangle theory can give only initial guesses for a feasible operating point. To find more detailed separation/regeneration regions and consequently to find the optimum flow rate in each section, the optimum section length, and the optimum switching time that guarantee the required purity and the highest yield, successive simulations should be performed using realistic mathematical models (see Section 2.1 in Chapter 2). In Figure 3.3, the effect of mass transfer coefficient on flow–rate ratio constraints of sections II and III is shown for the separation of chiral epoxide enantiomers in microcrystalline cellulose triacetate using methanol as eluent (Pais et al., 1998). When the mass transfer coefficient $k$ decreases (higher mass transfer resistance) the separation region becomes smaller. The take-home message is: the separation region is reduced when mass transfer resistances become important.

## 3.4 ANALYTICAL SOLUTION OF TRUE MOVING BED FOR LINEAR SYSTEMS

Let us consider a TMB system operating in steady-state. Model equations include the mass balance in the fluid phase of a volume element of section $j$ and the mass balance

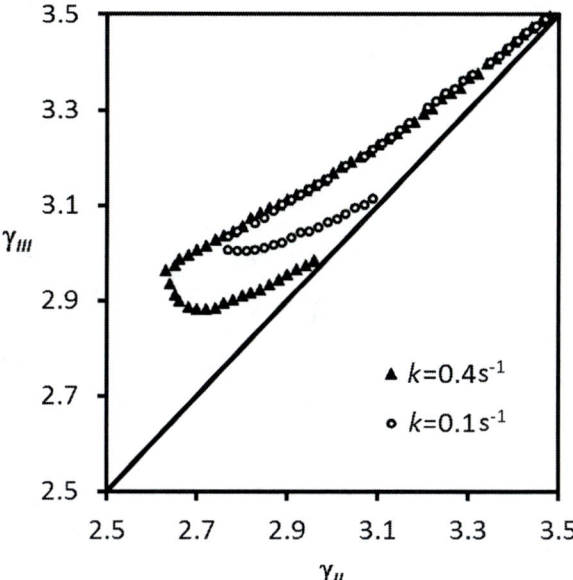

**Figure 3.3** Separation regions for 90% purity criteria in the presence of mass transfer resistances for the separation of chiral epoxide enantiomers (Pais et al., 1998). (Reprinted from *Journal of Chromatography A 827*, Pais, L.S., Loureiro, J.M., Rodrigues, A.E., Separation of enantiomers of a chiral epoxide by simulated moving bed chromatography, pp. 215–233, Copyright (1998), with permission from Elsevier.)

for the solid phase in the same volume element together with the adsorption equilibrium isotherm assumed to be linear.

Steady-state model equations assuming plug flow for the fluid and solid phases are:

$$\gamma_j \frac{dC_{i,j}}{dx} + \frac{1-\varepsilon}{\varepsilon} \alpha_i \left( q_{i,j}^* - q_{i,j} \right) = 0 \tag{3.30}$$

$$\frac{dq_{i,j}}{dx} + \alpha_i \left( q_{i,j}^* - q_{i,j} \right) = 0 \tag{3.31}$$

$$q_{i,j}^* = K_i C_{i,j} \tag{3.32}$$

where $C_{i,j}$ is the fluid-phase concentration of species $i$ in section $j$, $q_{i,j}^*$ is the solid concentration at the particle/fluid interface in equilibrium with $C_{i,j}$, $K_i$ is the slope of the adsorption equilibrium isotherm, $q_{i,j}$ is the average adsorbed concentration in the particle, $x = \frac{z}{L}$ is the dimensionless axial coordinate, $\varepsilon$ is the bed porosity. The dimensionless model parameters are $\gamma_j = \frac{v_j}{u_s}$, the ratio of interstitial velocities of fluid and solid phases, and $\alpha_i = \frac{L k_{h,i}}{u_s}$, the number of mass-transfer units with all sections with the same length $L$. We recall that $k_{h,i} = k_{p,i}/K_i$ and $k_{p,i} = 15 D_{p,i}/R_p^2$, where $D_{p,i}$ is the pore diffusion coefficient of component $i$.

The boundary conditions are:

$$Q_{IV} C_{i,IV,1} = Q_I C_{i,I,0} \tag{3.33}$$

$$C_{i,I,1} = C_{i,II,0} \tag{3.34}$$

$$Q_{II}C_{i,II,1} + Q_F C_{i,F} = Q_{III}C_{i,III,0} \tag{3.35}$$

$$C_{i,III,1} = C_{i,IV,0} \tag{3.36}$$

$$q_{i,j-1,1} = q_{i,j,0} \tag{3.37}$$

We should note that if $j = 1$ then $j - 1$ is section IV.

The node mass balances are:

$$Q_{IV} + Q_D = Q_I \quad \text{(Eluent or desorbent node)} \tag{3.38}$$

$$Q_I - Q_X = Q_{II} \quad \text{(Extract node)} \tag{3.39}$$

$$Q_{II} + Q_F = Q_{III} \quad \text{(Feed node)} \tag{3.40}$$

$$Q_{III} - Q_R = Q_{IV} \quad \text{(Raffinate node)} \tag{3.41}$$

Combining Eqns (3.30) and (3.31) and integrating from $x = 0$ up to a position $x$ in section $j$, we get the operating line connecting the average solid concentration at a position $x$ in section $j$ with the fluid-phase concentration at the same position, that is,

$$q_{i,j}(x) = q_{i,j,0} + \frac{\gamma_j}{v}\left(C_{i,j}(x) - C_{i,j,0}\right) \tag{3.42}$$

where $v = \frac{1-\varepsilon}{\varepsilon}$.

The fluid-phase concentration profile in section $j$ is then calculated from:

$$\frac{\gamma_j}{\alpha_i}\frac{dC_{i,j}}{dx} + M_{i,j}C_{i,j} = N_{i,j} \tag{3.43}$$

with $M_{i,j} = vK_i - \gamma_j$ and $N_{i,j} = vq_{i,j,0} - \gamma_j C_{i,j,0}$.

The solution is

$$C_{i,j}(x) = \frac{v\left(K_i C_{i,j,0} - q_{i,j,0}\right)}{M_{i,j}}e^{-\alpha_i M_{i,j}x/\gamma_j} + \frac{N_{i,j}}{M_{i,j}} \tag{3.44}$$

The concentration profiles for the fluid- and solid-phase concentrations are:

$$C_{i,j}(x) = C_{i,j,0} + v\frac{q_{i,j,0} - K_i C_{i,j,0}}{vK_i - \gamma_j}\left(1 - e^{-\alpha_i(vK-\gamma_j)x/\gamma_j}\right) \tag{3.45}$$

$$q_{i,j}(x) = q_{i,j,0} + \gamma_j\frac{q_{i,j,0} - K_i C_{i,j,0}}{vK_i - \gamma_j}\left(1 - e^{-\alpha_i(vK-\gamma_j)x/\gamma_j}\right) \tag{3.46}$$

To calculate the concentrations at the beginning of each section, we need to use the above equations at the end of each section and connect all sections by the node mass balances.

We have

$$C_{i,j,1} = A_{i,j}C_{i,j,0} + B_{i,j}q_{i,j,0} \tag{3.47}$$

$$q_{i,j,1} = q_{i,j,0} + D_j C_{i,j,1} - D_j C_{i,j,0} \tag{3.48}$$

where

$$A_{i,j} = \frac{v K_i e^{\alpha_i M_{i,j}/\gamma_j} - \gamma_j}{M_{i,j}} \tag{3.49}$$

$$B_{i,j} = v \frac{1 - e^{\alpha_i M_{i,j}/\gamma_j}}{M_{i,j}} \tag{3.50}$$

$$D_j = \frac{\gamma_j}{v} \tag{3.51}$$

The fluid and solid concentrations at the end of each section can be eliminated by using the node mass balances:

Section I

$$C_{i,I,1} = A_{i,I}C_{i,I,0} + B_{i,I}q_{i,I,0} = C_{i,II,0} \tag{3.52}$$

$$q_{i,I,1} = q_{i,I,0} + D_I C_{i,II,0} - D_I C_{i,I,0} = q_{i,II,0} \tag{3.53}$$

Section II

$$C_{i,II,1} = \frac{\gamma_{III}}{\gamma_{II}} C_{i,III,0} - \frac{\gamma_F}{\gamma_{II}} C_{i,F} = A_{i,II} C_{i,II,0} + B_{i,II} q_{i,II,0} \tag{3.54}$$

$$q_{i,II,1} = q_{i,III,0} = q_{i,II,0} + D_{II} \left( \frac{\gamma_{III}}{\gamma_{II}} C_{i,III,0} - \frac{\gamma_F}{\gamma_{II}} C_{i,F} \right) - D_{II} C_{i,II,0} \tag{3.55}$$

Section III

$$C_{i,IV,0} = C_{i,III,1} = A_{i,III} C_{i,III,0} + B_{i,III} q_{i,III,0} \tag{3.56}$$

$$q_{i,III,1} = q_{i,IV,0} = q_{i,III,0} + D_{III} C_{i,IV,0} - D_{III} C_{i,III,0} \tag{3.57}$$

Section IV

$$C_{i,IV,1} = \frac{\gamma_I}{\gamma_{IV}} C_{i,I,0} = A_{i,IV} C_{i,IV,0} + B_{i,IV} q_{i,IV,0} \tag{3.58}$$

$$q_{i,IV,1} = q_{i,I,0} = q_{i,IV,0} + D_{IV} \frac{\gamma_I}{\gamma_{IV}} C_{i,I,0} - D_{IV} C_{i,IV,0} \tag{3.59}$$

Equations (3.52)−(3.59) can be written in matrix form, $\mathbf{P} \cdot \mathbf{X} = \mathbf{Q}$ to get the variables $\mathbf{X}$, the fluid and solid concentrations at the beginning of each section:

$$
\begin{bmatrix}
A_{i,I} & -1 & 0 & 0 & B_{i,I} & 0 & 0 & 0 \\
0 & A_{i,II} & -\dfrac{\gamma_{III}}{\gamma_{II}} & 0 & 0 & B_{i,II} & 0 & 0 \\
0 & 0 & A_{i,III} & -1 & 0 & 0 & B_{i,III} & 0 \\
-\dfrac{\gamma_I}{\gamma_{IV}} & 0 & 0 & A_{i,IV} & 0 & 0 & 0 & B_{i,IV} \\
-D_I & D_I & 0 & 0 & 1 & -1 & 0 & 0 \\
0 & -D_{II} & D_{II}\dfrac{\gamma_{III}}{\gamma_{II}} & 0 & 0 & 1 & -1 & 0 \\
0 & 0 & -D_{III} & D_{III} & 0 & 0 & 1 & -1 \\
D_{IV}\dfrac{\gamma_I}{\gamma_{IV}} & 0 & 0 & -D_{IV} & -1 & 0 & 0 & 1
\end{bmatrix}
\cdot
\begin{bmatrix}
C_{i,I,0} \\
C_{i,II,0} \\
C_{i,III,0} \\
C_{i,IV,0} \\
q_{i,I,0} \\
q_{i,II,0} \\
q_{i,III,0} \\
q_{i,IV,0}
\end{bmatrix}
=
\begin{bmatrix}
0 \\
-C_{i,F}\dfrac{\gamma_F}{\gamma_{II}} \\
0 \\
0 \\
0 \\
D_{II}\dfrac{\gamma_F}{\gamma_{II}}C_{i,F} \\
0 \\
0
\end{bmatrix}
$$

From the solution of this system of algebraic equations, the liquid- and solid-phase concentrations at the beginning of each section are obtained, which are then used in Eqns (3.45) and (3.46) to determine the internal concentration profiles. The liquid-phase concentrations of the extract ($C_{i,X} = C_{i,II,0}$) and raffinate ($C_{i,R} = C_{i,IV,0}$) can also be used to calculate the SMB performance parameters (defined in Section 2.2 in Chapter 2).

The analytical solution was applied to model the separation of glucose and fructose (Silva et al., 2004). The model parameters and operating conditions used are given in Table 3.1. The obtained internal liquid-phase concentration profile is shown in Figure 3.4 in comparison with the one obtained using the TMB steady-state model (Section 2.1.2 in Chapter 2).

The separation regions for different extract and raffinate purities, 95% and 98%, determined using the analytical solution and the same operating parameters (Table 3.1) are shown

**Table 3.1** Glucose/Fructose SMB Model Parameters and Operating Conditions

| SMB Geometry | Model Parameters | Operating Conditions |
|---|---|---|
| $L_c = 29$ cm | $\varepsilon = 0.4$ | $T = 50\,°C$ |
| $d_c = 2.6$ cm | $k_{hGl} = 1.86$ min$^{-1}$ | $t^* = 3.3$ min |
| Number of columns: 12 | $k_{hFr} = 1.31$ min$^{-1}$ | $Q_D = 11.59$ cm$^3$/min |
| Configuration: 3-3-3-3 | $K_{Gl} = 0.27$ m$^3$/kg | $Q_X = 10.79$ cm$^3$/min |
|  | $K_{Fr} = 0.53$ m$^3$/kg | $Q_F = 6.95$ cm$^3$/min |
|  |  | $Q_R = 7.67$ cm$^3$/min |
|  |  | $Q_I^{SMB} = 36.47$ cm$^3$/min |

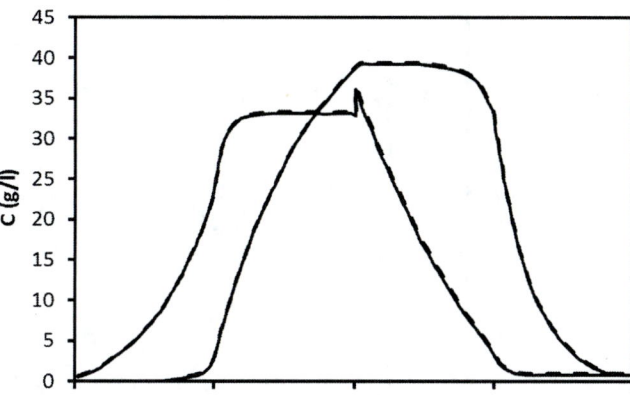

**Figure 3.4** Comparison of the liquid-phase concentration profiles. Solid line, calculated by the analytical solution; dashed line, simulated using the TMB steady-state model. (Reprinted from Industrial and Engineering Chemistry Research 43, Silva, V.M.T., Minceva, M., Rodrigues, A.E., Novel analytical solution for a simulated moving bed in the presence of mass-transfer resistance, pp. 4494–4502, Copyright (2004), with permission from American Chemical Society.)

in Figure 3.5. It can be seen that even for purities lower than 100% the separation region is reduced owing to mass transfer resistances. When the mass transfer resistances are negligible, lower purities are obtained within separation regions bigger than the original triangle from the equilibrium theory constructed to 100% purities. This is shown in Figure 3.6.

In the presence of mass transfer resistance, the minimum flow rate in section I will be higher than the one calculated from the equilibrium theory to be able to get a separation region. That region will increase when $\gamma_I$ increases and so will the productivity. This is

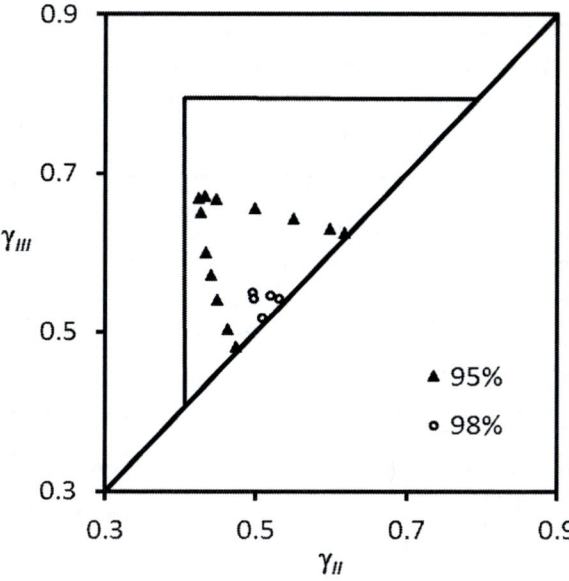

**Figure 3.5** Separation regions for different extract and raffinate purities in the case of mass transfer limitations: $\alpha_{Fr} = 12.95$ and $\alpha_{Gl} = 18.40$. (Reprinted from Industrial and Engineering Chemistry Research 43, Silva, V.M.T., Minceva, M., Rodrigues, A.E., Novel analytical solution for a simulated moving bed in the presence of mass-transfer resistance, pp. 4494–4502, Copyright (2004), with permission from American Chemical Society.)

**Figure 3.6** Separation regions for different extract and raffinate purities in the case of negligible mass transfer resistance. *(Reprinted from Industrial and Engineering Chemistry Research 43, Silva, V.M.T., Minceva, M., Rodrigues, A.E., Novel analytical solution for a simulated moving bed in the presence of mass-transfer resistance, pp. 4494–4502, Copyright (2004), with permission from American Chemical Society.)*

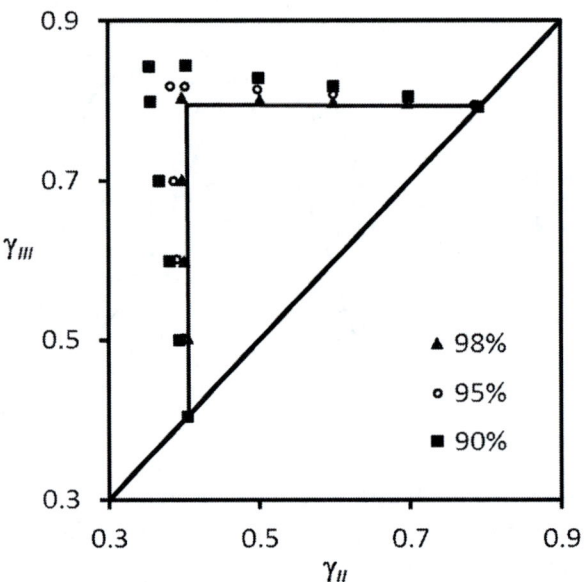

shown in Figure 3.7 and will be discussed in more detail in the next section in which the concept of "separation volume" is introduced.

## 3.5 THE CONCEPT OF SEPARATION VOLUME

The design of an SMB process based on the equilibrium theory (Section 3.1) results in the explicit definition of the boundaries of the separation region in terms of solid and fluid flow rates, for which complete separation is obtained. This theory predicts that,

**Figure 3.7** Influence of $\gamma_I$ on productivity for several values of $\gamma_{IV}$. *(Reprinted from Industrial and Engineering Chemistry Research 43, Silva, V.M.T., Minceva, M., Rodrigues, A.E., Novel analytical solution for a simulated moving bed in the presence of mass-transfer resistance, pp. 4494–4502, Copyright (2004), with permission from American Chemical Society.)*

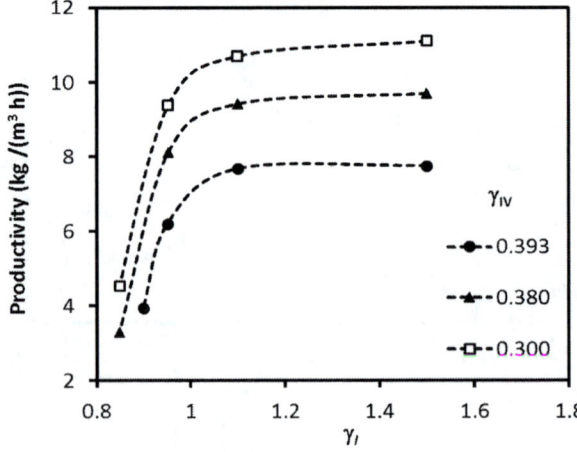

for any value of $\gamma_1$ and $\gamma_4$ (provided that Eqns (3.7) and (3.10) are obeyed), the separation region $\gamma_2 \times \gamma_3$ is constant. However, when mass transfer effects are present, the adsorbed-phase concentration can no longer be considered equal to the value in equilibrium with bulk-phase concentration. The simplifications applied in the equilibrium theory are not valid anymore and new constraints dependent on the mass transfer rate constants are required.

To study the influence of mass transfer resistance, not only on the restrictions of sections II and III, but also on the restrictions of sections I and IV, the "separation volume" methodology was developed (Azevedo and Rodrigues, 1999b), in which numerical simulation is performed to find a three-dimensional separation volume, instead of a two-dimensional separation area. Within this methodology two possibilities can be obtained: if the flow rate in section I ($\gamma_1$) is fixed, the design leads to a ($\gamma_2 \times \gamma_3 \times \gamma_4$) volume for a given separation requirement; if the flow rate in section IV ($\gamma_4$) is fixed, the design will result in a ($\gamma_2 \times \gamma_3 \times \gamma_1$) volume for a given separation requirement.

The "separation volume" was already applied for the design of an SMB for the separation of sugars (Azevedo and Rodrigues, 1999b) and p-xylene (Minceva and Rodrigues, 2002) and for chiral separation (Rodrigues and Pais, 2004).

The application of this methodology will be exemplified by the separation of fructose and glucose using a Dowex Monosphere cationic resin. A steady-state TMB model was employed, which assumes axially dispersed flow for the fluid phase, plug flow for the solid phase, external mass transfer resistance expressed by the film model, intraparticle mass transfer represented by a bilinear driving force model, and linear isotherms (Azevedo and Rodrigues, 1999a). The model equations are the following:

Mass balance to the bulk fluid phase:

$$\frac{\gamma_j}{Pe_j} \frac{\partial^2 C_{i,j}}{\partial x^2} - \gamma_j \frac{\partial C_{i,j}}{\partial x} - \frac{(1-\varepsilon)}{\varepsilon} \left[ \frac{Bi_{mj}}{5 + Bi_{mj}} \alpha_{pj} \left( C_{i,j} - \overline{C}_{pi,j} \right) \right] = 0 \qquad (3.60)$$

Mass balance to intraparticle fluid phase:

$$\frac{\partial \overline{C}_{pi,j}}{\partial x} + \frac{Bi_{mj}}{5 + Bi_{mj}} \frac{\alpha_{pj}}{\varepsilon_p} \left( C_{i,j} - \overline{C}_{pi,j} \right) + \frac{\alpha_{\mu j}}{\varepsilon_p} \rho_p \left[ \frac{K_i}{\rho_p} \overline{C}_{pi,j} - \overline{q}_{i,j} \right] = 0 \qquad (3.61)$$

Mass balance in intraparticle solid phase:

$$\frac{\partial \overline{q}_{i,j}}{\partial x} + \alpha_{\mu j} \rho_p \left[ \frac{K_i}{\rho_p} \overline{C}_{pi,j} - \overline{q}_{i,j} \right] = 0 \qquad (3.62)$$

In these equations $C_{i,j}$ is the fluid-phase concentration of species $i$ in section $j$, $\overline{C}_{pi,j}$ is the average concentration of component $i$ in the pores over the pellet in section $j$, $\overline{q}_{i,j}$ is the average solid concentration of component $i$ over the pellet in section $j$, $K_i$ is the slope of the adsorption equilibrium isotherm, $x = \frac{z}{L}$ is the dimensionless axial coordinate, $\varepsilon$ is the bed porosity, $\varepsilon_p$ is the pellet porosity, and $\rho_p$ is the pellet density.

Boundary conditions for section $j$:

$$x = 0 \quad C_{i,j,0} - \frac{1}{Pe_j} \frac{\partial C_{i,j,0}}{\partial x} = C_{i,j}^{in} \qquad (3.63)$$

$$x = 1 \quad \frac{\partial C_{i,j,1}}{\partial x} = 0 \qquad (3.64a)$$

$$\overline{C}_{pi,j,1} = \overline{C}_{pi,(j+1),0} \qquad (3.64b)$$

$$\overline{q}_{i,j,1} = \overline{q}_{i,(j+1),0} \qquad (3.64c)$$

Note that when $j = 1$, $j - 1$ is section IV.

The dimensionless model parameters are:

$$\text{Ratio between fluid and solid interstitial velocities} \quad \gamma_j = \frac{v_j}{u_s} \qquad (3.65a)$$

$$\text{Peclet number} \quad Pe = \frac{v_j L_j}{D_{L_j}} \qquad (3.65b)$$

$$\text{Mass Biot number} \quad Bi_{mj} = \frac{k_{fj} R_p}{D_{pe}} \qquad (3.65c)$$

$$\text{Number of macropore mass-transfer units} \quad \alpha_{pj} = \frac{k_p L_j}{u_s} \qquad (3.65d)$$

$$\text{Number of microparticle mass-transfer units} \quad \alpha_{\mu j} = \frac{k_\mu L_j}{u_s} \qquad (3.65e)$$

The inlet concentrations to each section are calculated from:

$$\text{Eluent node} \quad C_I^{in} = \frac{Q_{IV}}{Q_I} C_{IV,1} \qquad (3.66a)$$

$$\text{Extract node} \quad C_{II}^{in} = C_{I,1} \qquad (3.66b)$$

$$\text{Feed node} \quad C_{III}^{in} = \frac{Q_{II}}{Q_{III}} C_{II,1} + \frac{Q_F}{Q_{III}} C_F \qquad (3.66c)$$

$$\text{Raffinate node} \quad C_{IV}^{in} = C_{III,1} \qquad (3.66d)$$

Figure 3.8 shows two ($\gamma_2 \times \gamma_3 \times \gamma_1$) separation volumes for 90% purities, determined at 30 °C and 50 °C, by fixing $\gamma_4$ to 70% of the maximum value given by the equilibrium theory (Azevedo and Rodrigues, 2001). The model parameters employed in the simulations are shown in Table 3.2. The change in the size and shape of the separation region with the change in $\gamma_1$ can be observed.

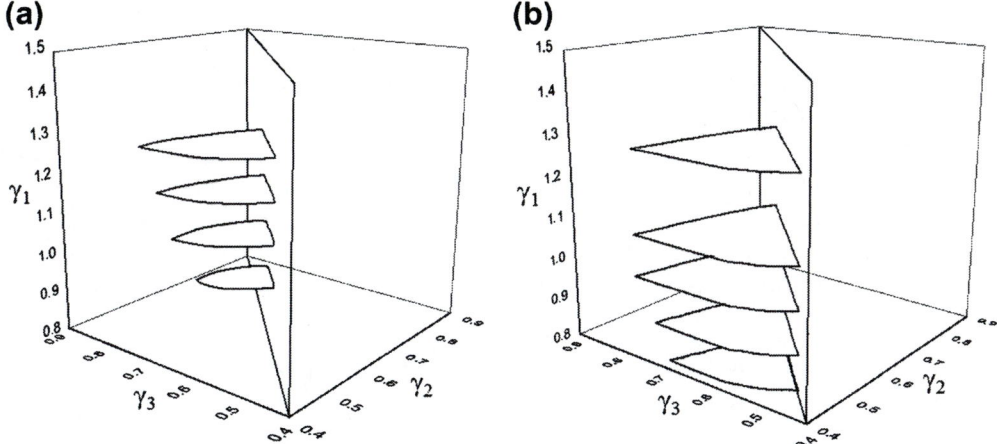

**Figure 3.8** Separation volumes (both product purities above 90%) at (a) 30 °C and (b) 50 °C for fructose–glucose separation on a LICOSEP 12–26 SMB pilot unit using Dowex Monosphere cationic resin. *(Reprinted from AIChE Journal 47, Azevedo, D.C.S., Rodrigues, A.E. Fructose-glucose separation in a SMB pilot unit: Modeling, simulation, design, and operation, pp. 2042–2051, Copyright (2001), with permission from John Wiley & Sons Inc.)*

**Table 3.2** Model Parameters for Fructose–Glucose Separation on a LICOSEP 12–26 SMB Pilot Unit Using Dowex Monosphere Cationic Resin

| Parameter | $T = 30\,°C$ | | $T = 50\,°C$ | |
|---|---|---|---|---|
| | Glucose | Fructose | Glucose | Fructose |
| $Pe$ | | 500 | | 500 |
| $Bi_m$ | | 200 | | 300 |
| $\varepsilon$ | | 0.4 | | 0.4 |
| $\varepsilon_p$ | | 0.1 | | 0.1 |
| $K$ | 0.18 | 0.5 | 0.17 | 0.43 |
| $k_p \times 10^2\ (s^{-1})$ | 3.33 | 3.33 | 4.17 | 4.17 |
| $k_\mu \times 10^2\ (s^{-1})$ | 1.33 | 1.33 | 2.17 | 2.17 |

## 3.6 OPTIMIZATION OF THE SIMULATED MOVING BED OPERATION

The optimization of an SMB unit considers the selection of appropriate operating conditions (flow rates in each zone, switching time, temperature) and/or geometric parameters (number of columns, number of columns per section, column dimensions) that minimize or maximize a given objective function, taking into account previously defined constraints (e.g., product purity and/or recovery).

Some studies have been carried out to develop optimization strategies for SMB operation (Storti et al., 1988; Gentilini et al., 1998; Karlsson et al., 1999; Beste et al., 2000;

Dunnebier et al., 2000; Klatt et al., 2000; Toumi et al., 2003; Minceva and Rodrigues, 2005).

Minceva and Rodrigues (2005) proposed a two-level optimization procedure based on the "separation volume" methodology presented in Section 3.5 and on an equivalent TMB modeling strategy. The employed mathematical model considers axial dispersed plug flow for the liquid phase and plug flow for the solid phase, linear driving force for the intraparticle mass transfer rate, and multicomponent adsorption equilibria described by the Langmuir isotherm. This procedure consists of two consecutive optimizations, each with a single objective function. In each level the product (extract and raffinate) purities and recoveries are imposed as constraints.

The objective function selected for the first level is the maximization of the SMB productivity ($PR$) and is given by:

$$\max J = PR = \frac{Q_F C_F}{V_{ads}} \tag{3.67}$$

To start the first-level optimization, the flow rate in section I ($\gamma_1$) is fixed and a low value is given for the flow rate of section IV ($\gamma_4$). The flow rates of sections II and III ($\gamma_2$, $\gamma_3$) are then optimized to obtain the maximum productivity for the fixed ($\gamma_1$, $\gamma_4$) pair. The maximum productivity is related to the difference between $\gamma_2$ and $\gamma_3$, that is, $\max(PR) = \max(QF) = \max(\gamma_3 - \gamma_2)$. The values are stored and the procedure is repeated keeping the same $\gamma_1$ and increasing successively $\gamma_4$ until the constraints can no longer be satisfied. Then, the value of $\gamma_1$ is increased and the same process is carried out. The flow sheet of the first-level optimization procedure is shown in Figure 3.9.

The objective function of the second level is the minimization of the desorbent consumption ($DC$) and is given by:

$$\min J = DC = \frac{Q_D}{Q_F C_F} \tag{3.68}$$

For the second-level optimization, some optimum ($\gamma_2$, $\gamma_3$) pairs of the first-level optimization are selected and the DC is minimized. The minimum desorbent flow rate is related to the difference between $\gamma_1$ and $\gamma_4$, that is, $\min(DC) = \min(QD) = \min(\gamma_1 - \gamma_4)$. The flow sheet of the second-level optimization procedure is shown in Figure 3.9.

The global optimization results in the definition of the flow rates on the four sections of the SMB that maximize the unit productivity with the minimum possible DC under the imposed constraints.

The effects of $\gamma_1$ and $\gamma_4$ on the SMB productivity and DC are summarized in Figure 3.10. For a constant $\gamma_1$, as $\gamma_4$ increases, the DC decreases, with minimum effect

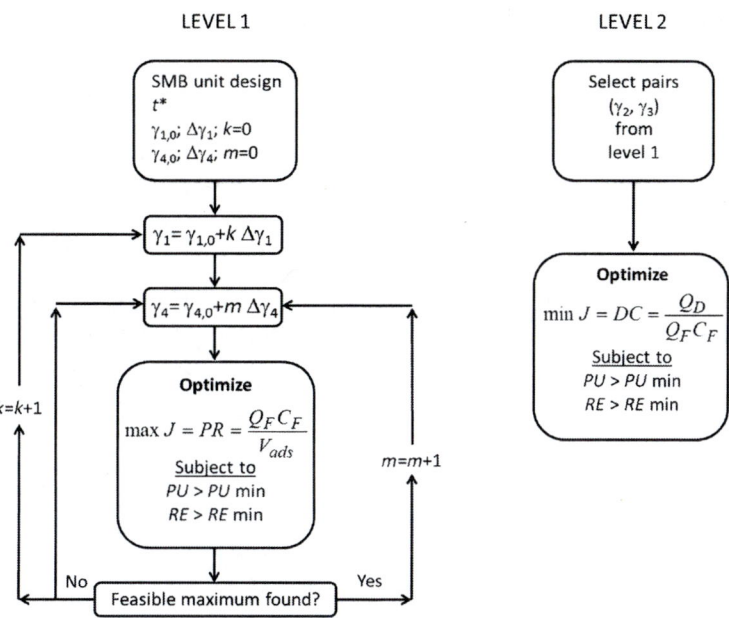

**Figure 3.9** Two-level optimization procedure. *(Reprinted from Computers & Chemical Engineering 29, Minceva, M., Rodrigues, A.E., Two-level optimization of an existing SMB for p-xylene separation, pp. 2215–2228, Copyright (2005), with permission from Elsevier.)*

on the productivity, until a value of $\gamma_4$ is reached for which the separation requirements cannot be obtained. For constant $\gamma_4$, as $\gamma_1$ increases, so does the productivity and DC.

The procedure was applied to the optimization of an existing SMB unit for the separation of *p*-xylene from a mixture of xylene isomers (*p*-xylene, *m*-xylene, *o*-xylene,

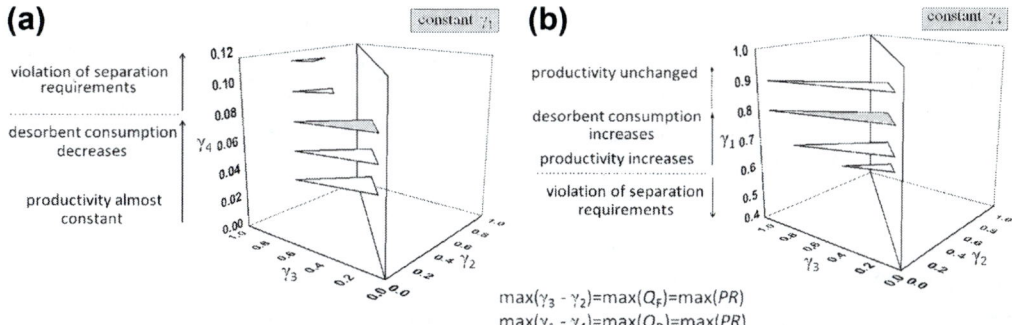

$$\max(\gamma_3 - \gamma_2) = \max(Q_F) = \max(PR)$$
$$\max(\gamma_1 - \gamma_4) = \max(Q_D) = \max(PR)$$

**Figure 3.10** (a) Influence of $\gamma_1$ on the separation region shown in the $\gamma_1 \times \gamma_2 \times \gamma_3$ plot, for fixed value of $\gamma_4$, and (b) influence of $\gamma_4$ on the separation region shown in the $\gamma_4 \times \gamma_2 \times \gamma_3$ plot, for fixed value of $\gamma_1$. Two-level optimization procedure. *(Reprinted from Computers & Chemical Engineering 29, Minceva, M., Rodrigues, A.E., Two-level optimization of an existing SMB for p-xylene separation, pp. 2215–2228, Copyright (2005), with permission from Elsevier.)*

**Table 3.3** SMB Model Parameters and Operating Conditions

| SMB Geometry | Model Parameters | Operating Conditions |
|---|---|---|
| $L_c = 1.14$ m<br>$d_c = 4.12$ m<br>Number of columns: 24<br>Configuration: 6-9-6-3 | $Pe = 1000$<br>$\varepsilon = 0.39$<br>$\varepsilon_p = 0.37$<br>$\rho_p = 1.48$ g/cm$^3$<br>$k_p$ (p-x; m-x; o-x; eb) $= 8.1$ min$^{-1}$<br>$k_p$ (p-DEB) $= 6.8$ min$^{-1}$ | $T = 180\,^\circ$C<br>$t^* = 1.15$ min<br>$Q_D^* = 133.06$ m$^3$/h<br>$Q_X^* = 88.65$ m$^3$/h<br>$Q_F^* = 87.00$ m$^3$/h<br>$Q_R^* = 131.41$ m$^3$/h<br>$Q_I^* = 637.38$ m$^3$/h |

| | Langmuir Isotherm Parameters | |
|---|---|---|
| **Component** | **$K$ (m$^3$/kg)** | **$q_m$ (kg/kg)** |
| *p*-Xylene | 1.9409 | 0.1024 |
| *m*-Xylene | 0.8884 | 0.0917 |
| *o*-Xylene | 0.8884 | 0.0917 |
| Ethylbenzene | 1.0263 | 0.0966 |
| *p*-Diethylbenzene | 2.3556 | 0.0847 |

ethylbenzene) using *p*-diethylbenzene as desorbent. *p*-Xylene purity and recovery higher than 99 and 98%, respectively, were imposed as constraints. The operating conditions and model parameters for this case study are shown in Table 3.3. Figure 3.11 shows the optimum points obtained after the first-level optimization, that is, the maximum productivity points as a function of $\gamma_4$ for different $\gamma_1$. Some optimum points from the first-level optimization were selected for the DC minimization (second-level optimization). The obtained optimum operation parameters are given in Table 3.4.

**Figure 3.11** Optimum points of the first-level optimization—maximum productivity points as a function of $\gamma_4$ for different $\gamma_1$. Two-level optimization procedure. (*Reprinted from Computers & Chemical Engineering 29, Minceva, M., Rodrigues, A.E., Two-level optimization of an existing SMB for p-xylene separation, pp. 2215–2228, Copyright (2005), with permission from Elsevier.*)

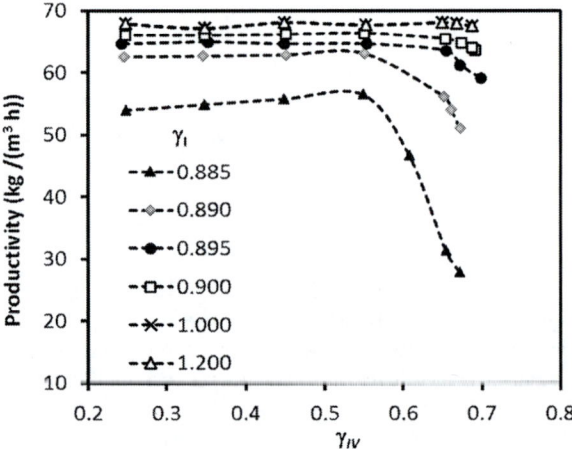

**Table 3.4** Optimum Operation Point

| $Q_F$ (m³/h) | $\gamma_1$ | $\gamma_2$ | $\gamma_3$ | $\gamma_4$ | $PU_{p-x}$ (%) | $RE_{p-x}$ (%) | $DC$ (m³/kg) | $PR$ (kg/(h m³)) |
|---|---|---|---|---|---|---|---|---|
| 88.04 | 0.949 | 0.800 | 1.089 | 0.661 | 99.00 | 98.00 | $5.839 \times 10^{-3}$ | 67.70 |

## 3.7 CONCLUDING REMARKS

The equilibrium theory is very useful for the initial design of SMB processes; the triangle theory allows the construction of the separation region in the plan ($\gamma_2 \times \gamma_3$) for linear and some nonlinear systems. An analytical solution is available for linear systems accounting for the effects of mass transfer resistances on the size of the separation region. When mass transfer is significant the flow rates estimated by the equilibrium theory for the regeneration regions are just minimum (for section I) and maximum (for section IV) values. The whole picture is given by the "separation volume" in the spaces ($\gamma_1 \times \gamma_2 \times \gamma_3$) or ($\gamma_4 \times \gamma_2 \times \gamma_3$), respectively. The two-level optimization of SMB processes is presented for productivity and desorbent consumption as practiced in industry.

## NOMENCLATURE

| | |
|---|---|
| $b_i$ | Langmuir isotherm parameter (m³/mol) |
| $Bi$ | Mass Biot number |
| $C$ | Liquid-phase concentration (mol/m³) |
| $\overline{C_p}$ | Average concentration in the particle pores over the pellet (mol/m³) |
| $d_c$ | Column diameter (m) |
| $DC$ | Desorbent consumption (m³/kg) |
| $D_L$ | Axial dispersion coefficient (m²/s) |
| $D_P$ | Pore diffusion coefficient (m²/s) |
| $H$ | Linear coefficient in the Linear + Langmuir isotherm |
| $k_f$ | Film mass transfer coefficient (m/s) |
| $k_h$ | Homogeneous mass transfer coefficient (s⁻¹) |
| $k_p$ | Mass transfer coefficient in the pores (s⁻¹) |
| $k_\mu$ | Mass transfer coefficient in the microparticle or crystal (s⁻¹) |
| $K$ | Linear isotherm parameter |
| $L$ | Column length (m) |
| $m$ | Ratio between fluid and solid flow rates |
| $Pe$ | Peclet number |
| $PR$ | Productivity (kg/(h m³)) |
| $PU$ | Purity |
| $q$ | Average adsorbed-phase concentration in the microparticle (mol/m³) |
| $q^*$ | Adsorbed phase concentration in equilibrium with $C$ (mol/m³) |
| $\overline{q}$ | Average adsorbed concentration over the pellet (mol/m³) |
| $q_m$ | Langmuir isotherm parameter (mol/m³) |
| $Q$ | Volumetric flow rate (m³/s) |

| | |
|---|---|
| $RE$ | Recovery |
| $R_p$ | Particle radius (m) |
| $t$ | Time variable (s) |
| $t^*$ | Switching time (s) |
| $T$ | Temperature (°C) |
| $u_s$ | Interstitial solid velocity (m/s) |
| $\nu$ | Fluid interstitial velocity in the TMB (m/s) |
| $V_{ads}$ | Adsorbent volume (m$^3$) |
| $x$ | Dimensionless axial coordinate |
| $z$ | Axial coordinate (m) |

## Greek Letters

| | |
|---|---|
| $\alpha$ | Number of mass-transfer units |
| $\beta$ | Security factor |
| $\varepsilon$ | Bed porosity |
| $\varepsilon_p$ | Particle porosity |
| $\gamma$ | Ratio between fluid and solid interstitial velocities |
| $\rho_p$ | Pellet density (kg/m$^3$) |

## Subscripts

| | |
|---|---|
| $0$ | Inlet |
| $D$ | Desorbent |
| $F$ | Feed |
| $i$ | Component $i$ ($i = A, B$) |
| $j$ | Section $j$ ($j = I, II, III, IV$ or 1, 2, 3, 4) |
| $p$ | Pore |
| $R$ | Raffinate |
| $s$ | Solid |
| $X$ | Extract |
| $\mu$ | Microparticle or crystal |

## REFERENCES

Azevedo, D.C.S., Rodrigues, A.E., 1999a. Bilinear driving force approximation in the modeling of a simulated moving bed using bidisperse adsorbents. Ind. Eng. Chem. Res. 38, 3519—3529.

Azevedo, D.C.S., Rodrigues, A.E., 1999b. Design of a simulated moving bed in the presence of mass-transfer resistances. AIChE J. 45, 956—966.

Azevedo, D.C.S., Rodrigues, A.E., 2001. Fructose—glucose separation in a SMB pilot unit: modeling, simulation, design, and operation. AIChE J. 47, 2042—2051.

Beste, Y.A., Lisso, M., Wozny, G., Arlt, W., 2000. Optimization of simulated moving bed plants with low efficient stationary phases: separation of fructose and glucose. J. Chromatogr. A 868, 169—188.

Chiang, A.S.T., 1998. Equilibrium theory for simulated moving bed adsorption processes. AIChE J. 44, 2431—2441.

Dunnebier, G., Fricke, J., Klatt, K.U., 2000. Optimal design and operation of simulated moving bed chromatographic reactors. Ind. Eng. Chem. Res. 39, 2290—2304.

Gentilini, A., Migliorini, C., Mazzotti, M., Morbidelli, M., 1998. Optimal operation of simulated moving-bed units for non-linear chromatographic separations - II. Bi-Langmuir isotherm. J. Chromatogr. A 805, 37—44.

Helfferich, F.G., 1967. Multicomponent ion exchange in fixed beds: generalized equilibrium theory for systems with constant separation factors. Ind. Eng. Chem. Fundam. 6, 362–364.

Karlsson, S., Pettersson, F., Westerlund, T., 1999. A MILP-method for optimizing a preparative simulated moving bed chromatographic separation process. Comput. Chem. Eng. 23, S487–S490.

Kaspereit, M., Seidel-Morgenstern, A., Kienle, A., 2007. Design of simulated moving bed processes under reduced purity requirements. J. Chromatogr. A 1162, 2–13.

Klatt, K.U., Hanisch, F., Dunnebier, G., Engell, S., 2000. Model-based optimization and control of chromatographic processes. Comput. Chem. Eng. 24, 1119–1126.

Klein, G., Tondeur, D., Vermeulen, T., 1967. Multicomponent ion exchange in fixed beds: general properties of equilibrium systems. Ind. Eng. Chem. Fundam. 6, 339–351.

Ma, Z., Wang, N.H.L., 1997. Standing wave analysis of SMB chromatography: linear systems. AIChE J. 43, 2488–2508.

Mazzotti, M., 2006. Equilibrium theory based design of simulated moving bed processes for a generalized Langmuir isotherm. J. Chromatogr. A 1126, 311–322.

Mazzotti, M., Storti, G., Morbidelli, M., 1994. Robust design of countercurrent adsorption separation processes: 2. Multicomponent systems. AIChE J. 40, 1825–1842.

Mazzotti, M., Storti, G., Morbidelli, M., 1996. Robust design of countercurrent adsorption separation: 3. Nonstoichiometric systems. AIChE J. 42, 2784–2796.

Mazzotti, M., Storti, G., Morbidelli, M., 1997a. Optimal operation of simulated moving bed units for nonlinear chromatographic separations. J. Chromatogr. A 769, 3–24.

Mazzotti, M., Storti, G., Morbidelli, M., 1997b. Robust design of countercurrent adsorption separation processes: 4. Desorbent in the feed. AIChE J. 43, 64–72.

Migliorini, C., Mazzotti, M., Morbidelli, M., 2000. Robust design of countercurrent adsorption separation processes: 5. Nonconstant selectivity. AIChE J. 46, 1384–1397.

Minceva, M., Rodrigues, A.E., 2002. Modeling and simulation of a simulated moving bed for the separation of p-xylene. Ind. Eng. Chem. Res. 41, 3454–3461.

Minceva, M., Rodrigues, A.E., 2005. Two-level optimization of an existing SMB for p-xylene separation. Comput. Chem. Eng. 29, 2215–2228.

Minceva, M., Silva, V.M.T., Rodrigues, A.E., 2005. Analytical solution for reactive simulated moving bed in the presence of mass transfer resistance. Ind. Eng. Chem. Res. 44, 5246–5255.

Pais, L.S., Loureiro, J.M., Rodrigues, A.E., 1998. Separation of enantiomers of a chiral epoxide by simulated moving bed chromatography. J. Chromatogr. A 827, 215–233.

Rajendran, A., 2008. Equilibrium theory-based design of simulated moving bed processes under reduced purity requirements. Linear isotherms. J. Chromatogr. A 1185, 216–222.

Rodrigues, A.E., Pais, L.S., 2004. Design of SMB chiral separations using the concept of separation volume. Sep. Sci. Technol. 39, 245–270.

Ruthven, D.M., Ching, C.B., 1989. Counter-current and simulated counter-current adsorption separation processes. Chem. Eng. Sci. 44, 1011–1038.

Silva, V.M.T., Minceva, M., Rodrigues, A.E., 2004. Novel analytical solution for a simulated moving bed in the presence of mass-transfer resistance. Ind. Eng. Chem. Res. 43, 4494–4502.

Storti, G., Masi, M., Morbidelli, M., 1989. In: Rodrigues, A.E., LeVan, M.D., Tondeur, D. (Eds.), Optimal Design of Simulated Moving Bed Adsorption Separation Units through Detailed Modeling and Equilibrium Theory, 158. Adsorption Science and Technology, London, Kluwer, pp. 357–381.

Storti, G., Masi, M., Paludetto, R., Morbidelli, M., Carra, S., 1988. Adsorption separation processes - countercurrent and simulated countercurrent operations. Comput. Chem. Eng. 12, 475–482.

Storti, G., Mazzotti, M., Morbidelli, M., Carra, S., 1993. Robust design of binary countercurrent adsorption separation processes. AIChE J. 39, 471–492.

Tondeur, D., Klein, G., 1967. Multicomponent ion exchange in fixed beds: constant-separation-factor equilibrium. Ind. Eng. Chem. Fundam. 6, 351–361.

Toumi, A., Engell, S., Ludemann-Hombourger, O., Nicoud, R.M., Bailly, M., 2003. Optimization of simulated moving bed and Varicol processes. J. Chromatogr. A 1006, 15–31.

Xie, Y., Wu, D.J., Ma, Z.D., Wang, N.H.L., 2000. Extended standing wave design method for simulated moving bed chromatography: linear systems. Ind. Eng. Chem. Res. 39, 1993–2005.

# CHAPTER 4

# Process Development for Liquid-Phase Simulated Moving Bed Separations: Methodology and Applications

## 4.1 METHODOLOGY

The optimization of a binary or pseudo-binary separation is based on the selection of a proper combination of a stationary-phase (solid) and a mobile-phase (solvent) composition. It is a much more challenging task when we are dealing with a preparative separation instead of a traditional analytical separation problem. The complexity will increase if we consider using a simulated moving bed (SMB), in which high feed concentrations are normally used to improve the process performance. In these situations, the desired high concentrations of the different solutes inside the chromatographic columns will enhance significantly the mutual competition between solutes for adsorption with the stationary phase. From a preparative point of view, and when considering the choice of the solvent composition, a high selectivity of the enantiomers should not be the only goal, as is frequently the case at the analytical scale. In addition to the choice of a stationary phase with high loading capacity, a high solubility of the solutes in the solvent and low retention times should also be taken into account, to improve the process performance. A scheme of the methodology of process development is presented in Figure 4.1, and a description of the different tasks is provided beneath.

The screening of the solvent composition is always the first experimental step if facing a new mixture separation problem. In these "preliminary studies" the main goal is to measure the mixture solubility covering the wider range of pure solvents and solvent

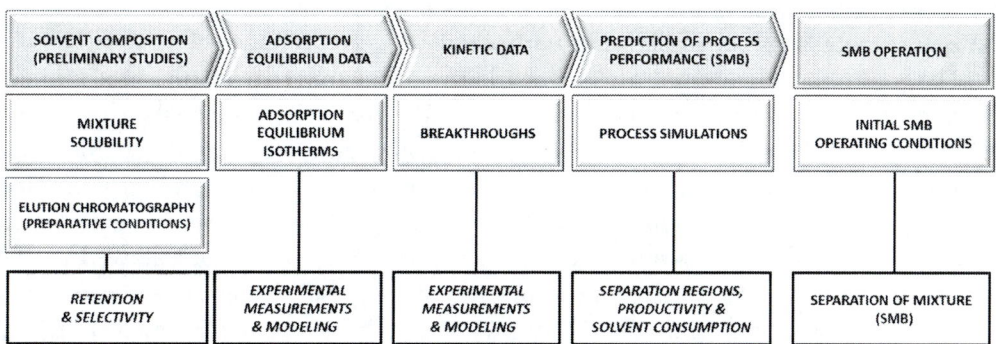

**Figure 4.1** Scheme of the methodology for preparative separation using the SMB technology.

*Simulated Moving Bed Technology*
ISBN 978-0-12-802024-1

mixtures. Other helpful data are the retention and the selectivity obtained by elution chromatography. At this point in the process development, experimental work is normally performed using various types of stationary phases and increasing feed concentrations, toward preparative conditions. The solvent screening is normally carried out using chromatographic columns with analytical dimensions (4.6 mm i.d. × 250 mm L) and analytical particle size diameters (3 or 5 μm). After these preliminary studies, the solvents or solvent mixtures that present a better trade-off between solubility, retention, and selectivity are selected to measure the adsorption equilibrium isotherms and the kinetic data. These data are collected using chromatographic columns with preparative dimensions (20 mm i.d. × 100 mm L) and preparative particle size diameters (10 and 20 μm) that are the same as those of the other columns installed in the SMB unit. Generally, the bulk stationary phase is commercially available and stainless steel columns are packed with the prepared slurry. The hydrodynamic characterization of each "SMB column" is also a relevant step because the common modeling of the SMB operation assumes that all columns are identical in terms of dimensions and that they contain equal amounts of stationary phase. The experimental packing procedure can lead to local fluctuations in the packing density of each SMB column. Therefore, characterization of each SMB column must be done to minimize the differences between columns: non-retained-compound pulses (retention), HETP (Height Equivalent to Theoretical Plate) experiments (total porosity), and/or pressure drop (bulk porosity). After column packing and characterization, the most promising solvent compositions are checked by performing elution pulse experiments using one column with the preparative dimensions, the "SMB column". Then, the equilibrium adsorption isotherms are experimentally determined. This task is very critical and an accurate experimental method is advised to ensure the quality of the equilibrium data. Frontal analysis and perturbation methods are commonly the most used methods because they are less time-consuming. However, the adsorption/desorption method will provide accurate and reliable adsorption equilibrium experimental data. Thereafter, there are several adsorption equilibrium models that can be used to fit experimental data. Breakthrough experiments are then used to collect the kinetic data and develop a dynamic fixed-bed adsorption model, validating at the same time the adsorption equilibrium isotherm data.

The adsorption equilibrium and kinetic data are mandatory to perform a prediction of the separation process performance using the SMB technology. This task is based on several simulation runs in which the separation region is constructed and the performance parameters are estimated: purity, recovery, solvent consumption, and productivity. Finally, based on these performance parameters, the decision of the better combination between solvent/solvent mixture and stationary phase must be made, followed by the definition of the initial SMB operating conditions, and experimental separation performed using the real pilot/laboratory scale unit. The experimental results obtained from the separation of a mixture are used to validate the SMB model.

## 4.2 CHIRAL SEPARATIONS

Chiral separations are, nowadays, one of the most relevant issues for the pharmaceutical industry. This can be confirmed by the increase in the number of racemic drugs that are being separated and marketed as single enantiomer drugs. Notwithstanding the ongoing use of the organic enantioselective synthesis of single enantiomer molecules, the preparative liquid separation of racemic chiral drugs by means of the SMB or multicolumn technologies has become an attractive alternative. Developments in terms of more compatibility and reliability of the manufactured stationary phases and improvements in the systems based on SMB technologies, with the rise of new and more efficient modes of operation, led to the success of this technology (Pais and Rodrigues, 2003; Sá Gomes and Rodrigues, 2007; Seidel-Morgenstern et al., 2008; Rajendran et al., 2009). In this section, the separation of two different racemic pharmaceutical drugs will be used to illustrate the application of the methodology described in the previous section (Ribeiro et al., 2008, 2009, 2011a,b). Ketoprofen, (R,S)-2-(3-benzoylphenylpropionic acid), and flurbiprofen, (R,S)-2-(2-fluoro-4-biphenyl)propionic acid (Figure 4.2), belong to the family of profens, which are known as a major group of the nonsteroidal anti-inflammatory drugs.

The medical prescription of profens is related mainly to the relief of inflammatory diseases such as rheumatoid arthritis or osteoarthritis or to promote an antipyretic, analgesic, or antidysmenorrheic action. However, the use of this type of drug is also related to some severe risks, such as gastrointestinal, hepatic, or renal toxicity and even cardiovascular

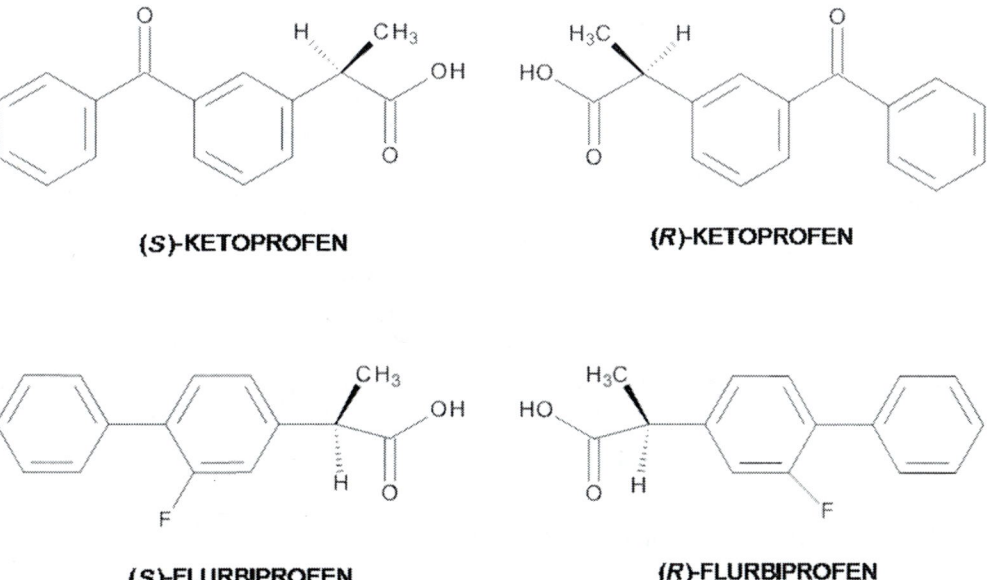

Figure 4.2 Molecular structures of ketoprofen (top) and flurbiprofen enantiomers (bottom).

failure (Panico et al., 2005). Several studies suggest that this problem is related, or could be related, to the fact that these drugs are still marketed as racemic mixtures. Another important aspect is that when their enantiomers are separated and administered as single enantiomer drugs, they can display quite different behaviors or even new therapeutic actions. The S-enantiomers of both ketoprofen and flurbiprofen are associated with the anti-inflammatory and analgesic therapeutic action of these drugs, whereas the R-enantiomer of ketoprofen is associated with periodontal disease prevention and analgesic or antipyretic actions, and the R-enantiomer of flurbiprofen has been reported to have some anticarcinogenic potential as well as slowing down the progress of Alzheimer disease. The separation of chiral drugs is necessary to perform studies to better understand their therapeutic action and/or their pharmacological behavior as single enantiomer drugs.

## 4.2.1 Solubility of Racemic Mixtures

Following the methodology proposed in the previous section, the first experimental task is to measure the solubility of each racemic mixture in different pure solvents and different types of solvent mixtures. For this purpose a gravimetric method (Granberg and Rasmuson, 1999) is applied using the equation

$$S = \frac{(w_{VR} - w_V)}{(w_{VS} - w_{VR})} \times 10^3 \tag{4.1}$$

where $S$ is the solubility of the racemic mixture expressed in mass (g) of solute per unit mass (kg) of solvent (on a solute-free basis), $w_V$ is the previously weighed glass vial, $w_{VS}$ is the mass of the vial plus the saturated solution, and $w_{VR}$ is the mass of vial and residue. The solubility measurements results are presented in Figure 4.3.

Using pure solvents, we can observe that both racemic mixtures are "almost" insoluble in pure n-hexane and they have high solubility in pure methanol or ethanol. Thus, to better study the solubility, the first choices were methanol/n-hexane and ethanol/n-hexane mixtures. However, owing to the large range of immiscibility between methanol and n-hexane (6—60% at 25 °C), this type of mixture was not studied. In the two graphs on the right, the effects of increasing the ethanol content in the mixture ethanol/n-hexane on the solubility of the flurbiprofen and ketoprofen racemic mixtures are represented. We can observe that high hydrocarbon content represents a low solubility value for both profens. We can also observe a solubility increase with the increase in the ethanol content. For pulse experiments the selected solvent compositions are pure methanol, pure ethanol, and mixtures of ethanol in n-hexane.

## 4.2.2 Selection of the Chiral Stationary Phase

Before performing any liquid chromatography experiments, the selection of a proper chiral stationary phase (CSP) must be done. From an analytical point of view this decision

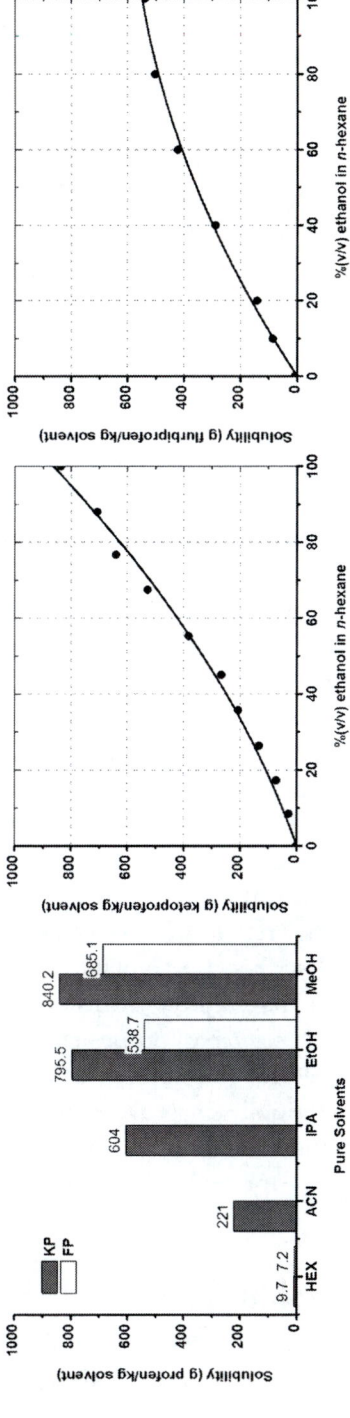

**Figure 4.3** Solubility of ketoprofen and flurbiprofen racemic mixtures in pure solvents (left) and of ketoprofen in ethanol/*n*-hexane mixtures (middle) and of flurbiprofen in ethanol/*n*-hexane mixtures (right) at 23 °C. *(Reprinted from Separation and Purification Technology 61, Ribeiro, A.E., Graca, N.S., Pais, L.S., Rodrigues, A.E., Preparative separation of ketoprofen enantiomers: choice of mobile-phase composition and measurement of competitive adsorption isotherms, pp. 375–383, Copyright (2008), with permission from Elsevier.)*

is normally made based on the complete resolution of all enantiomers and with the lower analysis time. However, in a preparative separation, and using the SMB technology, it must be pointed that the adsorbent must have high loading capacity because it will allow high feed concentrations. Therefore, the selection of the CSP is done from the commercially available CSPs and establishing a compromise between resolution, retention, and high loading capacity. Particularly, high chiral recognition is provided using the phenyl-carbamate derivatives of polysaccharides (cellulose- and amylose-based) as CSPs. The amylose 3,5–dimethylphenylcarbamate (e.g., Chiralpak® AD) is the most used for the separation of profen racemates (Maier et al., 2001; Tachibana and Ohnishi, 2001; Yashima, 2001) and is used for these two case studies.

### 4.2.3 Elution Chromatography

To provide a global overview of the ketoprofen and flurbiprofen selectivity in the three previously selected solvent compositions under preparative conditions (high concentrations), a set of preliminary pulse experiments was done. These experiments consisted of several pulse injections of various concentrations for both profen racemic drugs in a column with analytical dimensions and a particle size of 20 μm and using two different injection loops, 100 μl and 1 ml. A flow rate of 1 ml/min was used. To better compare the two case studies, only the most representative results are presented (Figure 4.4 for ketoprofen, and Figure 4.5 for flurbiprofen).

The elution chromatography experimental results obtained for ketoprofen enantiomers (Figure 4.4) show that the hydrocarbon mobile phase leads to important chromatographic tails (see 20E/80C6, 1 ml), which is an indication of strong nonlinear behavior not desired in preparative separations. The 20% ethanol/80% hydrocarbon composition presents also considerably higher retention times than the pure ethanol or methanol compositions. From all these pulses it can be noted that the increase in the amount injected leads to a decrease in the retention of both enantiomers, which is a well-known behavior for systems described by favorable isotherms. For pulses using pure methanol, it can be concluded that, although this composition presents higher solubility, it does not allow acceptable selectivity values and, hence, ketoprofen enantiomer separation. The results obtained for flurbiprofen enantiomers (Figure 4.5) show that the 10% ethanol/90% n-hexane mobile phase represents a reasonable compromise between selectivity, retention time and solubility. For pure methanol and, despite high solubility, the use of such mobile phase presents low selectivity, as can be observed in the 100 M, 1 ml curve.

### 4.2.4 Packing and Testing One SMB Column

As soon as the promising solvent compositions are identified, the next experimental step is to collect adsorption equilibrium and kinetic data using a column with the

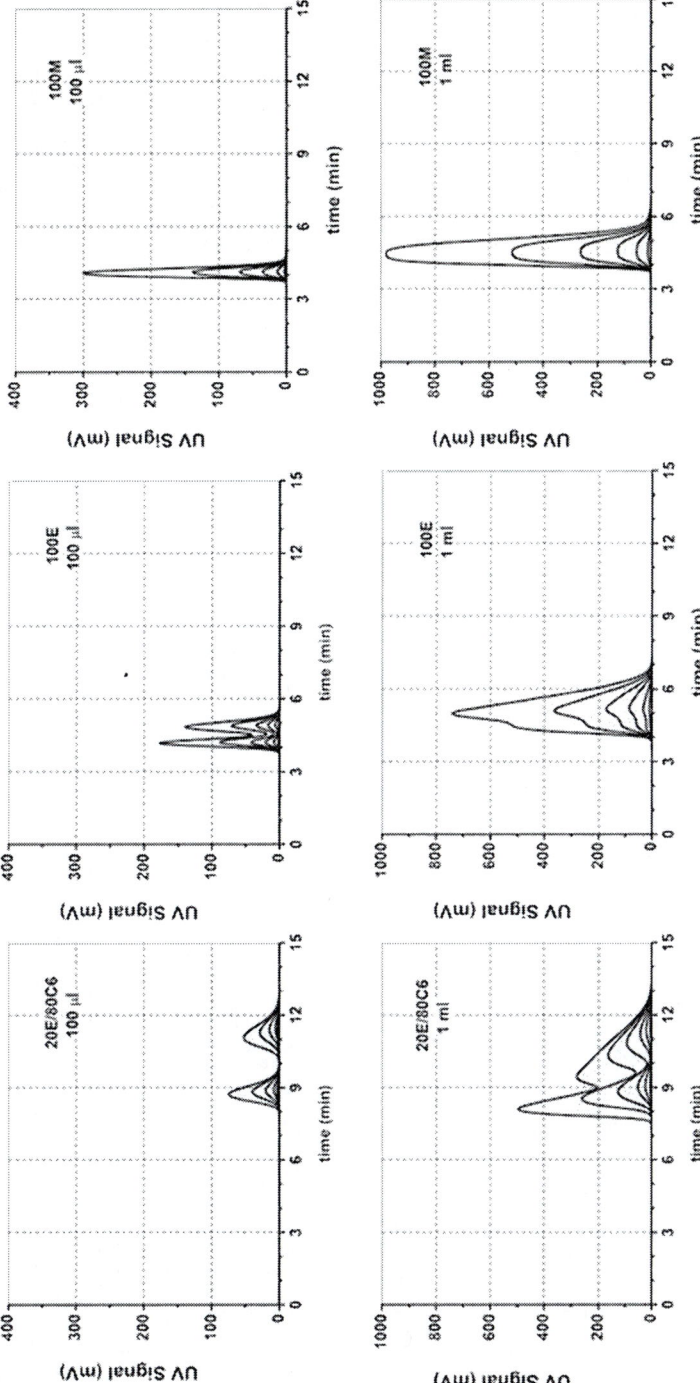

**Figure 4.4** Experimental elution profiles of ketoprofen enantiomers in three different solvent compositions: 20% ethanol/80% *n*-hexane, 100% ethanol, and 100% methanol. Racemic ketoprofen concentrations at six levels, 0.05, 0.2, 0.5, 1.0, 2.0, and 4.0 g/l; injection volumes of 100 µl and 1 ml; preparative column (20 µm); UV detection at 260 nm; flow rate of 1 ml/min; temperature of 23 °C. *(Reprinted from Separation and Purification Technology 61, Ribeiro, A.E., Graca, N.S., Pais, L.S., Rodrigues, A.E., Preparative separation of ketoprofen enantiomers: choice of mobile-phase composition and measurement of competitive adsorption isotherms, pp. 375–383, Copyright (2008), with permission from Elsevier.)*

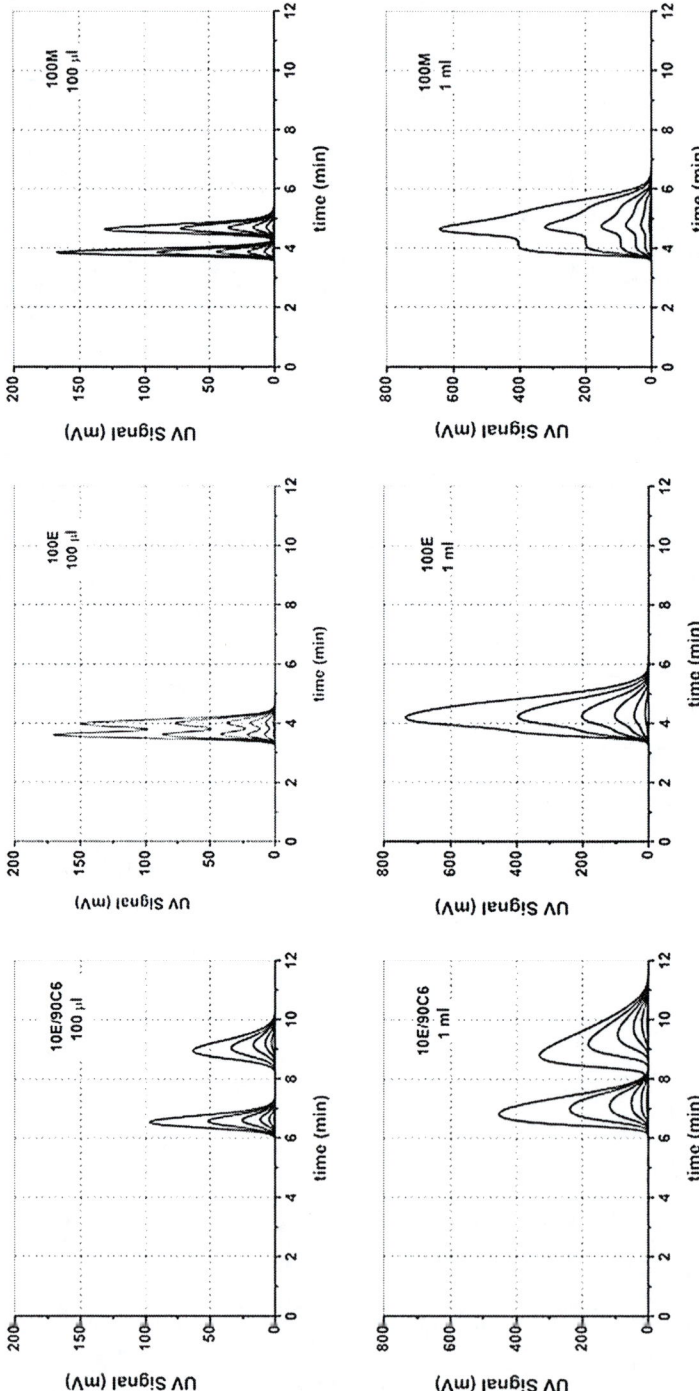

**Figure 4.5** Experimental elution profiles of flurbiprofen enantiomers in three different solvent compositions: 10% ethanol/90% *n*-hexane, 100% ethanol, and 100% methanol. Racemic flurbiprofen concentrations at six levels, 0.05, 0.2, 0.5, 1.0, 2.0, and 4.0 g/l; injection volumes of 100 μl and 1 ml; preparative column (20 μm); UV detection at 260 nm; flow rate of 1 ml/min; temperature of 23 °C. *(Reprinted from Separation and Purification Technology 68, Ribeiro, A.E., Graça, N.S., Pais, L.S., Rodrigues, A.E., Optimization of the mobile-phase composition for preparative chiral separation of flurbiprofen enantiomers, pp. 9–23, Copyright (2009), with permission from Elsevier.)*

same characteristics as those installed in the SMB unit. To do so, a stainless steel column must be packed with the selected CSP and some pulse tests with a nonretained compound must be carried out to characterize the column hydrodynamics. In Figure 4.6 the column packing apparatus (left), an experimental pulse and modeling (middle), and a breakthrough experiment and modeling (right) in the SMB column are presented. These tests are also repeated using both racemic profen mixtures to estimate the Peclet number ($Pe = 2500$), the total porosity ($\varepsilon_T = 0.69$ and $\varepsilon_T = 0.71$ for the less polar and more polar solvent composition, respectively), and the mass transfer coefficient, as will be presented in the following section for frontal chromatography. Small differences were found between model and experimental adsorption profiles (elution and breakthrough) that can be related to a small dead volume in the SMB column.

## 4.2.5 Elution Chromatography Using the SMB Column

Several pulse experiments were performed with the SMB column to study the column dynamics for the most promising selected solvent compositions, 100% ethanol for ketoprofen and 10% ethanol/90% n-hexane for flurbiprofen. A racemic mixture concentration of 20 g/l and a flow rate of 5 ml/min were used in these studies. The experimental results are presented in Figure 4.7 for ketoprofen and flurbiprofen enantiomers.

These experimental results show that for ketoprofen enantiomers, with a 100% ethanol composition, it is possible to achieve lower retention times compared with the 10% ethanol/90% n-hexane composition (nearly four times lower). It can be also seen that selectivity is significantly higher with the 10% ethanol/90% n-hexane composition. However, the selectivity obtained in pure ethanol shows some potential for preparative separation, because an acceptable separation of the two enantiomers is observed. For flurbiprofen enantiomer separation, the results show that a strong nonpolar solvent composition enables a notably higher resolution compared with the polar mobile-phase composition. Furthermore, a stronger nonpolar solvent represents higher retention values and such behavior is not convenient for the SMB separation, because it represents longer cycle times. Separation obtained using 100% ethanol, despite showing very low resolution values for linear chromatography (analytical), may be enough for the preparative separation.

## 4.2.6 Experimental Measurement and Modeling of the Adsorption Equilibrium Isotherms

After elution chromatography the competitive adsorption isotherms were determined in the two mobile-phase compositions. These experiments were performed using the adsorption—desorption method. To fit the experimental data several different models were used, such as the Langmuir model, the Linear + Langmuir model and the bi-Langmuir model. The model that best fitted the experimental data was the modified

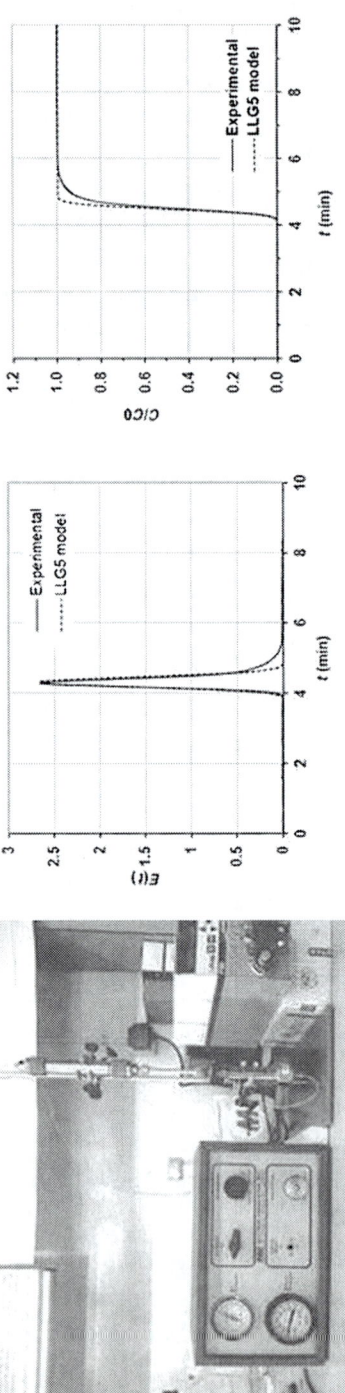

**Figure 4.6** SMB column characterization: packing system (left), experimental and modeling of pulses (middle), and breakthroughs (right) of the nonretained compound.

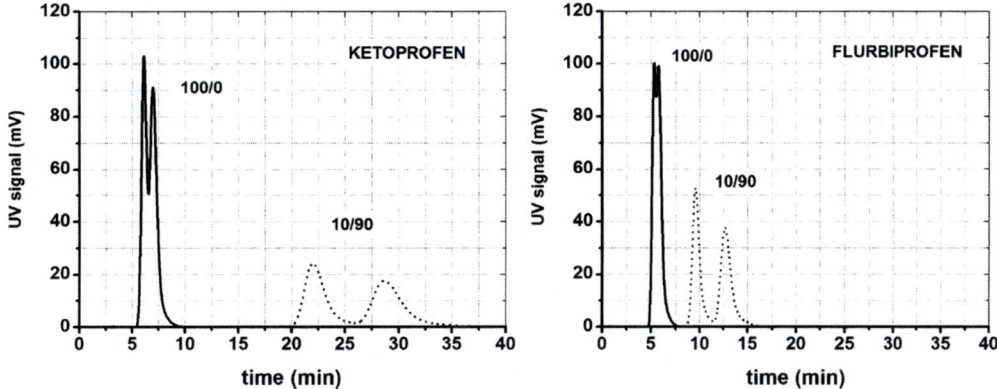

**Figure 4.7** Experimental elution profiles of ketoprofen (left) and flurbiprofen (right) enantiomers in the SMB column, using two different mobile-phase compositions: pure ethanol (100/0) and 10% ethanol/90% *n*-hexane (10/90). For both racemic mixtures: concentration 20 g/l; UV detector set at 260 nm; $Q = 5$ ml/min; loop 100 µl.

**Table 4.1** Estimated Linear + Langmuir Model Parameters for Ketoprofen and Flurbiprofen Adsorption Isotherms in the Two Mobile-Phase Compositions Using the SMB Column

| | $M$ | $N$ | $m_{L1}$ | $m_{L2}$ | $Q_{sat}$ | $b_1$ | $b_2$ |
|---|---|---|---|---|---|---|---|
| **10% Ethanol/90% *n*-hexane** | | | | | | | |
| Ketoprofen | 14 | 5 | 1.700 | 1.475 | 20.63 | $1.725 \times 10^{-1}$ | $2.707 \times 10^{-1}$ |
| Flurbiprofen | 14 | 5 | 0.2372 | $3.945 \times 10^{-7}$ | 100.0 | $1.635 \times 10^{-2}$ | $2.724 \times 10^{-2}$ |
| **100% Ethanol** | | | | | | | |
| Ketoprofen | 18 | 5 | 0.6613 | 0.6680 | 9.122 | $4.613 \times 10^{-2}$ | $7.981 \times 10^{-2}$ |
| Flurbiprofen | 14 | 5 | 0.1760 | $1.243 \times 10^{-5}$ | 87.09 | $7.028 \times 10^{-3}$ | $1.085 \times 10^{-2}$ |

$M$ is the number of points; $N$ is the number of parameters; $m_{L1}$, $m_{L2}$, $b_1$, and $b_2$ are the estimated model parameters.

Linear + Langmuir model (LLG5) using five parameters, as presented in Table 4.1. The equations for this model are:

LLG5 model for the less retained enantiomer:

$$q_1^* = m_{L1} C_1 + \frac{Q_{sat} b_1 C_1}{1 + b_1 C_1 + b_2 C_2} \tag{4.2}$$

LLG5 model for the more retained enantiomer:

$$q_2^* = m_{L2} C_2 + \frac{Q_{sat} b_2 C_2}{1 + b_1 C_1 + b_2 C_2} \tag{4.3}$$

Figures 4.8 and 4.9 show the adsorption equilibrium isotherms and the related experimental selectivities obtained for the ketoprofen and flurbiprofen enantiomers, respectively. The 10% ethanol/90% *n*-hexane composition is represented by the dashed lines and 100% ethanol is represented by the solid lines. For all the isotherms and selectivity values, it can be noted that there is a good agreement between experimental points and the predicted adsorption behavior by the selected LLG5 model. Figure 4.8 also shows that the decrease in selectivity for high feed concentrations is much more pronounced for

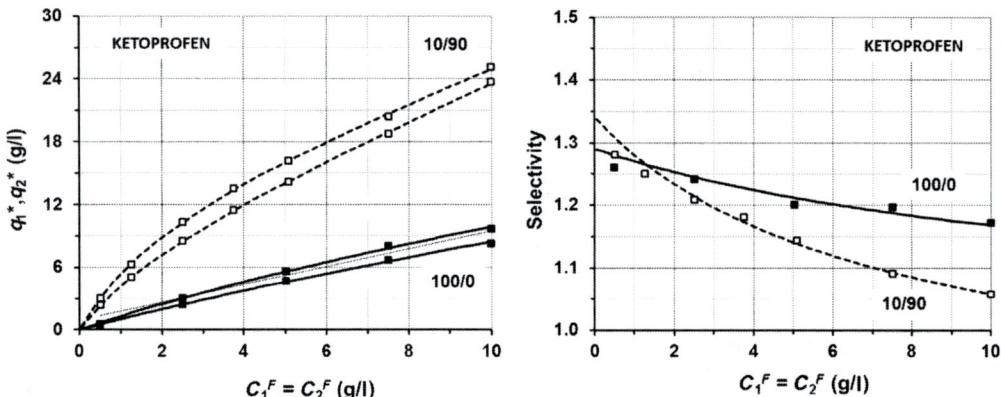

**Figure 4.8** Comparison between model and experimental results for the equilibrium adsorption isotherms of ketoprofen enantiomers using the modified Linear + Langmuir model on two different solvent compositions: 10% ethanol/90% *n*-hexane and 100% ethanol. Experimental and model adsorption isotherms at the left and the experimental and model selectivity at the right.

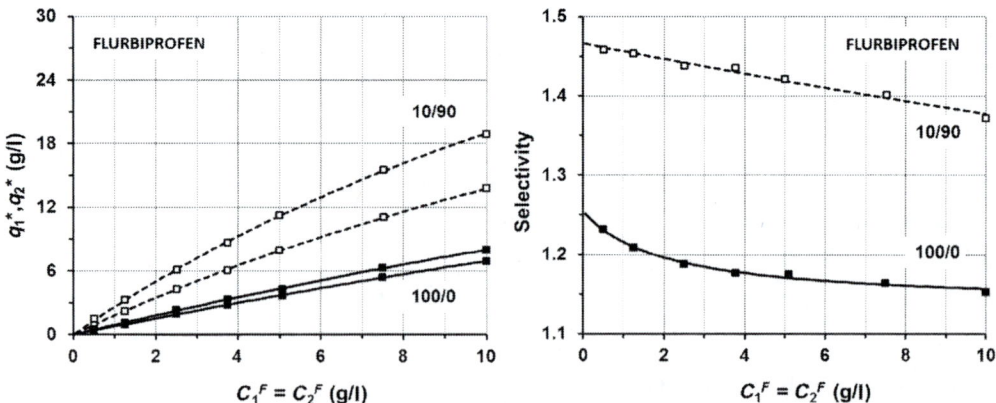

**Figure 4.9** Comparison between model and experimental results for the equilibrium adsorption isotherms of flurbiprofen enantiomers using the modified Linear + Langmuir model on two different solvent compositions: 10% ethanol/90% *n*-hexane and 100% ethanol. Experimental and model adsorption isotherms at the left and the experimental and model selectivity at the right.

the 10% ethanol/90% *n*-hexane composition. This is an important issue, because the 100% ethanol composition shows better selectivity at preparative conditions (high feed concentrations). For flurbiprofen, when a 10% ethanol composition is used the obtained selectivity is much higher than when a 100% ethanol composition is used. For ketoprofen and for preparative conditions a 100% ethanol mobile-phase composition represents higher selectivity values. Figures 4.8 and 4.9 also show how the selectivity depends on the enantiomer concentration. The decrease in selectivity with the increase in the enantiomer concentration and the fact that the selected LLG5 adsorption isotherm model can properly predict it can be clearly seen.

### 4.2.7 Experimental Determination and Modeling of Breakthroughs

After the measurement of the equilibrium data, different experiments of frontal chromatography were carried out with the purpose of testing the selected competitive adsorption isotherm model. The fixed-bed model is characterized by a convective and a dispersive term and the mass transfer resistance inside the adsorbent particle is described by a linear driving force model. The simulation of the fixed-bed column behavior requires knowledge of the Peclet number and mass transfer coefficient ($k$). The mass transfer coefficient was estimated by fitting the model to the experimental data using both profen racemic mixtures and in both mobile-phase compositions.

A concentration of 20 g/l of each profen mixture was used and was considered high enough to characterize the nonlinear region. Several samples were collected and analyzed at different times for the saturation and regeneration steps. These results are presented in Figures 4.10 and 4.11 for ketoprofen and flurbiprofen, respectively. The results show a good agreement between the experimental points and the model used for both solvent compositions. A nonretained compound was used for experimental determination of the Peclet number and the racemic profen mixture was used for the experimental determination of the mass transfer coefficient. The estimated value of the Peclet number ($Pe = 2500$) was the same for both compositions, whereas the estimated mass transfer coefficient was $k = 0.3 \text{ s}^{-1}$ for 10% ethanol/90% *n*-hexane and double that, $k = 0.6 \text{ s}^{-1}$, for 100% ethanol.

### 4.2.8 Predictions of Separation Regions and Performance Parameters

The estimated values of Peclet and mass transfer coefficient were used to predict the performance parameters of the SMB separation process. This study is normally done by constructing the separation region and estimation of the productivity and of the solvent consumption. This is done to support the final decision on the selection of the solvent composition to perform the real separation of each profen racemic mixture. Figure 4.12 (ketoprofen) and Figure 4.13 (flurbiprofen) present the simulation results obtained for both racemic mixtures and for the two studied solvent compositions: separation regions

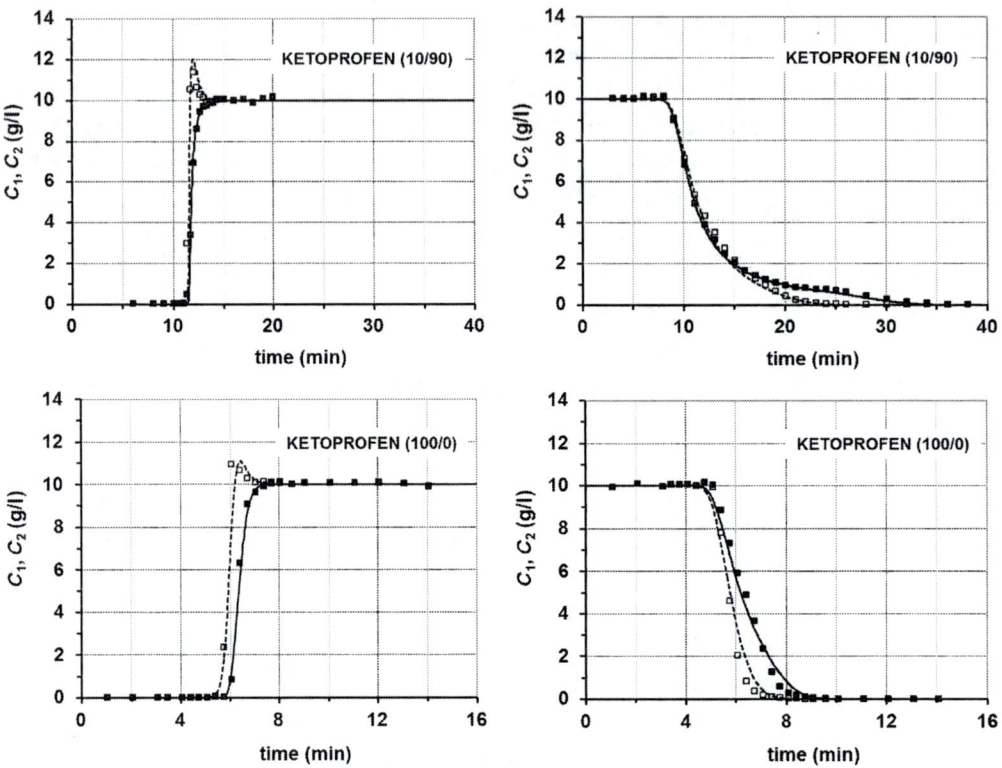

**Figure 4.10** Saturation (adsorption) and regeneration (desorption) curves for 20 g/l ketoprofen racemic feed concentration. Comparison between experimental (symbols) and simulation (lines) results for two different solvent compositions: (top) 10/90 and (bottom) 100/0 of ethanol/n-hexane. $Q = 5$ ml/min. Model parameters: $\varepsilon = 0.4$; $Pe = 2500$; $k = 0.3$ s$^{-1}$ for 10/90 and $k = 0.6$ s$^{-1}$ for 100/0 composition, and modified Linear + Langmuir model parameters (see Table 4.1).

(left), solvent consumption (middle), and productivity (right). In these figures the solvent consumption and productivity were calculated using the following equations:

Solvent consumption (l/g),
$$SC = \frac{Q_E + Q_F}{Q_F\left(C_A^X + C_B^X\right)} = \frac{1}{\left(C_A^F + C_B^F\right)}\left[1 + \frac{\gamma_I - \gamma_{IV}}{\gamma_{III} - \gamma_{II}}\right]$$
(4.4)

Productivity (g/(h l$_{\text{bed}}$)),
$$PR = \frac{Q_F\left(C_A^X + C_B^X\right)}{V_T} = \frac{\varepsilon}{N_C t^*}(\gamma_{III} - \gamma_{II})\left(C_A^F + C_B^F\right)$$
(4.5)

The definition of the separation region considered purities in the extract and raffinate outlet streams above 99.0% and an SMB classic mode of operation with six columns

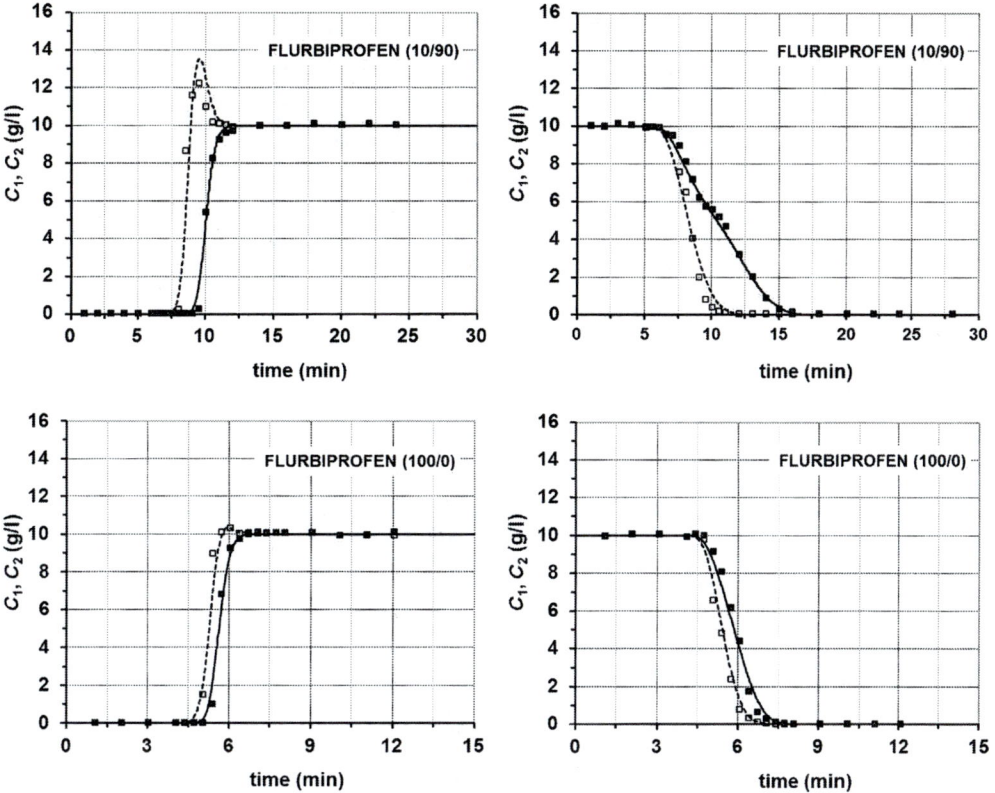

**Figure 4.11** Saturation (adsorption) and regeneration (desorption) curves for 20 g/l flurbiprofen racemic feed concentration. Comparison between experimental (symbols) and simulation (lines) results for two different solvent compositions: (top) 10/90 and (bottom) 100/0 of ethanol/n-hexane. $Q = 5$ ml/min. Model parameters: $\varepsilon = 0.4$; $Pe = 2500$; $k = 0.3\,\text{s}^{-1}$ for 10/90 and $k = 0.6\,\text{s}^{-1}$ for 100/0 composition, and modified Linear + Langmuir model parameters (see Table 4.1).

(1-2-2-1 configuration). Further predictions and final SMB separation results will be presented in Section 4.3.3.

We can observe that for ketoprofen, and despite the very narrow separation regions, better performance parameters are obtained using 100% ethanol composition: lower solvent consumption and higher productivity values for preparative (high) feed concentrations. For the separation of flurbiprofen enantiomers, the 10% ethanol composition predicts better performance results: bigger separation regions, less solvent consumption, and higher productivity values for high feed concentrations (nearly four times higher). Therefore, for the separation of ketoprofen enantiomers the final choice of solvent composition is 100% ethanol, whereas for the separation of flurbiprofen enantiomers the solvent composition is 10% ethanol/90% n-hexane.

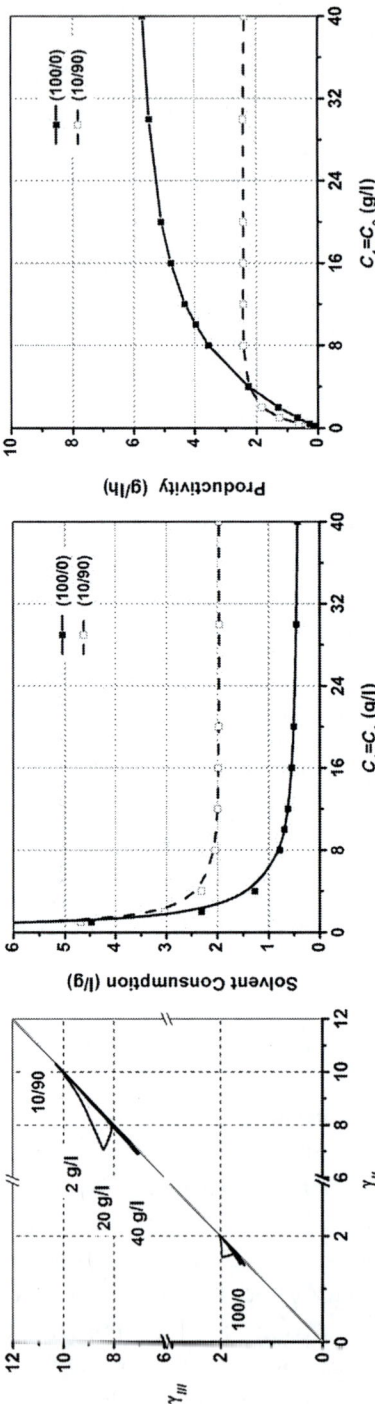

**Figure 4.12** SMB separation regions for the two mobile-phase compositions (10/90 and 100/0 ethanol/*n*-hexane) and three different racemic feed concentrations of ketoprofen enantiomers (2, 20, and 40 g/l) (left). Prediction of SMB: solvent consumption (middle) and productivity (right) for the two mobile-phase compositions, as a function of the feed concentration (racemic mixtures). Dashed line for the 10/90/0.01 and solid line for the 100/0/0.01 composition.

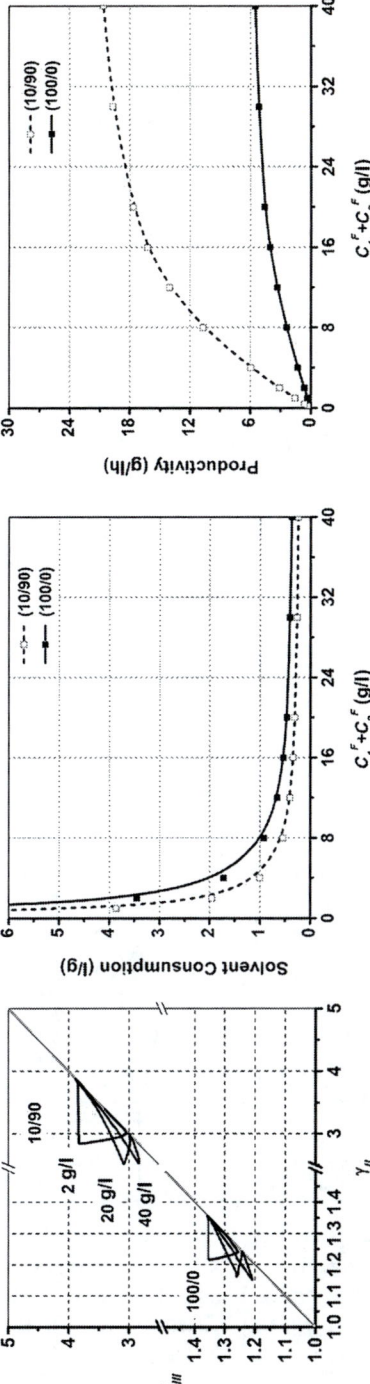

**Figure 4.13** SMB separation regions for the two mobile-phase compositions (10/90 and 100/0 ethanol/n-hexane) and three different racemic feed concentrations of flurbiprofen enantiomers (2, 20, and 40 g/l) (left). Prediction of SMB: solvent consumption (middle) and productivity (right), for the two mobile-phase compositions, as a function of the feed concentration (racemic mixtures). Dashed line for the 10/90/0.01 and solid line for the 100/0/0.01 composition. *(Reprinted from Chirality 23, Ribeiro, A.E., Gomes, P.S., Pais, L.S., Rodrigues, A.E., Chiral separation of flurbiprofen enantiomers by preparative and simulated moving bed chromatography, pp. 602–611, Copyright (2011), with permission from John Wiley & Sons Inc.)*

## 4.3 THE FLEXSMB-LSRE® UNIT

### 4.3.1 Design and Construction

The FlexSMB-LSRE® is a unit that was built taking into account its flexibility as the main key objective. The unit design was directed for study and research purposes and, in such situations, the operation of several different systems under different specifications is an important advantage (Chin and Wang, 2004). The unit was also constructed with the objective of being able to use "nonconventional" modes of operation (Sá Gomes et al., 2010; Sá Gomes and Rodrigues, 2012). The FlexSMB-LSRE® unit is presented in Figure 4.14 when the configuration is adapted for a classic (1–2–2–1) mode of operation.

#### 4.3.1.1 Valves and Pumps

Typically, there are two different classifications for the design of SMB valves: central or distributed valve arrangement. The central valve arrangement normally uses a single

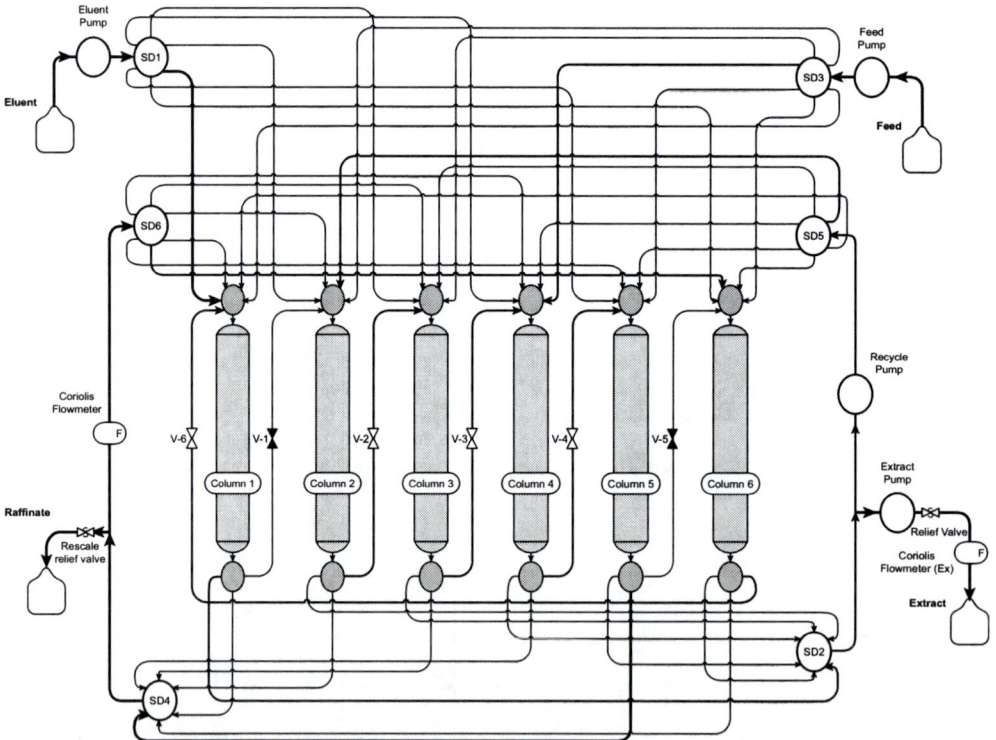

**Figure 4.14** The FlexSMB-LSRE® scheme of pumps and valves at the first step of the (1-2-2-1) classic mode of SMB operation. Bold lines are the active connections; thin lines are stagnant volumes. (Reprinted from *Chemical Engineering & Technology* 35, Sá Gomes, P., Rodrigues, A.E., *Simulated moving bed chromatography: from concept to proof-of-concept*, pp. 17–34, Copyright (2011), with permission from John Wiley & Sons Inc.)

rotary valve, as is the case of the UOP rotary valve or the ISEP®/CSEP systems. This type of valve design allows easy scale-up, reliability, and reasonably low dead volumes, but is quite limited to adaptation for the nonconventional SMB operating modes such as Varicol, JO, etc. The other option is the distributed valve SMB configuration, which uses several generic valves: two-way valves (on—off; with one inlet and one outlet) and/or rotary (multiposition) valves. This second type of valve arrangement provides increased unit flexibility, but with comparatively higher costs because of the need for a considerable number of valves and, in some cases, considerable extra column dead volumes (particularly relevant for small units).

The selected valve system configuration for the FlexSMB unit is similar to the one proposed by Negawa and Shoji in U.S. Patent 5 456 825 (1995). The unit is based on two SD (select-dead-end flow path) valves per stream in the extract and raffinate internal lines, one SD per stream in the feed and eluent lines, and one two-way valve per column. The presented valve and pump scheme allows the unit to use nonconventional modes of operation. This can be easily achieved through the computer interface by simply changing some parameters in the automation routines (power feed, Varicol, Outlet Streams Swing, partial-feed, partial-withdrawal) or by making other modifications in the automation routines (JO process, Intermittent-SMB). The SD2 valve fully withdraws the stream that leaves section I, in this case from column 1. Part of this stream is collected as extract and the remainder is pumped to valve SD5, which delivers the stream to the inlet of section II, in this case, the inlet of column 2. Similarly, valve SD4 withdraws the complete stream from column 5 and splits part of it into the raffinate outlet, and the remainder follows to valve SD6, which returns the stream to the inlet of column 6. The feed inlet flows to column 4 by means of valve SD3 and the eluent/desorbent current with valve SD1 to column 1. Between each two columns is placed a two-way valve (acting as an on—off valve) that is closed to direct all fluid to either extract or raffinate lines. Just four pumps are used: one pump, positioned next to the extract node (in the beginning of section II), allows the fluid to recycle within the system; another pump, placed in the extract line, withdraws the extract current; and two pumps are used to feed the system (one for the feed itself and the other for the eluent/desorbent current).

### 4.3.1.2 Equipment Assembly and Automation

The choice of arrangement in the assembly of an SMB unit is a very critical decision, because several issues may introduce, directly or indirectly, strong limitations to the unit performance owing to the introduction of dead volumes. For instance, tubing and other equipment will introduce dead volumes and the arrangement of valves and pumps may also introduce dead volume asymmetries (Migliorini et al., 2000; Minceva and Rodrigues, 2003; Sá Gomes et al., 2010; Lim and Bhatia, 2011). To minimize the effect of dead volumes on the unit performance, distinct types of equipment assemblage, namely tubing with different lengths and inner diameters, were studied. Dead volumes were minimized by using tubing with 1-mm inner diameter and short dead-end valves. The symmetry of the assembling was maximized by placing all columns in a carousel and by using the same length for tubes that are used for the same function. To reduce the pressure fluctuation around the unit due to the pump flow rates (inlets, outlets, and internal flow rates), four HPLC pumps were used and assisted by two Coriolis flowmeters. The total pressure of the system can be manually adjusted by means of two purge valves, one installed next to the extract pump and other installed at the outlet of the raffinate stream. The purge valves also work as pressure safety devices protecting the HPLC pump heads and SD valves. It is possible to operate the FlexSMB unit using up to 12 columns because all columns are interconnected with 12-way SD valves. For the specific case of a six-column configuration (the conventional mode of operation), the columns are connected to the SD valves using just the odd-valve connecting positions (1–11). This distributed connection scheme will provide reduced variations owing to the port switching velocity asymmetries. Finally, the unit is protected from electrical fluctuations and possible current discharges by connecting all equipment to an integrated power supply.

The automation and the control routines of the SMB unit are critical for the operating flexibility and ensured by using a LabVIEW platform-developed software. With this platform it is possible for the operator to program asynchronous port shifts to all the inlets/outlets or even change the flow rate inside any of the four sections, both changes with a time discretization of 1/5 of the time switch (Sá Gomes, 2009).

### 4.3.1.3 Number and Geometry of Columns

A lower number of columns will be profitable in terms of less complex tubing and valve scheme and also a reduced effort in the column packing procedure. However, the efficiency of an SMB unit will increase with an increasing number of columns used because of the difference decrease between real SMB and theoretical true moving bed (TMB) operation (Pais and Rodrigues, 2003; Silva et al., 2012). The SMB units that use a large number of columns are usually found in large industrial separations, as is the case of the Parex units for $p$-xylene separation (24 columns) (Broughton et al., 1970). The design of SMB units that are usually applied for high added value products is based on a reduced number of columns (five or six). The reduced number of columns could be an advantage by reducing relevant problems associated with extra column dead volumes and the packing and repacking of the columns owing to the considerable aging rate of the adsorbent (around 10% per year). In an SMB unit with a lower number of columns, other more complex modes of operation, such as the Varicol operation, are normally used to compensate for the lower efficiency related to the aging of the adsorbent.

The FlexSMB-LSRE® unit is mostly directed to high added value products separation and so it was assembled with only six standard lab-scale columns. The geometry of the columns is determined by minimizing the maximum pressure drop allowed inside the system, which is directly imposed by both the type and amount of adsorbent and the type (viscosity) of the solvent. Typically, a diameter-to-column ratio of 10 is established for both larger and smaller SMB units (Sá Gomes et al., 2006).

### 4.3.1.4 Dead Volume Compensation

There are several issues that affect the prediction of the performance parameters of an SMB unit by simulation using common separation process models. These issues are related to the uncertainty in model parameters, such as the equilibrium isotherm data (model parameters), the kinetic data (mass transfer coefficient), the hydrodynamics data (diffusivity and axial dispersion coefficients), packing asymmetries (channeling and dead volumes within the columns), extra dead volumes due to other SMB equipment (tubing, valves, and pumps), variations in the port switching velocity (asymmetries and delays), and fluctuations in pump flow rates (fluid inlets, outlets, and internal), among others. Normally, all these issues are not accounted for in the most commonly used SMB models. The uncertainty in the equilibrium, kinetic, and hydrodynamic data is efficiently reduced by experimental determination using adequate methodologies. Other types of issues must be considered before the construction of the SMB unit, such as the equipment dead volume, the tubing, and the valve asymmetries. However, and despite all the efforts spent to minimize these issues, any SMB unit will remain with several points that must be taken into account. The FlexSMB-LSRE® unit, even after the effort spent to minimize these issues during its construction, still has 11.5% dead volume and the port switching velocity variation (delay) is near 0.8 s. There are several compensation strategies for these discrepancies to get simulation predictions close to the experimental results. A simple switching time-compensating measure was developed for the FlexSMB-LSRE®. This measure accounts both for equipment dead volumes and for the switching time delay or asymmetry (Sá Gomes et al., 2010):

$$t^*|_{FlexSMB} = t^*|_{SMB} + \frac{\sum_{j=1}^{IV} V_j^D}{\overline{Q_j^*}} \frac{1}{\sum_{j=I}^{IV} n_j} + t_{subswitch} \tag{4.6}$$

where $t^*|_{SMB}$ is the switching time for an SMB unit with zero dead volumes, $V_j^D$ is the total dead volume in section $j$, $\overline{Q_j^*}$ is the unit's average flow rate, and $t_{subswitch}$ is the switching time delay. The effect of this compensation strategy (Sá Gomes, 2009) in the separation regions of the SMB can be clearly observed in Figure 4.15.

## 4.3.2 Detailed Modeling

### 4.3.2.1 Fixed-Bed Chromatography

For SMB modeling, it must be first noted that an SMB unit can be considered as a sequence of several fixed-bed columns. There are several different models that can be

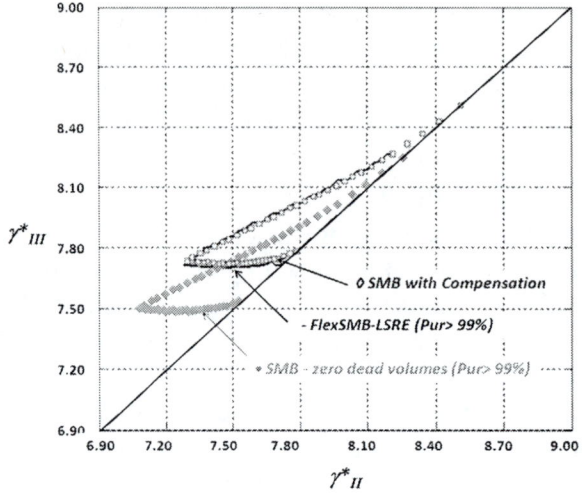

**Figure 4.15** Separation regions according to the equilibrium theory, SMB zero dead volumes, and the FlexSMB-LSRE®, the last two obtained for a minimum purity requirement of 99.0% in both extract and raffinate outlet streams. *(Reprinted from AIChE Journal 56, Sá Gomes, P., Zabkova, M., Zabka, M., Minceva, M., Rodrigues, A.E., Separation of chiral mixtures in real SMB units: the FlexSMB-LSRE®, pp. 125–142, Copyright (2009), with permission from John Wiley & Sons Inc.)*

used to this purpose, such as the ideal model, the equilibrium-dispersive model, the lumped kinetic model, the lumped pore model, or the general rate model. Separations of high added value products (such as chiral separations) using SMB technology are normally well described using a simplification of the general rate model written in terms of the particle pore concentration. Additionally, a linear driving force model approximation is used to describe the overall particle mass transfer resistances, $a_i = \frac{3}{R_p} k_{ov} t^*$, and an axial dispersion term is considered by means of the dimensionless Peclet number. The model is based on the bulk fluid mass balance equations,

$$\frac{\partial C_{b_{i,k}}}{\partial \theta} = \gamma_j^* \left( \frac{1}{Pe_k} \frac{\partial^2 C_{b_{i,k}}}{\partial x^2} - \frac{\partial C_{b_{i,k}}}{\partial x} \right) - \frac{(1 - \varepsilon_b)}{\varepsilon_b} a_i (C_{b_{i,k}} - C_{P_{i,k}}) \tag{4.7}$$

the mass balance in the particle,

$$\varepsilon_P \frac{\partial C_{P_{i,k}}}{\partial \theta} + \frac{\partial q_{i,k}}{\partial \theta} = a_i (C_{b_{i,k}} - C_{P_{i,k}}) \tag{4.8}$$

the initial conditions,

$$\theta = 0: \quad C_{b_{i,k}}(x, 0) = C_{P_{i,k}}(x, 0) = q_{i,k}(x, 0) = 0 \tag{4.9}$$

the boundary conditions,

$$x = 0: \quad C_{b_{i,k}}^0 = C_{b_{i,k}}\Big|_{x=0} - \frac{1}{Pe_x}\frac{\partial C_{b_{i,k}}}{\partial x}\Big|_{x=0} \quad (4.10)$$

$$x = 1: \quad \frac{\partial C_{b_{i,k}}}{\partial x}\Big|_{x=1} = 0 \quad (4.11)$$

and adsorption equilibrium isotherms,

$$q_{i,k} = q_{i,k}^* = f(C_{P_{i,k}}) \quad (4.12)$$

where $\gamma_j^* = \frac{u_j^*}{u_s}$ is the ratio between the SMB fluid and the TMB solid interstitial velocities, $Pe_k = \frac{u_j^* L_c}{D_{L_k}}$ is the column $k$ Peclet number, $C_{P_{i,k}}$ is the particle pore concentration, and $q_{i,k}$ is the adsorbed phase concentration of species $i$ in column $k$.

### 4.3.2.2 SMB Modeling and Design

The modeling of the SMB unit is done as described in Chapter 2. The first step of the design of the SMB unit is the estimation of the maximum flow rate at the inlet of section I, and with that value, it is possible to determine the critical values for $\gamma_I$ and $\gamma_{IV}$ using a safety margin ($\beta$) and the following equations:

$$\gamma_I = \frac{1 - \varepsilon}{\varepsilon}(m_{L2} + Q_{sat}b_2)\beta \quad (4.13)$$

$$\gamma_{IV} = \frac{1 - \varepsilon}{\varepsilon}(m_{L1} + Q_{sat}b_1)/\beta \quad (4.14)$$

where $m_{L1}$, $m_{L2}$, $Q_{sat}$, $b_1$, and $b_2$ are the estimated parameters of the Linear + Langmuir adsorption isotherm model. The safety margin was considered here as 25%, that is, $\beta = 1.25$. Afterward, the switching time was fixed using

$$t^* = \frac{\varepsilon V_c}{Q_I^*}(\gamma_I + 1) \quad (4.15)$$

The definitions of $Q_I^*$, $\gamma_I$, $\gamma_{IV}$, and $t^*$ for pure ethanol and 10% ethanol/90% $n$-hexane, the SMB operating conditions, were obtained by evaluating, through simulation, the ($\gamma_{II} \times \gamma_{III}$) separation regions. The experimental operation of the SMB unit is then carried out by choosing a pair of ($\gamma_{II} \times \gamma_{III}$) values inside the separation region evaluated before. The corresponding SMB $\gamma_j^*$ values are then evaluated by using

$$\gamma_j^* = \gamma_j + 1 \quad (4.16)$$

and then the internal flow rates inside the other sections, by using

$$Q_j^* = \frac{\varepsilon V_c}{t^*}\gamma_j^* \quad (j = II, II, IV) \quad (4.17)$$

Finally, the inlet, outlet, and recycle SMB flow rates are estimated by

$$Q_E = Q_I^* - Q_{IV}^* \qquad (4.18)$$

$$Q_X = Q_I^* - Q_{II}^* \qquad (4.19)$$

$$Q_F = Q_{III}^* - Q_{II}^* \qquad (4.20)$$

$$Q_R = Q_{III}^* - Q_{IV}^* \qquad (4.21)$$

$$Q_{Rec} = Q_{IV}^* \qquad (4.22)$$

## 4.3.3 Test and Operation

### 4.3.3.1 Predictions of Separation Regions and Initial Operating Conditions

The estimated values of Peclet and mass transfer coefficient were used to predict the performance parameters of the SMB separation process. This study is normally done by constructing the separation region and calculating the productivity and the solvent consumption values. These estimations are used to define the final decision on the selection of the solvent composition to perform the real separation of each profen racemic mixture. As examples, the two case studies presented in Section 4.2 will be used to perform the final separation and test the FlexSMB unit.

The maximum pressure allowed in the FlexSMB-LSRE® unit is near 50 bar owing to a trade-off between maximum adsorbent pressure drop and hardware limitations. In view of these limitations, it was decided to work with a maximum system pressure around 35 bar, measured at the beginning of section I. This pressure limit implies that, for a 100% ethanol solvent composition, it is possible to operate with an internal flow rate in section I of $Q_I^* = 20$ ml/min, whereas, for a 10% ethanol/90% $n$-hexane composition, the corresponding maximum flow rate at section I is $Q_I^* = 30$ ml/min. These two maximum internal SMB flow rates were fixed for the evaluation of all separation regions. Additionally, a classic SMB (1-2-2-1) configuration and a safety margin of 25% for sections I and IV was used and a 99.0% purity criterion in both outlet streams was fixed to estimate SMB separation regions.

### 4.3.3.2 Experimental Operation

The initial operating conditions (switching time, eluent, extract, feed, and recycle flow rates) are set on the FlexSMB-LSRE® control and automation routine, written in LabView platform (National Instruments). After the SMB unit is initialized, the extract and raffinate outlet streams are continuously collected and, at the end of a complete cycle ($6t^*$), both samples are weighed to evaluate and correct, if necessary, the flow rates. In addition, the extract and the recycle flow rates are also checked by two Coriolis flowmeters. The collected extract and raffinate samples are then analyzed to evaluate the enantiomer concentrations in the outlet streams. The cyclic steady state is considered

**Table 4.2** Experimental Operating Conditions Used in SMB Operation for the Separation of the Ketoprofen and Flurbiprofen Enantiomers, Using a Feed Concentration of 20 g/l

| Flow Rates (ml/min) | | | | Performance |
| --- | --- | --- | --- | --- |
| SMB Operation | SMB Internals | Pressure (Bar) | TMB Ratios | Parameters |
| **Ketoprofen: $C_1 + C_2 = 20.1$ g/l; $T = 24$ °C; 36 cycles** | | | | |
| $Q_E = 9.14$ | $Q_I^* = 19.56$ | $\Delta P_F = 8$ | $\gamma_I = 3.047$ | $C_2^X = 0.667$ g/l |
| $Q_F = 0.51$ | $Q_{II}^* = 12.45$ | $\Delta P_E = 22$ | $\gamma_{II} = 1.576$ | $C_1^R = 1.95$ g/l |
| $Q_X = 7.11$ | $Q_{III}^* = 12.96$ | $\Delta P_{Rec} = 24$ | $\gamma_{III} = 1.681$ | $PU_X = 98.7\%$ |
| $Q_R = 2.54$ | $Q_{IV}^* = 10.42$ | $\Delta P_X = 36$ | $\gamma_{IV} = 1.156$ | $PU_R = 100.0\%$ |
| $Q_{Rec} = 10.42$ | | | | $SC = 0.94$ l/g |
| | | | | $PR = 3.26$ g/(l h) |
| **Flurbiprofen: $C_1 + C_2 = 20.1$ g/l; $T = 24$ °C; 23 cycles** | | | | |
| $Q_E = 15.90$ | $Q_I^* = 31.00$ | $\Delta P_F = 16$ | $\gamma_I = 5.414$ | $C_2^X = 1.22$ g/l |
| $Q_F = 1.45$ | $Q_{II}^* = 19.00$ | $\Delta P_E = 23$ | $\gamma_{II} = 2.931$ | $C_1^R = 2.72$ g/l |
| $Q_X = 12.00$ | $Q_{III}^* = 20.45$ | $\Delta P_{Rec} = 24$ | $\gamma_{III} = 3.231$ | $PU_X = 99.6\%$ |
| $Q_R = 5.35$ | $Q_{IV}^* = 15.10$ | $\Delta P_X = 34$ | $\gamma_{IV} = 2.124$ | $PU_R = 99.7\%$ |
| $Q_{Rec} = 15.10$ | | | | $SC = 0.59$ l/g |
| | | | | $PR = 9.29$ g/(l h) |

to be obtained when both the extract and the raffinate concentrations do not change more than 3% during five successive cycles. At that moment, several samples are collected to obtain the internal concentration profiles, using a six-port valve installed at the outlet of column 6. The samples are collected at 25%, 50%, 75%, and 95% of a switch time and then analyzed by HPLC using an analytical column. The main experimental results obtained for both separations (Ribeiro, 2010) are presented in Table 4.2.

Figure 4.16 presents the evolution of both enantiomer concentrations in the extract and raffinate outlets (note that, at the beginning of these SMB runs, the unit is not free of profen enantiomers; it corresponds to the final state of a previous run). In this figure is also presented the experimental and simulated internal concentration profiles at the cyclic steady state for both profen racemic mixtures. The internal concentration profiles show two small plateaus near the extract and raffinate outlets, due to dead volumes of the FlexSMB-LSRE® unit and taken into account in the simulation and control software. Figure 4.16 shows a good agreement between experimental data and simulation predictions using the SMB model. For ketoprofen and a feed concentration of 20 g/l, a productivity of $PR = 3.26$ g/(l h) and a solvent consumption of $SC = 0.94$ l/g are obtained, with purities of 98.7% and 100.0% at the extract and raffinate streams, respectively. For flurbiprofen and a feed concentration of 20 g/l, a productivity of $PR = 9.29$ g/(l h) and a solvent consumption of $SC = 0.59$ l/g are obtained, with purities of 99.6% and

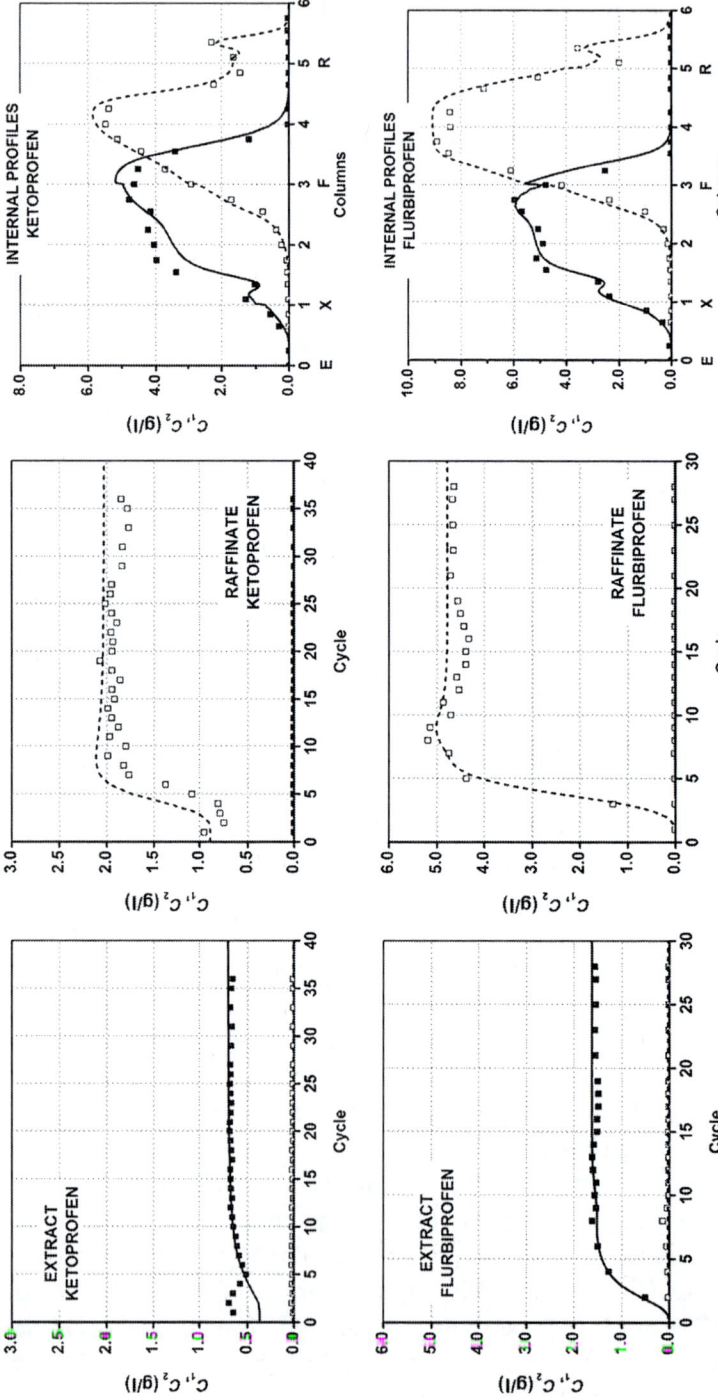

**Figure 4.16** Experimental profiles of enantiomers concentration in the extract outlet stream (left), raffinate (middle) and the concentration profiles inside the 6 SMB columns at the cyclic steady state (right) for both enantiomers and for the two racemic mixtures separation (ketoprofen, at the top and flurbiprofen, at the bottom). Experimental results are presented as points, and predicted adsorption behavior is presented by lines. Total feed concentration of 20 g/l for both racemic mixtures.

99.7% at the extract and raffinate streams, respectively. As main conclusions for these two profen chiral drugs, it can be noted that the "best" mobile-phase compositions for two chiral systems of the same family (both are profens) are different. These results show that individual studies must be carried out for each chiral separation system, because different profen drugs may have different adsorption behaviors.

## 4.4 CONCLUSIONS

This chapter stresses the methodology for process development using SMB technology for chiral separations as practiced in the LSRE group. The methodology must be explored for each racemic mixture and screening of the wider possible range of solvents using a stationary phase suitable for preparative separations. Each step of the methodology is illustrated using the chiral separation of two nonsteroidal anti-inflammatory drugs, ketoprofen and flurbiprofen. These two drugs are prescribed worldwide and are still marketed as racemic mixtures and frequently associated with several side effects. It should be noted that the "best" mobile-phase compositions for two chiral systems of the same family are different. In the last part of this chapter a general description of the design and construction of the FlexSMB-LSRE® is given, as well as its detailed modeling, design, and operational testing.

## NOMENCLATURE

| | |
|---|---|
| $b$ | Adsorption isotherm parameter ($dm^3/g$) |
| $C$ | Liquid-phase concentration ($g/dm^3$) |
| $D_L$ | Axial dispersion coefficient ($m^2/s$) |
| $k$ | Mass transfer coefficient ($s^{-1}$) |
| $k_{ov}$ | Global mass transfer coefficient ($m/s$) |
| $L_c$ | Column length (m) |
| $m_L$ | Adsorption isotherm parameter |
| $N_C$ | Number of columns |
| $Pe$ | Peclet number |
| $PR$ | Productivity ($g/(h\ l_{bed})$) |
| $PU$ | Purity |
| $q$ | Average adsorbed phase concentration ($g/dm^3$) |
| $q^*$ | Adsorbed phase concentration in equilibrium with $C$ ($g/dm^3$) |
| $Q_{sat}$ | Adsorption isotherm parameter ($g/dm^3$) |
| $Q$ | Volumetric flow rate (ml/min) |
| $\overline{Q_j^*}$ | The unit's average flow rate (ml/min) |
| $R_p$ | Particle radius (m) |
| $S$ | Racemic mixture solubility ($g_{solute}/kg_{solvent}$) |
| $SC$ | Solvent consumption (l/g) |
| $t^*$ | Switching time (s) |
| $u_s$ | Interstitial solid velocity (m/s) |
| $u^*$ | Fluid interstitial velocity in the SMB (m/s) |

| | |
|---|---|
| $V_j^D$ | Dead volume in section $j$ (m³) |
| $V_T$ | Total volume (m³) |
| $w$ | Mass (g) |
| $x$ | Dimensionless axial coordinate |

## Greek Letters

| | |
|---|---|
| $\varepsilon$ | Bed porosity |
| $\varepsilon_T$ | Total porosity |
| $\gamma$ | Ratio between fluid (TMB) and solid interstitial velocities |
| $\gamma^*$ | Ratio between fluid (SMB) and solid interstitial velocities |
| $\theta$ | Dimensionless time |

## Subscripts and Superscripts

| | |
|---|---|
| 0 | Inlet |
| $b$ | Bed |
| $c$ | Column |
| $E$ | Eluent |
| $F$ | Feed |
| $i$ | Component $i$ ($i = 1, 2$ or $A, B$) |
| $j$ | Section $j$ ($j = I, II, III, IV$ or $1, 2, 3, 4$) |
| $k$ | Column ($k = 1, 2, \ldots, N_c$) |
| $p$ | Particle |
| $R$ | Raffinate |
| $REC$ | Recycle |
| $s$ | Solid |
| $VR$ | Vial plus residue |
| $VS$ | Vial plus saturated solution |
| $V$ | Vial |
| $X$ | Extract |

## REFERENCES

Broughton, D.B., Neuzil, R.W., Pharis, J.M., Brearley, C.S., 1970. Parex process for recovering paraxylene. Chem. Eng. Prog. 66, 70—75.

Chin, C.Y., Wang, N.H.L., 2004. Simulated moving bed equipment designs. Sep. Purif. Rev. 33, 77—155.

Granberg, R.A., Rasmuson, A.C., 1999. Solubility of paracetamol in pure solvents. J. Chem. Eng. Data 44, 1391—1395.

Lim, Y.I., Bhatia, S.K., 2011. Effect of dead volume on performance of simulated moving bed process. Adsorption 17, 109—120.

Maier, N.M., Franco, P., Lindner, W., 2001. Separation of enantiomers: needs, challenges, perspectives. J. Chromatogr. A 906, 3—33.

Migliorini, C., Mazzotti, M., Morbidelli, M., 2000. Robust design of countercurrent adsorption separation processes: 5. Nonconstant selectivity. AIChE J. 46, 1384—1399.

Minceva, M., Rodrigues, A.E., 2003. Influence of the transfer line dead volume on the performance of an industrial scale simulated moving bed for p-xylene separation. Sep. Sci. Technol. 38, 1463—1497.

Negawa, M., Shoji, F., 1995. Simulated Moving Bed Separation System. US Patent 5 456 825.

Pais, L.S., Rodrigues, A.E., 2003. Design of simulated moving bed and Varicol processes for preparative separations with a low number of columns. J. Chromatogr. A 1006, 33—44.

Panico, A.M., Cardile, V., Gentile, B., Garufi, F., Avondo, S., Ronsisvalle, S., 2005. "In vitro" differences among (R) and (S) enantiomers of profens in their activities related to articular pathophysiology. Inflammation 29, 119—128.

Rajendran, A., Paredes, G., Mazzotti, M., 2009. Simulated moving bed chromatography for the separation of enantiomers. J. Chromatogr. A 1216, 709—738.

Ribeiro, A.E., Gomes, P.S., Pais, L.S., Rodrigues, A.E., 2011a. Chiral separation of flurbiprofen enantiomers by preparative and simulated moving bed chromatography. Chirality 23, 602—611.

Ribeiro, A.E., Gomes, P.S., Pais, L.S., Rodrigues, A.E., 2011b. Chiral separation of ketoprofen enantiomers by preparative and simulated moving bed chromatography. Sep. Sci. Technol. 46, 1726—1739.

Ribeiro, A.E., Graca, N.S., Pais, L.S., Rodrigues, A.E., 2008. Preparative separation of ketoprofen enantiomers: choice of mobile phase composition and measurement of competitive adsorption isotherms. Sep. Purif. Technol. 61, 375—383.

Ribeiro, A.E., Graca, N.S., Pais, L.S., Rodrigues, A.E., 2009. Optimization of the mobile phase composition for preparative chiral separation of flurbiprofen enantiomers. Sep. Purif. Technol. 68, 9—23.

Ribeiro, A.M.E., 2010. Influência do solvente no desempenho de separações quirais em leito móvel simulado (Ph.D thesis). University of Porto.

Sá Gomes, P., 2009. Advances in Simulated Moving Bed (Ph.D thesis). University of Porto.

Sá Gomes, P., Minceva, M., Rodrigues, A.E., 2006. Simulated moving bed technology: old and new. Adsorption 12, 375—392.

Sá Gomes, P., Rodrigues, A.E., 2007. Outlet Streams Swing (OSS) and MultiFeed operation of simulated moving beds. Sep. Sci. Technol. 42, 223—252.

Sá Gomes, P., Rodrigues, A.E., 2012. Simulated moving bed chromatography: from concept to proof-of-concept. Chem. Eng. Technol. 35, 17—34.

Sá Gomes, P., Zabkova, M., Zabka, M., Minceva, M., Rodrigues, A.E., 2010. Separation of chiral mixtures in real SMB units: the FlexSMB-LSRE (R). AIChE J. 56, 125—142.

Seidel-Morgenstern, A., Kessler, L.C., Kaspereit, M., 2008. New developments in simulated moving bed chromatography. Chem. Eng. Technol. 31, 826—837.

Silva, V.M.T.M., Sá Gomes, P., Rodrigues, A.E., 2012. Use of ion exchange resins in continuous chromatography for sugar processing. In: Inamuddin, D., Luqman, M. (Eds.), Ion Exchange Technology II. Springer, Netherlands, pp. 109—135.

Tachibana, K., Ohnishi, A., 2001. Reversed-phase liquid chromatographic separation of enantiomers on polysaccharide type chiral stationary phases. J. Chromatogr. A 906, 127—154.

Yashima, E., 2001. Polysaccharide-based chiral stationary phases for high-performance liquid chromatographic enantioseparation. J. Chromatogr. A 906, 105—125.

# CHAPTER 5

# The Parex Process for the Separation of *p*-Xylene

## 5.1 *p*-XYLENE PRODUCTION

Worldwide, the production of aromatics is obtained from three supply sources: reformates (~68%), pyrolysis gasoline (~29%), and coke-oven light oil (~3%). As such, aromatics are obtained as co-products in refinery catalytic reformers, olefin plants, or from coal-tar processing. However, some of these streams have low xylene content and, consequently, are not suitable for economic production of *p*-xylene.

*Mixed xylenes* (a mixture of the three dimethylbenzenes and ethylbenzene, with the general formula $C_8H_{10}$), from which individual isomers are recovered, are usually obtained from reformate, pyrolysis gasoline, or after transalkylation or toluene disproportionation. *Mixed xylenes* are frequently used as solvents (e.g., paint industry), and they are large contributors to the gasoline pool due to their high octane number (in limited quantity due to the environmental and public health problems that they may cause).

The main applications of *p*-xylene, *m*-xylene, and *o*-xylene involve their oxidation to terephthalic acid, isophthalic acid (precursor of polyesters), and phthalic anhydride (precursor of plasticizers), respectively. In the case of *p*-xylene, terephthalic acid is reacted with ethylene glycol to give polyester for fibers, films, and molding resins (Wittcoff et al., 2004). In the case of ethylbenzene, it is dehydrogenated to styrene, which is then converted to polystyrene and other polymers. Among the xylene isomers, *p*-xylene is the chemical with the greatest market demand.

Generally, about 75% of the *mixed xylenes* are extracted from reformate. The reformate composition has a strong dependence on the operating conditions of the reformer and on the naphtha feedstock.

Pyrolysis gasoline (or pygas) is obtained in the production of olefins by steam cracking (in which naphtha or gas oil is used as feed). This stream has lower *mixed xylenes* content than reformate, and frequently up to 40% of the *mixed xylenes* fraction obtained is ethylbenzene. Consequently, pyrolysis gasoline is commonly used for the extraction of benzene instead of xylenes. The $C_8$ fraction obtained from pyrolysis gasoline is frequently sent to the gasoline pool instead of being used for the individual separation of isomers. When the steam cracker is part of a larger aromatics complex, the extraction of this xylene fraction becomes more relevant.

Aromatics are obtained either from coal or crude oil; nevertheless, the production of *mixed xylenes* from coal is of minor importance since it represents only around 1% of the

*Simulated Moving Bed Technology*
ISBN 978-0-12-802024-1

xylenes produced (Locating and estimating air emissions from sources of xylene, 1994) and less than 5% of the total global supply of aromatics. Coke oven light oil contains only between 4% and 8% of *mixed xylenes*.

The production of *p*-xylene from the mixture of xylenes involves mostly two steps: *p*-xylene separation (from its isomers and ethylbenzene) and isomerization to reestablish the xylenes equilibrium toward the formation of *p*-xylene after separation. Recently, hybrid processes have been developed.

Typically, separations in the refining and petrochemical industry are made by distillation through the exploitation of the different compound boiling points. In the case of the separation of the xylene isomers and ethylbenzene, the only feasible separation through distillation is the separation of *o*-xylene from the other components, since *o*-xylene has a boiling point at least 5 °C above the others. The separation of the other $C_8$ isomers through distillation is barely feasible. Although it is possible to separate *o*-xylene with a minimum purity of 98% and 95% yield, it requires 120–150 effective plates and a reflux ratio of 10 to 15 (Fabri et al., 2012), which leads to an expensive separation. To overcome the difficulty of separation through distillation, *p*-xylene is separated essentially through crystallization, adsorption, or by using hybrid combinations of both processes.

Since the development of the simulated moving bed (SMB) technology for the separation through adsorption, it has been preferred over crystallization, and almost all new *p*-xylene production capacity uses adsorption. A baseline production of *p*-xylene using crystallization is maintained based on sites that existed before the adsorptive separation technology was established.

Furthermore, among the adsorption separation technologies, the Parex process (UOP Sorbex trademark processes) produces the majority of *p*-xylene, with approximately 21.9 million metric tons per annum of *p*-xylene in 2010 (Millard, 2010).

### 5.1.1 Crystallization Processes

The first implemented process for the separation of *p*-xylene from its isomers and ethylbenzene was crystallization at low temperature.

*p*-Xylene has a higher freezing point than *o*-xylene, *m*-xylene, and ethylbenzene. In the crystallization processes, the mixture of xylenes and ethylbenzene is refrigerated to precipitate *p*-xylene as a crystalline solid. In commercial processes, *p*-xylene crystallization is carried out at a temperature just above the eutectic point. *p*-Xylene crystals usually form at about −4 °C, and the *p*-xylene/*m*-xylene eutectic is reached at about −68 °C (Cannella, 2007). At this temperature, *p*-xylene is still soluble in the remaining liquid containing $C_8$ aromatics (mother liquor), which means that crystallizers are limited to the eutectic composition restricting *p*-xylene recovery to about 60–65% per pass.

After separation from the mother liquor by centrifugation or filtration, a washing step follows using toluene or a portion of the *p*-xylene product. In commercial processes, *p*-xylene purity higher than 99% is obtained after two or more crystallization steps separated by centrifuges.

The several commercial processes can be separated by the type of refrigeration: indirect (Krupp—Koppers (Fabri et al., 2012) and the Phillips Petroleum (McKay and Goard, 1965)) processes or direct (Arco (Desideri et al., 1974), Chevron (Laurich, 1969), Sohio/BP (Fabri et al., 2012), and Maruzen Oil (Ockerbloom, 1971)) processes.

## 5.1.2 Adsorption Processes

The first process licensed for the separation of *p*-xylene through adsorption was the Parex process (Broughton et al., 1970). This process was first patented in 1961 by Broughton and Gerhold (US Patent 2985589), assigned by Universal Oil Products (UOP), and introduced the SMB technology for the continuous separation of *p*-xylene through adsorption.

When in 1971, UOP started commercializing the Parex process for the separation of *p*-xylene (the first unit to be put on stream was the U.R.B.K. plant in Wesseling, West Germany) using the SMB technology, this process started to be preferred over crystallization, since the advantages of this new adsorptive process are significant. In only 4 years, UOP licensed the first 15 Parex units (Broughton et al., 1975).

The Parex process is an efficient, reliable, and simple way to recover more than 97 wt% of *p*-xylene in the feed per pass (99.9 wt% purity), which is considerably higher than the recovery achieved through crystallization. A higher *p*-xylene recovery provides two other advantages to the overall process. Firstly, the recycle flow is lower, so the capacity of the isomerization reactor (located downstream) is smaller. In addition, since the isomerization reaction is limited by the equilibrium concentration of *p*-xylene, any quantity of *p*-xylene in the feed of the isomerization reactor reduces the *p*-xylene production per pass.

In the SMB technology, the differences in affinity of the several components of the feed for the adsorbent are explored by simulating the true countercurrent movement of the solid and the liquid. To avoid the problems of maintaining plug flow of the solid, abrasion of the equipment, and formation of fines due to friction between the adsorbent particles, in the SMB the solid does not move, and its movement is simulated by switching the four ports (raffinate, extract, feed, and desorbent) periodically and synchronously in the direction of fluid flow. In the UOP Sorbex (Johnson, 2004) processes (Molex, Parex, MX Sorbex, Olex, Cresex, Cymex, and Sarex), port switching is controlled by a rotary valve (Carson and Purse, 1962).

In this selective adsorption process, *p*-xylene is the component with higher affinity to the adsorbent (faujasite-type zeolite) and it is collected in the extract stream. The other xylene isomers and ethylbenzene are collected in the raffinate stream.

In the Parex process, separation is accomplished at 177 °C and around 9 bars.

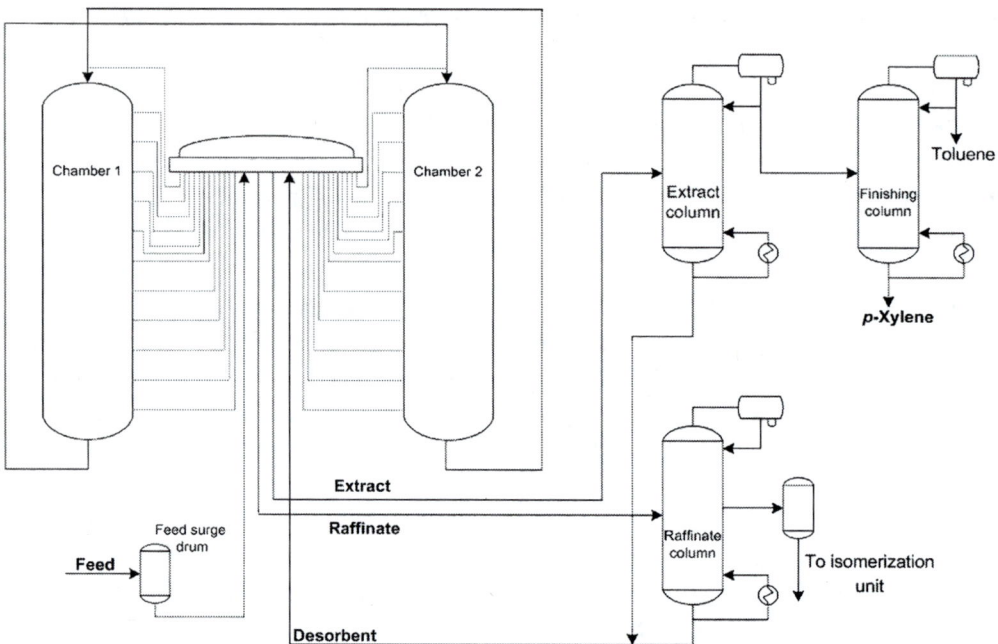

**Figure 5.1** Parex process.

The extract and raffinate are obtained diluted in the desorbent (*p*-diethylbenzene), which is separated through distillation and recycled back to the adsorbent chambers/rotary valve (Figure 5.1). *p*-Diethylbenzene has a higher boiling point than xylenes and, therefore, it is obtained in the bottom of the extract and raffinate distillation columns. The extract is sent to a finishing column in which *p*-xylene is obtained in the bottom as a final product. The finishing column is used to separate toluene and nonaromatics from *p*-xylene.

Table 5.1 summarizes the earliest patents from UOP for the SMB technology associated with the separation of *p*-xylene.

Over the years, the *p*-xylene production capacity of the Parex units commissioned has increased significantly due to advances in the operation of the Parex unit and improvement of adsorbents.

Focusing on the process operation and only on the patents issued in the last few years, advances are related mainly to flushing streams and solutions to avoid the effect of dead volumes created by bed lines that connect the ports in the adsorbent chambers to the rotary valve (Frey, 2007; Hotier, 2009; Porter, 2011).

In recent years, some other solutions related to the utilization of SMB units in series (Lee and Shin, 2009) and parallel (Leflaive et al., 2011a) have also emerged.

**Table 5.1** First Patents of the Simulated Moving Bed (SMB) Technology for the Separation of *p*-Xylene Assigned by UOP

| Patent No. Title Date | Inventors Assignee | Invention |
|---|---|---|
| **US 2985589** *Continuous sorption process employing fixed bed of sorbent and moving inlets and outlets* 23/05/1961 | **Broughton, D.B.** **Gerhold, C.G.** Universal Oil Products Company | Continuous process for separating components of a mixture by contacting with an elongated bed or a plurality of fixed beds of selective solid sorbent. Through the simulated countercurrent circulation of the solid and liquid, the component with higher affinity to the adsorbent is collected in the extract and the others are collected in the raffinate. The points of introducing the feedstock and the desorbent and collecting the extract and raffinate are constantly shifting downstream in equal increments. |
| **US 3040777** *Rotary valve* 26/06/1962 | **Carson, D.B.** **Purse, F.V.** Universal Oil Products Company | Rotary multiport valve for distributing fluid streams simultaneously to a plurality of process lines according to a periodic sequence determined by the valve construction. |
| **US 3214247** *Fluid distributing means for packed chambers* 26/10/1965 | **Broughton, (1965)** Universal Oil Products Company | Fluid distributing means for packed chambers, which comprises a combination of perforate distributor pipe and a baffle plate. This combination is adapted for use in a fluid–solid contacting chamber to improve mixing and uniformity of flow through the contact beds. |
| **US 3558730** *Aromatic hydrocarbon separation by adsorption* 26/01/1971 | **Neuzil, R.W.** Universal Oil Products Company | Process for separating *p*-xylene from a mixture of isomers and ethylbenzene using a faujasite type adsorbent. |
| **US 3789989** *Flow distribution apparatus* 05/02/1974 | **Carson, (1974)** Universal Oil Products Company | Baffling and flow distribution apparatus for use in a system in which a fluid is desired to be passed from one bed of particles to another bed of particles with addition or withdrawal of fluids between the beds. |

Lee and Shin (2011) presented the idea of injecting each raw material (coming from different units) at different points of the unit according to the composition of each stream. Further developing this idea, Porter and Pilliod (2012) suggest the injection of streams with different p-xylene concentrations in the SMB unit by having two independent rotary valves in the same SMB system.

The major advances in the Parex process over the years are related to the development of new adsorbents. Improved adsorbents are responsible for a considerable capacity increase in the plants (Wantanachaisaeng and O'Neil, 2007). Among the commercial adsorbents, namely, ADS-3, ADS-7, ADS-27 (UOP ADS-27™ Adsorbent, 2007), and recently, ADS-37 (UOP ADS-37™ Adsorbent, 2009) and ADS-47 (Brookes, 2012; Werba, 2011), a capacity increase of 15% from ADS-7 to ADS-27 (Cannella, 2007; Wantanachaisaeng and O'Neil, 2007) and of 6% from ADS-27 to ADS-37 (UOP ADS-37™ Adsorbent, 2009) has been observed. The patents concerning new adsorbents for the separation of p-xylene in recent years (Priegnitz et al., 2012; Hurst et al., 2012; Cheng and Hurst, 2012; Cheng and Johnson, 2009; Kulprathipanja et al., 2009) have been dedicated mainly to binderless adsorbents.

Besides UOP's Parex, other processes for the separation of p-xylene through adsorption have emerged, such as IFP's Eluxyl (Ash et al., 1994; Wolff and Leflaive, 2004) and Toray's Aromax[1] (Otani et al., 1973, 1977). Both processes claim to achieve the p-xylene market specifications in terms of purity.

The Aromax process was developed in the 1970s by Toray Industries, Inc. (Japan). The process operates in the liquid phase at 140 °C and consists of a horizontal series of independent chambers containing fixed-bed adsorbent. The process does not operate with a rotary valve, and instead, specially designed on—off valves controlled by computer are used to open and close inlet/outlet ports around the beds.

The Eluxyl process was introduced in 1994 by IFP, and also uses a series of on—off valves controlled by computer to move the inlet/outlet ports along the fixed-bed columns. The Eluxyl process uses an adsorbent designated SPX 3000. This process continues to be improved, and several patents have been published in the last few years. As an example, the valve system has been improved and the number of valves reduced (Hotier et al., 2009).

Some patents of recent years regarding p-xylene separation through the SMB technology are associated with the simultaneous production of compounds (Wolff et al., 2009; Hotier and Il, 2009; Leflaive et al., 2005, 2004). Examples include production of p-xylene and styrene by dehydrogenation of ethylbenzene to produce styrene; production of p-xylene and benzene by combining selective disproportionation, crystallization, and SMB in a process; and the production of p-xylene and m-xylene or o-xylene by sending the raffinate to another SMB unit.

---

[1] Toray's Aromax—It should not be confused with Chevron's Aromax process for reforming naphtha into aromatics.

In the production of p-xylene through adsorption, technology for the separation through pressure swing adsorption (PSA) has also emerged. The main goal of this new PSA application is to maintain the whole p-xylene separation process in the gas phase and to eliminate the expensive compression step before the isomerization reactor. As an example, in the patent of Williams et al. (2007) by BP Corporation North America, Inc., several possible embodiments of applying a nondecreasing total-pressure/swinging partial-pressure method are described to adsorb p-xylene from a gaseous feed of xylene isomers, desorb, collect a p-xylene-enriched product, and isomerize an unadsorbed p-xylene-depleted portion of the feed to produce more p-xylene.

### 5.1.3 Adsorption/Crystallization Hybrid Processes

Hybrid processes using crystallization and adsorption have generally appeared to increase the capacity of existing crystallization plants (Cannella, 2007). However, the combination of both operations can be useful for eliminating eutectic constraints and enabling single-stage crystallization above 90% (Commissaris, 2004).

The idea of combining adsorption and crystallization was first announced by IFP and Chevron in 1994 to further purify the extract stream from the SMB (with a p-xylene purity of 90—95%) in a single stage crystallizer (the filtrate is recycled back to the SMB unit). The patents of Mikitenko and Hotier (1999) and Ash et al. (1999) are examples of patents granted to both assignees using adsorption and crystallization combined for the production of p-xylene.

A recent patent by Lee and Shin (2010) also claims this arrangement of having crystallization after SMB units to further purify the extract stream.

### 5.1.4 Separation/Isomerization Hybrid Processes

Isomerization is part of the p-xylene production processes to maximize the process yield in p-xylene. Isomerization reactors are commonly located after the p-xylene separation unit to reestablish the xylene equilibrium in the raffinate stream (p-xylene-depleted stream obtained in the separation unit). In the typical UOP aromatics complex, the stream obtained from the isomerization unit (Isomar unit) is recycled to the Parex process (Parex—Isomar loop; Par—Isom™ Process, 2007), as is shown in Figure 5.2.

The xylene isomerization process converts m-xylene and o-xylene to p-xylene in fixed-bed catalytic reactors, according to the thermodynamic equilibrium at a certain process temperature (Perego and Pollesel, 2009). Isomerization catalysts are designed to also convert ethylbenzene to xylenes, benzene, and lights, or benzene and diethylbenzene. Xylene-isomerization feedstock usually contains more than 10% of ethylbenzene. If ethylbenzene is not converted, it will accumulate in the recycle streams. By

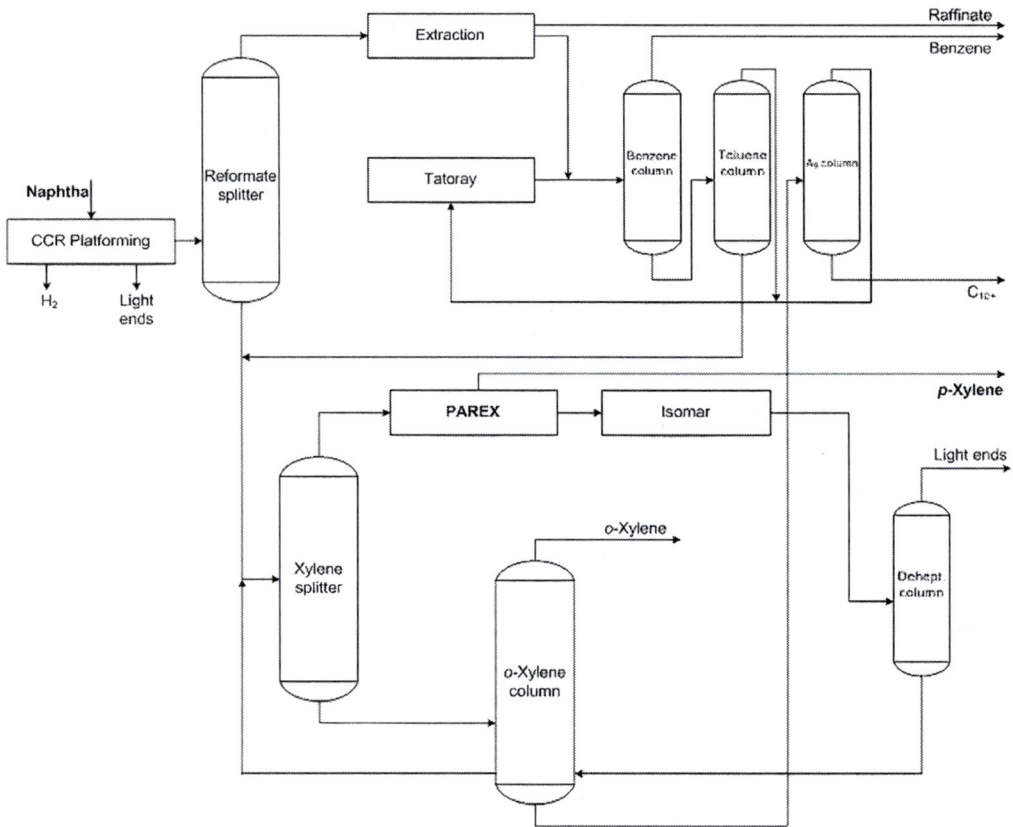

**Figure 5.2** Integrated UOP aromatics complex (Johnson, 2004).

converting ethylbenzene to easily removable products by distillation, this problem is avoided.

Recently, the combination of the separation through adsorption using the SMB technology and reaction (SMB reactor) in one unit was studied by Minceva et al. (2008) and patented by Bergeot et al. (2009).

Using combinations of SMB and isomerization, new ideas related to the separation of ethylbenzene from the raffinate stream using membranes, or including a second withdrawal of raffinate in the SMB unit, have appeared, so that separated ethylbenzene can be treated in a different isomerization reactor.

In the patent of Leflaive et al. (2011b), three withdrawals from the SMB unit are considered: an extract, an intermediate raffinate that contains essentially ethylbenzene, and a second raffinate stream that contains essentially *m*- and *o*-xylene. The intermediate raffinate will be treated in an isomerization step operating in the vapor phase, and the second raffinate stream will be treated in another isomerization step operating in the vapor or liquid phase. If *p*-xylene purity is not achieved, a crystallization step is proposed.

The isomerization of *m*- or *o*-xylene requires only mild isomerization conditions, whereas for ethylbenzene more stringent conditions are needed (leading to a certain deal-kylation of *m*-xylene/*o*-xylene). To overcome this problem, Leflaive et al. (2009) proposed to separate ethylbenzene by means of a membrane module. This procedure purges the ethylbenzene and avoids its accumulation in the process, whereas the single isomerization only treats a stream depleted in ethylbenzene. The patent presents two process configurations, considering one or two raffinate stream withdrawals from the SMB unit, but in both cases only one isomerization unit is used.

## 5.2 THE PAREX PROCESS

The Parex process (Broughton and Gerhold, 1961) operates according to the SMB principle; however, a seven-zone design is used. The objective is to separate *p*-xylene from a mixture containing: *m*-xylene, *o*-xylene, ethylbenzene, a nonaromatic fraction, and vestigial quantities of toluene.

### 5.2.1 Configuration

To obtain the extract stream, rich in *p*-xylene, and the raffinate stream, depleted in *p*-xylene, the commercial Parex unit has 24 adsorbent beds divided into two chambers (Figure 5.3). Each adsorbent chamber has 12 beds and each one is connected to a rotary valve. The rotary valve switches periodically and synchronously the inlet and outlet ports of the unit in the direction of fluid flow.

The liquid circulates between the two adsorbent chambers by means of pumps, which are located, respectively, between bed 12 of column 1 and bed 13 of column 2 (push-around pump), and bed 24 of column 2 and bed 1 of column 1 (pumparound pump) (Silva, 2013). In the pumparound line, a pumparound controller keeps each zone flow rate constant as the zone moves around the adsorption circuit. The pumparound control is based on zone equations written as a function of the rotary valve position at any given time. The position of the rotary valve is usually defined by the feed-stream location (feed-inlet position, *FIP*). The system is full of practically incompressible liquid, therefore fixing the flow rate in one zone necessarily fixes the flow rates in all zones, since at each inlet or outlet point, the zone flow rate downstream of the point is equal to the flow rate upstream, plus or minus whatever volume of liquid is added or removed at that point. In the push-around line, a flow meter permits an independent check of the constancy of the zone flow rates.

The rotary valve, patented by UOP in 1962 (Carson and Purse, 1962), is a device for distributing the seven streams entering and leaving the adsorbent chambers. The valve rotates periodically by 15° at a constant rate; the time interval between two consecutive

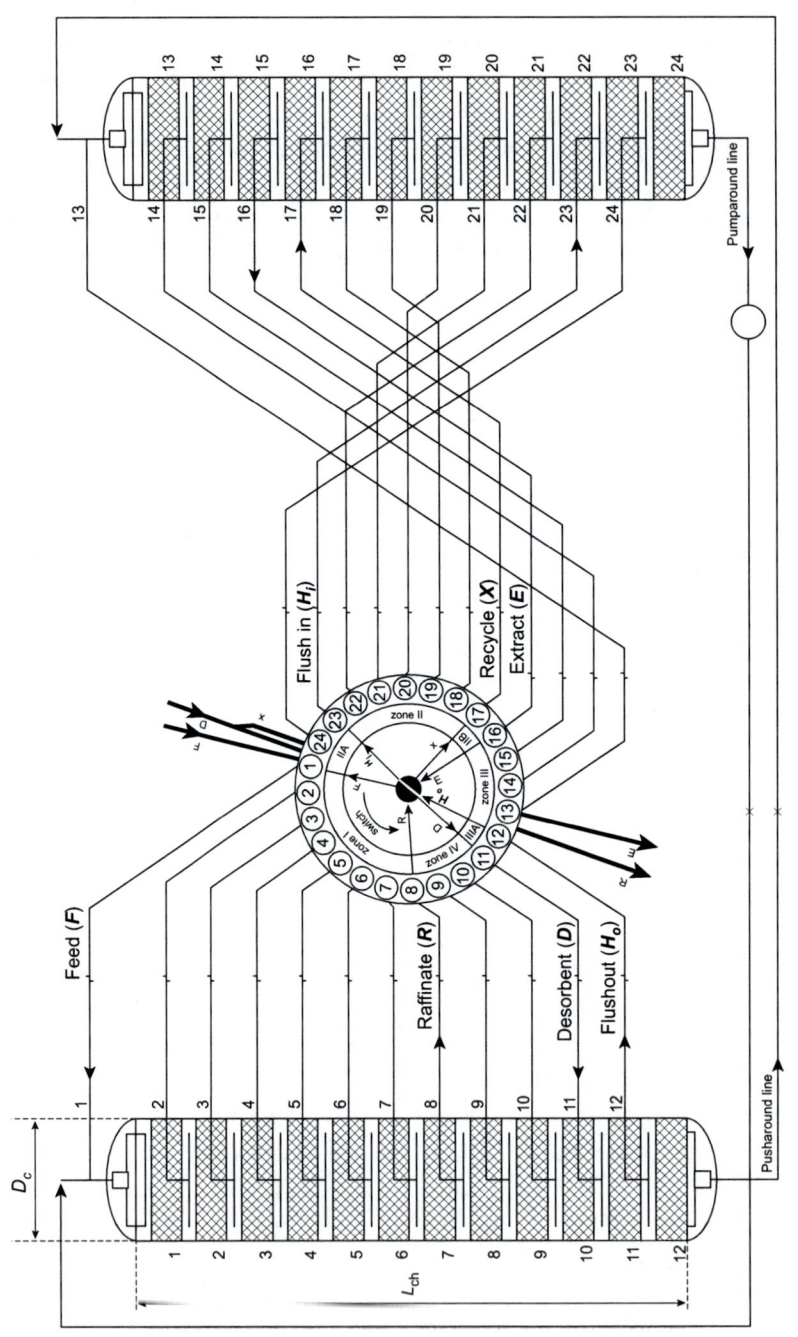

**Figure 5.3** Diagram of the Parex unit and configuration of the rotary valve (Silva, 2013, 2015). *(Reprinted from AIChE Journal 61, Silva, M.S.P., Rodrigues, A.E., Mota, J.P.B., Modeling and simulation of an industrial-scale parex process, 1345–1363, Copyright (2015), with permission from John Wiley & Sons Inc.)*

valve switches is termed the switching time, and is within the range of $1-2$ min, depending on the operating conditions.

In the commercial Parex unit, the adsorbent beds are divided into seven zones, instead of four, named zones *I, IIA, II, IIB, III, IIIA,* and *IV*. The existence of extra three zones is linked to the fact that the same bed lines are used to feed and withdraw the streams, generating dead volumes that contaminate the streams passing in the lines in subsequent switching-time periods. To avoid these problems, three ports are added in the rotary valve for flushing streams (Figure 5.3).

For example, the zone configuration of the Parex unit can be 7-2-6-1-4-1-3 (*I, IIA, II, IIB, III, IIIA,* and *IV* ) (in four zones, 7-9-5-3). Commonly, in the literature, zones *I* and *III* are named, *III* and *I*, respectively.

The primary flush-out stream, $H_o$, is located between the extract and the desorbent ports (one bed before the desorbent injecting port) and has the function of removing the liquid volume with the composition of the extract that was held back in the line previously used to withdraw the extract. If this line is not washed before injecting the desorbent, the *p*-xylene of the extract will be injected in a regeneration zone in which *p*-xylene should not be found. The volume removed through the flush-out line is directly injected in the primary flush-in line, $H_i$, which is located two positions ahead of the feed port in the adsorption zone of *p*-xylene (zone *II*). The injection of this stream after the feed also pushes the liquid with the feed composition that was held in the line into the adsorption chamber (avoiding a loss of feed). A secondary flush (or recycle), *X*, is positioned after the primary flush-in stream and cleans with desorbent the line that is going to be used to remove the extract stream. Depending on the flow rate of the primary flush stream, the secondary flush pushes into the adsorbent chamber all feed or extract that was still in the line, leaving the line that is going to be used to remove the extract full with desorbent.

The zone between the primary flush-out port and the desorbent port is named zone *IIIA*, the injection of the primary flush-in creates the separation between zones *II* and *IIA*, and the secondary flush port creates the separation between zones *II* and *IIB*.

## 5.2.2 Adsorbent

The separation of *p*-xylene from *m*-xylene, *o*-xylene, and ethylbenzene through the SMB principle has been implemented using, in general, faujasitic zeolites X and Y (Broughton, 1982; Neuzil, 1971; Anderson, 1992; Oroskar et al., 1993; Kulprathipanja, 1996; Hotier et al., 1999; Priegnitz et al., 2012; Cheng and Hurst, 2012; Silva et al., 2012a,b). However, this separation has also been studied on other types of zeolites, such as MFI (Guo et al., 2000), ZSM-5 (Yan, 1989) or mordenite (Luna et al., 2010), and other types of materials such as $AlPO_4$-*n* (Chiang et al., 1991; Cavalcante et al., 2000) or MOFs (Gu et al., 2009; Bárcia et al., 2011; Moreira et al., 2011).

### 5.2.3 Operation

The operation of the SMB plant is dependent on the control of the stream flow rates (feed, desorbent, extract, raffinate, flush in/out, and secondary flush) and on the switching time of the ports. These parameters are inputs controlled by the operators.

The calculation of the stream flow rates, $Q$ (m³/h), and switching time, $t^*$ (h), is based on fixed and variable parameters (Figure 5.4).

The fixed parameters are the selective and nonselective volume of the unit (which is directly connected to the volume of the unit and the volume of adsorbent), $V_s$ (m³) and $V_w$ (m³), respectively, and the volume of the bed lines, $V_{line}$ (m³).

The other parameters are directly changed according to the operator's instructions. The feed flow rate enters the unit's control system as a percentage in relation to the project value, $f_F^{project}$, which has a fraction of C$_8$ aromatic hydrocarbons associated, $f_F^a$. The project flow rate of the industrial unit is presented as $Q_F^{project}$. $A/Q_F^a$ is a design parameter of the unit related to the simulated velocity of the solid and the velocity of the liquid, and can be translated by the ratio between the selective volume of the solid contained in the unit and the aromatic liquid volume of the feed injected in the unit during one complete cycle. $A$ (m³/h) is the rate of simulated circulation of the selective adsorbent pore volume and $Q_F^a$ is the volumetric rate of the feed stream containing the mixture of C$_8$ alkylaromatic hydrocarbons.

The fraction of the bed lines washed during flush in/out ($H$) and recycle or secondary flush ($X$), respectively, $f_H^{line}$ and $f_X^{line}$, are also input parameters of the unit. The flow rates of zones II, III, and IV ($LII$, $LIII$, and $LIV$) are input parameters; in the case of the operation of the industrial unit, they are input related to the rate of simulated circulation of the selective adsorbent pore volume, $A$, that is, $L_2/A$, $L_3/A$, and $L_4/A$, and with the nonselective volume of the unit, as given by Eqns (5.9)–(5.11).

The calculations performed in the plant to determine operational parameters of the process are presented below. According to the percentage of feed flow rate in relation to the project value, $f_F^{project}$, and the percentage of aromatics in the feed stream, $f_F^a$, the feed flow rate of aromatics, $Q_F^a$, is determined.

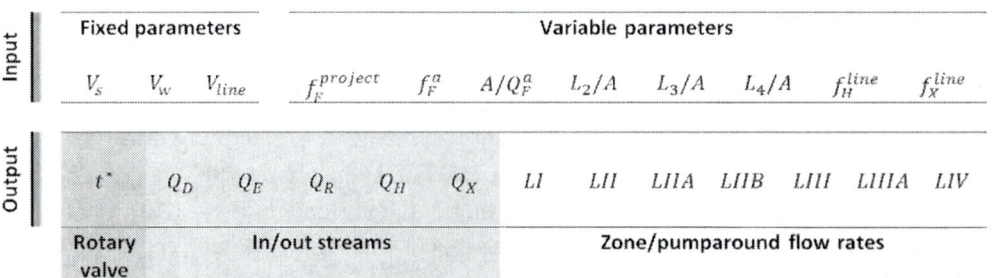

**Figure 5.4** Calculation of operational parameters—inputs introduced by operators and outputs to process (Silva, 2013, 2015). *(Reprinted from AIChE Journal 61, Silva, M.S.P., Rodrigues, A.E., Mota, J.P.B., Modeling and simulation of an industrial-scale parex process, 1345–1363, Copyright (2015), with permission from John Wiley & Sons Inc.)*

$$Q_F = Q_F^{project} \times f_F^{project} \tag{5.1}$$

$$Q_F^a = Q_F \times f_F^a \tag{5.2}$$

The feed flow rate of aromatics is then used to calculate the rate of simulated circulation of the selective adsorbent pore volume, $A$, to have an effective separation,

$$A = \varepsilon_\mu \, Q_{solid}^{TMB} = \left( A/Q_F^a \right) \times Q_F^a \tag{5.3}$$

which can be specified over a complete cycle, $t_c$:

$$A = \frac{V_s}{t_c} \tag{5.4}$$

in which $V_s$ is the selective volume of the adsorbent.

With the cycle time, $t_c$ (h), obtained in Eqn (5.4), the switching time of the ports, $t^*$ (h), is determined by dividing by the number of columns, in this case, 24.

$$t^* = \frac{t_c}{24} \tag{5.5}$$

The switching time, $t^*$, is used directly in the calculation of the flow rates of the flush-in/-out streams, $H$ ($H_i$ or $H_o$), and of the recycle stream (secondary flush), $X$.

$$Q_H = \frac{f_H^{line} \times V_{line}}{t^*} \tag{5.6}$$

$$Q_X = \frac{f_X^{line} \times V_{line}}{t^*} \tag{5.7}$$

The calculation of all flow rates of the unit (inlet/outlet stream flow rates and zone flow rates) is made from the other three input parameters: $L_2/A$, $L_3/A$, $L_4/A$. In the flow rate calculation, the term is introduced corresponding to the nonselective volume of the unit, $V_W$, which includes the nonselective voids of the bed, grids, and circulation piping. In a true moving-bed approach, $W$ is the rate of simulated circulation of the nonselective volume and corresponds to the amount of liquid that should pass in a countercurrent direction to the adsorbent to displace the fluid present in the nonselective void volume.

$$W = \frac{V_W}{t_c} \tag{5.8}$$

$$LII = (L_2/A) \times A + W \tag{5.9}$$

$$LIII = (L_3/A) \times A + W \tag{5.10}$$

$$LIV = (L_4/A) \times A + W \tag{5.11}$$

$$LIIA = LII + Q_H \tag{5.12}$$

$$LI = LIIA + Q_F \tag{5.13}$$

$$LIIB = LII - Q_X \tag{5.14}$$

$$LIIIA = LIII + Q_H \tag{5.15}$$

$$Q_D = LIIIA - LIV \tag{5.16}$$

$$Q_E = LIII - LIIB \tag{5.17}$$

$$Q_R = LI - LIV \tag{5.18}$$

The raffinate stream can be used to check the balances, because $Q_R = Q_F + Q_D + Q_X - Q_E$.

The zone flow rates ($LI$, $LIIA$, $LII$, $LIIB$, $LIII$, $LIIIA$, $LIV$) are equal to the flow rates regulated in the pump-around circulation.

As the dead volumes of the elliptical heads of the adsorbent chambers can be a source of contamination, small desorbent streams purged after the desorbent filters are injected in the top and bottom of the adsorbent chambers. These streams are then mixed with the circulating fluid inside the adsorbent chambers and are collected in the outlet streams of the unit with the rest of the desorbent. The flow rate of these streams is equal to the minimum volume that avoids contamination. The adsorbent chamber heads should be filled with at least 99% $p$-diethylbenzene. The plant calculations presented here neglect the influence of the bed heads in the unit operation, because the injection of desorbent in the elliptical heads is very small.

For evaluating the performance of the Parex unit during operation, two types of information are usually used: pump-around surveys and extract-out surveys. When the pressure drop through the Parex adsorbent chambers exceeds the maximum allowable design value, problems related to damage of the grid internals, loss of sieve containment, and reduction of performance in the unit and/or maldistribution of the fluid can be identified. Thus, bed pressure-drop surveys can also be used as a tool for solving troubleshooting problems in the Parex unit. In a bed pressure-drop survey, pressure drop across individual beds is monitored and the nonuniformity of the values across the 24 beds is analyzed. The pressure-drop survey is done at the highest desired throughput (when zone $I$ is passing in each bed). The bed pressure-drop survey is also a tool to identify the ultimate capacity of the adsorbent chambers.

This type of information is frequently used to identify problems affecting the performance of the unit.

The pump-around survey is a series of 24 samples taken from the pumparound circulation line at different points in the rotary valve cycle. Analysis of these samples shows the composition profile of each zone as it passes through the pumparound loop. The results of the pumparound survey are useful in troubleshooting purity and/or recovery problems and in optimizing the unit.

The survey of the extract stream is used to evaluate the purity of p-xylene obtained during one complete cycle. It is represented as function of the Feed Inlet Position, FIP, or Extract Outlet Position, EOP (EOP = FIP − 9 for FIP > 9; EOP = FIP + 15 for FIP < 10), and ideally p-xylene purity should be uniform along the 24 positions of the cycle. When p-xylene purity in the extract stream is not constant, problems in separation zone II are confirmed.

## 5.3 PERFORMANCE

### 5.3.1 Effect of the Flush Streams

To better understand the operation of the unit, it is important to assess the impact of the flow rates of the three flushing streams (primary, H, and secondary, X, flush streams) on the performance of the separation.

Figure 5.5 shows the effect of $f_X^{line}$ and $f_H^{line}$ on the performance of the Parex process in terms of p-xylene purity, $PUX_{p-x}^E$, and recovery, $REX_{p-x}^E$. When $f_X^{line}$ and $f_H^{line}$ are equal to 1, the flow rates of the washing streams are just enough to displace all the volume of the line once, that is, the concentration front moves from one end of the bed line, $z = 0$, to the other, $z = L$, during one switching time interval.

It is shown that $PUX_{p-x}^E$ increases with the increase of $f_H^{line}$ for a constant $f_X^{line}$. When $f_H^{line}$ is large enough to displace all the liquid, with the feed composition filling the flush in line at the beginning of the switching time interval, for any value of $f_X^{line}$ the contamination of the final product due to bed lines is avoided and purity becomes dependent only on the impurities of the extract (and possible impurities of the desorbent collected in zone III). If one neglects the effect of the axial dispersion, for any value of $f_X^{line}$, a value of $f_H^{line} = 1$ should be enough to avoid contamination of the extract by the liquid with the feed composition held back in the bed lines.

It should be noted that, for $f_H^{line} > 1$, the line used by the recycle stream has at the beginning of the switching time interval a certain quantity of p-diethylbenzene collected in zone III and therefore might have small amounts of contaminants. Depending on $f_X^{line}$, this fraction of p-diethylbenzene can be collected with the extract stream and, consequently, if the contaminants of p-diethylbenzene are only p-xylene, the recovery will increase; on the other hand, if some other components are present, the purity of the final product will decrease.

When the flush-out stream flow rate allows cleaning the whole line, the desorbent will not push extract into a regeneration zone. When the line used to inject the desorbent is not clean at the beginning of the switching time interval, p-xylene will be injected into the regeneration zone and will contaminate the raffinate stream. The contamination of the raffinate leads to a loss in recovery, and consequently, p-xylene will be sent to the isomerization reactor located downstream of the Parex unit, which will also reduce its conversion.

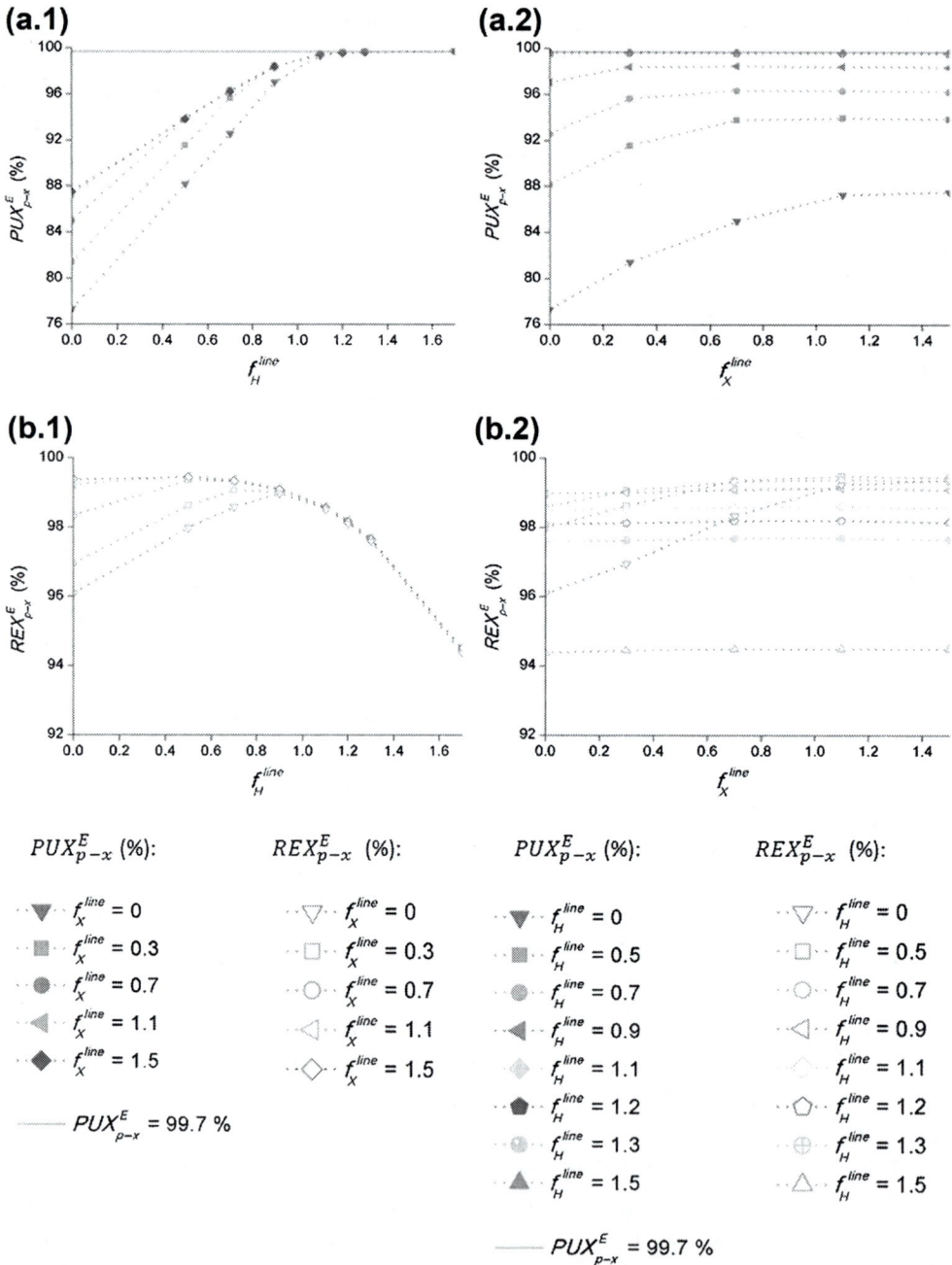

**Figure 5.5** $p$-Xylene purity, $PUX_{p-x}^E$ (a), and recovery, $REX_{p-x}^E$ (b), as a function of $f_H^{line}$ (1—for constant values of $f_X^{line}$) and $f_X^{line}$ (2—for constant values of $f_H^{line}$) (Silva, 2013).

Figure 5.5(b.1) shows that for each $f_X^{line}$ value a certain $f_H^{line}$ value maximizes the recovery of the unit. As $f_H^{line}$ increases, the fraction of the extract line filled with extract (at the beginning of the switching time interval) increases until a maximum is reached; after that, the fraction of extract in the line reduces until the line is completely filled with p-diethylbenzene. The maximum of extract in the line at the beginning of the switching time interval is related to the recovery obtained in the unit. For $f_X^{line} > 1$, the extract line is initially filled with p-diethylbenzene diverted from the desorbent stream (from the desorbent recovery units) and the recovery is constant for small values of $f_H^{line}$. For $f_X^{line} < 1$, the recovery increases with $f_H^{line}$ because $f_H^{line}$ reduces the quantity of feed in the extract line (at the beginning of the switching time interval) and the quantity of p-xylene lost through the raffinate (after being pushed into the unit by the desorbent stream). It is also observed that the recovery is maximized when the injection of the streams is done in a feed inlet position at which the concentration of the components in the internal concentration profile is close to the concentration of the liquid being injected.

The purity of p-xylene increases with $f_X^{line}$ for a fixed value of $f_H^{line}$; however, for $f_H^{line} > 1$ the purity is not enhanced by increasing $f_X^{line}$. Overall, the p-xylene recovery decreases with $f_X^{line}$. For high values of $f_H^{line}$, $f_X^{line}$ does not influence significantly the recovery of the unit, since the liquid with the feed composition is no longer held back in the extract line at the beginning of the switching time interval. However, considering plug flow in the lines, for $f_H^{line} > 1$ the flush-in stream injects not only feed but also extract (it is injected in $FIP = 3$ instead of being injected only in $FIP = 9$) and the recovery of the unit consequently starts decreasing.

Figure 5.6 shows how the extract history and the internal concentration profile of the unit are affected by different values of $f_H^{line}$ and $f_X^{line}$. By increasing $f_H^{line}$, the quantity of p-xylene injected in $FIP = 15$ (desorbent line) decreases, since for $f_H^{line} > 1$ the desorbent stream does not push extract into the adsorbent and desorbent regeneration zones.

When the flow rates of the primary flush streams are increased, the quantity of feed injected by the flush in stream ($FIP = 3$) increases, and for $f_H^{line} > 1$ the flush-in stream starts to inject extract; therefore, an increase of p-xylene concentration is observed in zone II as $f_H^{line}$ increases.

Analyzing the extract history for $f_X^{line} = 0$ and $f_H^{line} = 0.5$, a fraction of extract is first collected (transferred by the flush-out line to the flush-in line), then a fraction with the composition of the feed stream (from the flush in line) is collected and, finally, extract obtained from the column during that switching time interval starts to be collected. In the case of $f_X^{line} = 0$ and $f_H^{line} = 1.7$, at the beginning of the switching time interval, p-diethylbenzene withdrawn by the flush-out stream is collected and then extract starts to be obtained.

For $f_X^{line} = 1.5$, it is observed that by increasing $f_H^{line}$, the quantity of p-xylene collected in the extract drops, which is translated into a reduction of recovery.

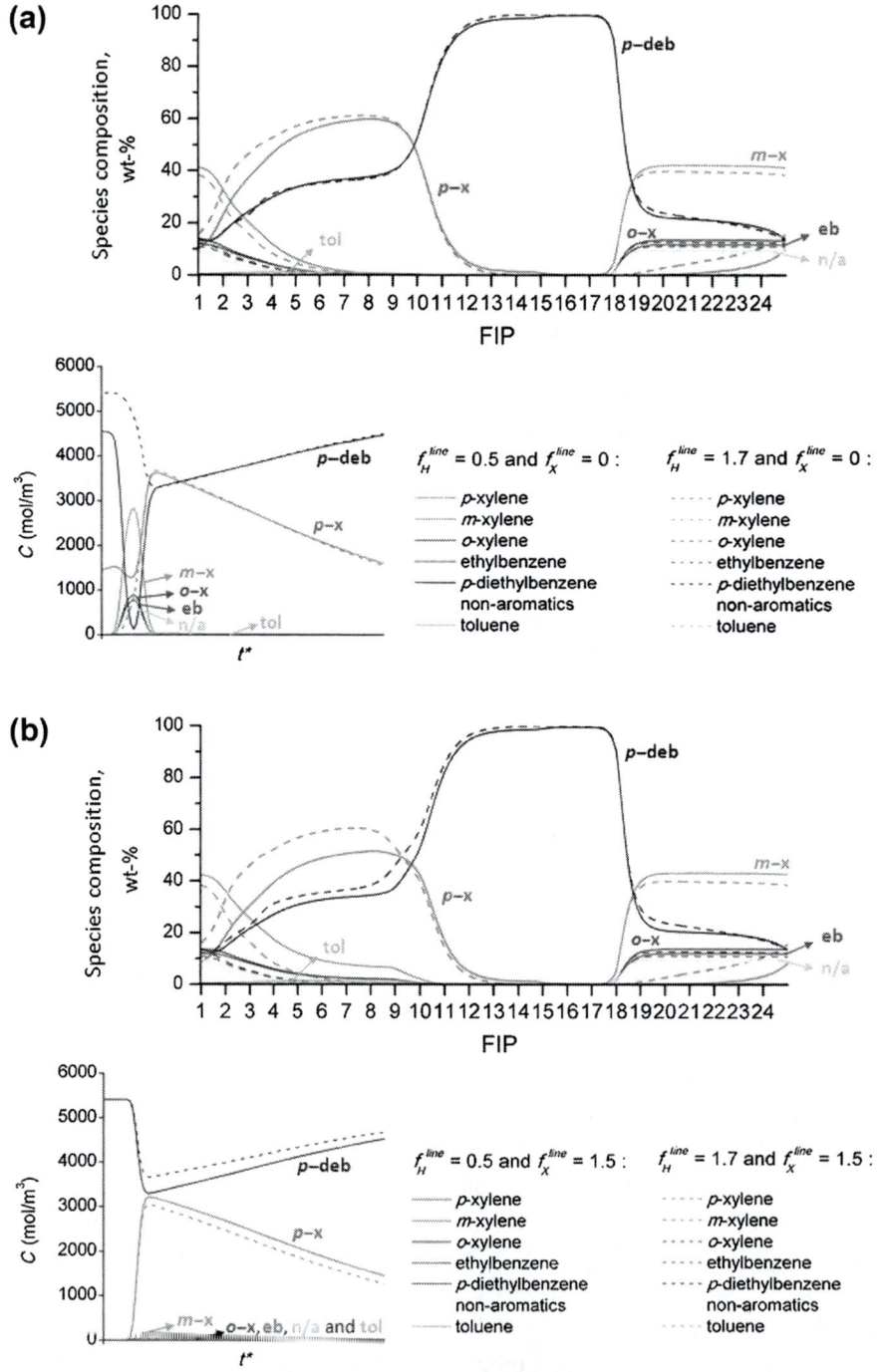

**Figure 5.6** Internal composition profiles and extract histories during one switching time interval imposing different $f_X^{line}$ and $f_H^{line}$ values: (a) effect of $f_H^{line}$ when $f_X^{line} = 0$; (b) effect of $f_H^{line}$ when $f_X^{line} = 1.5$ (Silva, 2013).

When the cases with different $f_X^{line}$ values are compared, a reduction of *p*-xylene concentration in zone *II* with the increase of $f_X^{line}$ is observed for low $f_H^{line}$ values. At the same time, an increase of impurities is observed in the extract due to the increasing concentrations of *m*-xylene, *o*-xylene, and ethylbenzene in zone *II*. For these operating conditions, the recycle (*FIP* = 9) located just one column ahead of the extract injects the feed stream causing a spread of extract contaminants over the *p*-xylene separation zone (zone *II*).

After analyzing the results presented, it is concluded that to achieve the purity requirements of the Parex process, one should impose $f_H^{line} > 1.2$ as is shown in Figure 5.5(a). As demonstrated, $f_H^{line} = 1.2$ leads to unit performance near the purity requirement limit. For a certain value of $f_X^{line}$, if a higher value for $f_H^{line}$ is chosen, the recovery will be reduced.

In relation to $f_X^{line}$, Figure 5.5 shows that recovery increases with $f_X^{line}$. If $f_X^{line} < 1$ then the extract will contain a fraction of *p*-diethylbenzene taken from the regeneration zone through the flush-out stream; consequently, if the regeneration of the desorbent is affected in some way and it is not complete during the operation of the unit, the desorbent impurities will be directly withdrawn with the extract stream. To avoid this possible loss of purity, for safety, the extract line can be filled with pure *p*-diethylbenzene, that is, $f_X^{line} > 1$. In this case, the recovery of the unit per pass will be reduced, but purity problems associated with the contamination of the desorbent over time can be prevented.

The results demonstrate that to minimize the effects of the bed lines, the flush streams should not alter the internal concentration profile; therefore, the streams should be injected through a port in which the internal concentrations are similar to the concentrations of the liquid being injected.

## 5.3.2 Optimization at a Fixed-Feed Flow Rate

The correct operation of the flushing streams (primary and secondary flushes) does not depend on other operating conditions of the unit. However, for a fixed feed flow rate, the switching time can be used to maximize the unit's performance. Usually, separation regions are determined for a fixed flow rate, and as such, it is important to understand its influence on the performance of the unit.

Through simulation, it is possible to understand how the performance of the unit is optimized for a certain feed flow rate and composition. For a certain value of $t^*$ (and corresponding $A/Q_F^a$), the value of $L_2/A$ is changed until the separation zone is found. By changing $L_2/A$, the operating point moves parallel to the $\gamma_I = \gamma_{II}$ line ($\gamma$ defines the ratio between the liquid and solid velocities). For each $t^*$, there is an interval of $L_2/A$ (or $\gamma_{II}$) in which the separation is accomplished with $(PUX_{p-x}^E)_{min} \geq 99.7\%$ and a recovery higher than a predefined value, for example $(REX_{p-x}^E)_{min} = 90\%$.

Figure 5.7 shows a separation region obtained for a fixed feed flow rate in terms of $\gamma_I$ versus $\gamma_{II}$ and $A/Q_F^a$ versus $L_2/A$, along with the variation of purity, recovery, desorbent consumption, and productivity in the separation region.

Figure 5.7 shows that the limit of the separation region $PUX_{p-x}^E = 99.7\%$ in the graph $\gamma_I$ versus $\gamma_{II}$ becomes linear when it is represented as $A/Q_F^a$ versus $L_2/A$.

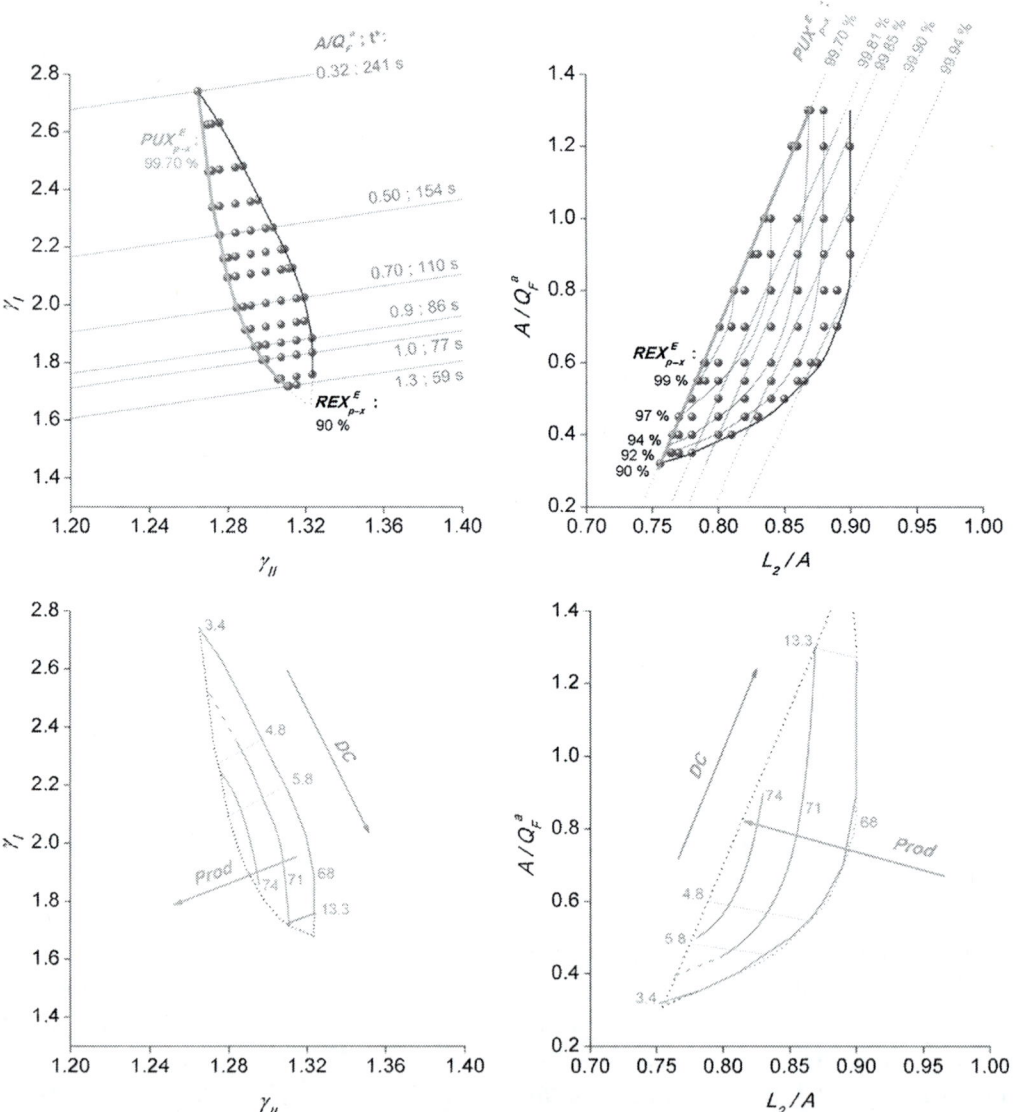

**Figure 5.7** Example of separation region, represented as $\gamma_I$ versus $\gamma_{II}$ and $A/Q_F^a$ versus $L_2/A$. Variation of p-xylene purity, $PUX_{p-x}^E$ (%), recovery, $REX_{p-x}^E$ (%), desorbent consumption, DC (kg p-diethylbenzene/kg p-xylene in extract), and productivity, Prod (kg p-xylene/h m³ adsorbent), in the separation zone. Definition of the $PUX_{p-x}^E = 99.70\%$ operation line for these conditions (Silva, 2013).

For a certain $A/Q_F^a$ within the feasible region of separation, the purity increases with $L_2/A$ whereas the recovery and productivity decrease. In relation to the desorbent consumption, lower $t^*$ values lead to an increase of desorbent flow rate for the same amount of $p$-xylene separated.

Each $A/Q_F^a$ defines an operating curve of the Parex unit. The variation of $L_2/A$ allows the operator to move the operating point along the operating curve until the $PUX_{p-x}^E = 99.7\%$ point is found. As $L_2/A$ increases, the purity and the desorbent consumption increase, whereas the recovery and productivity decrease. On each operating curve, a well-defined operating point exists for which the purity specification for the $p$-xylene is met. The comparison between the $PUX_{p-x}^E = 99.7\%$ points for several $A/Q_F^a$ values allows the determination of the pair $A/Q_F^a$ versus $L_2/A$ that leads to a maximum recovery. In the $PUX_{p-x}^E = 99.7\%$ line, when $A/Q_F^a$ is lower than the optimum value (higher $t^*$), the recovery increases with $A/Q_F^a$; however, when $A/Q_F^a$ is higher than the optimum value (lower $t^*$), the recovery decreases with $A/Q_F^a$.

Figure 5.8 shows the effect of $L_2/A$ on the performance parameters ($PUX_{p-x}^E$ and $REX_{p-x}^E$) and on the internal concentration profile.

For a fixed value of $A/Q_F^a$, increasing $L_2/A$ leads to an increase of purity and reduction of recovery. The variation of $L_2/A$ induces the variation of the internal flow rates of zones $I$ and $II$ (with $L_2/A$, $LI$, $LIIA$, $LII$ and $LIIB$ increase) and of the flow rates of extract and raffinate (with $L_2/A$, $Q_E$ decreases and $Q_R$ increases). Zones $III$ and $IV$ do not suffer any variation in their internal flow rates or inlet/outlet streams. The internal concentration profiles show that $L_2/A$ increases the concentration of $p$-xylene desorbed in zone $II$, and it decreases the concentration of the extract impurities. The analysis of the concentration profile shows that the concentration of $p$-xylene in the raffinate increases with $L_2/A$.

In the right-hand side of Figure 5.8, $A/Q_F^a$ is changed for a certain $L_2/A$ and purity is observed to decrease and recovery to increase with $A/Q_F^a$. In this situation, the switching time of the ports is altered, which causes an increase of all internal flow rates and inlet/outlet streams (for a constant feed flow rate and, in this case, equal to the project value). Consequently, the internal concentration profile shows alterations in all zones.

By varying the switching time interval and for a fixed feed flow rate, an optimum operation point in which the recovery and productivity of the unit are maximized can be found (Figure 5.9). When the switching time interval is below the optimum point (higher $A/Q_F^a$ values), the contact time between the solid and liquid phases is not enough to ensure the mass transfer of the components between the two phases. For switching times higher than the optimum, the simulated solid velocity is small and the solid does not have capacity to adsorb the $p$-xylene injected by the feed stream (zone $I$). If one considers the ratio between the internal liquid flow rate in zone $I$ (zone of adsorption of $p$-xylene) and the simulated solid flow rate, it is observed to increase with the

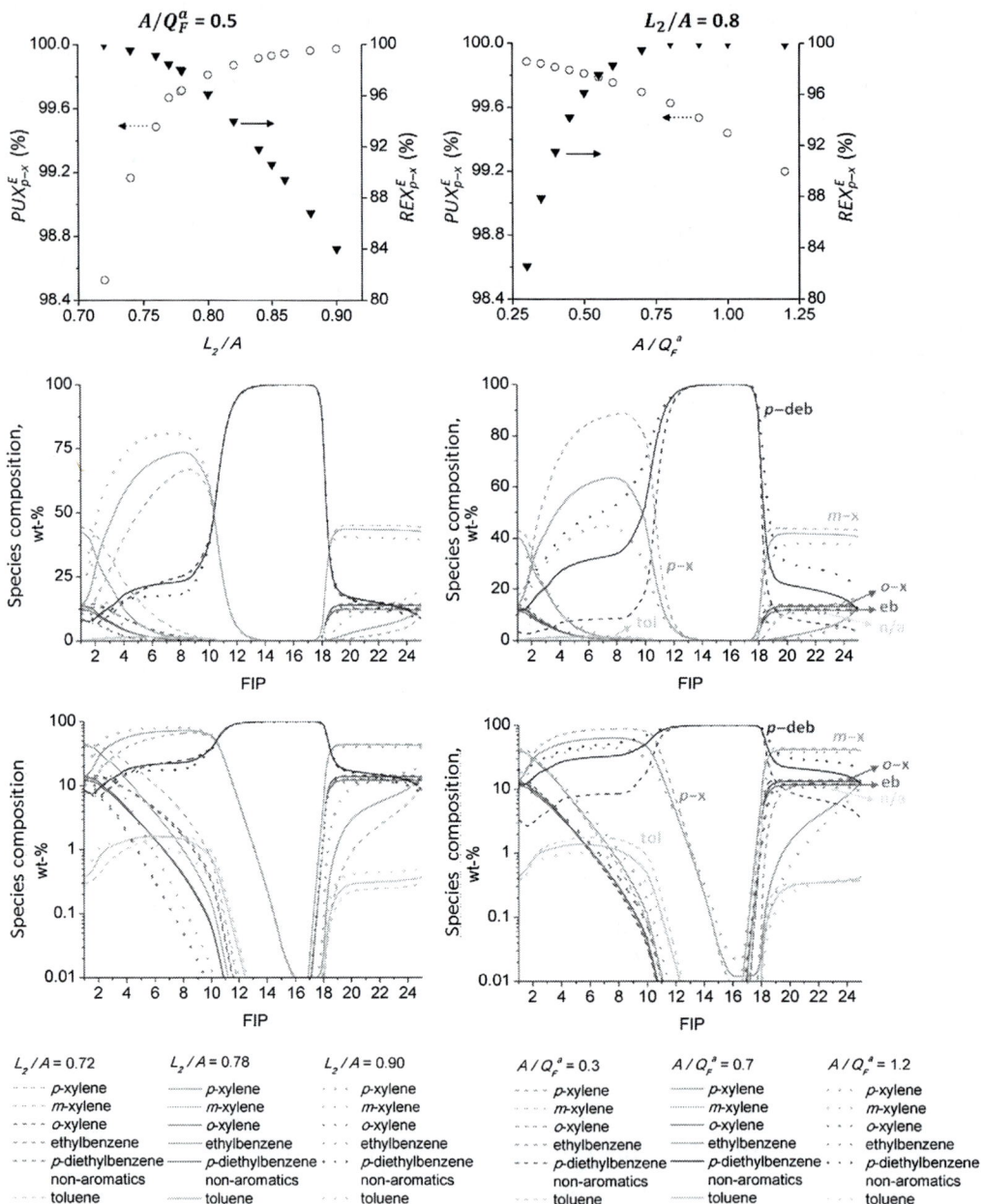

**Figure 5.8** $p$-Xylene purity, $PUX_{p-x}^{E}$ (%), and recovery, $REX_{p-x}^{E}$ (%): effect of $L_2/A$ when $A/Q_F^a = 0.5$ (left-hand side) and effect of $A/Q_F^a$ when $L_2/A = 0.8$ (right-hand side) (Silva, 2013).

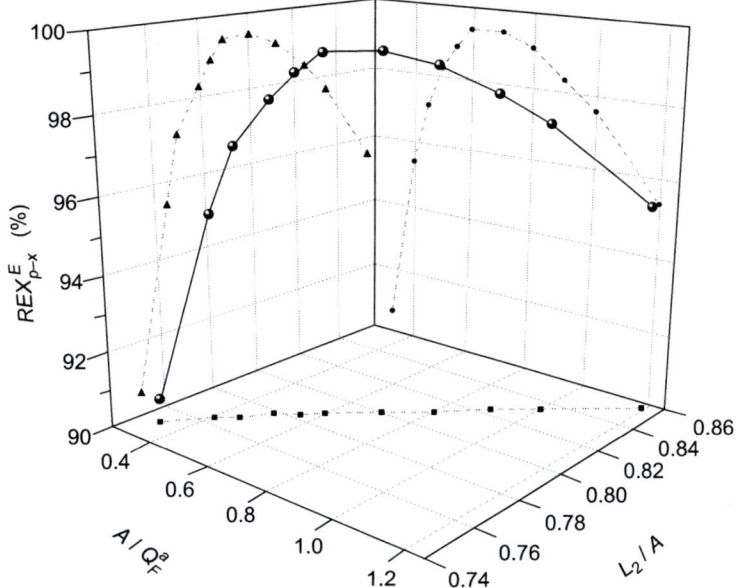

**Figure 5.9** Unit's recovery, $REX_{p-x}^{E}$ (%), as function of $A/Q_F^q$ versus $L_2/A$ (99.7% purity) (Silva, 2013).

switching time interval. That is, for higher switching time intervals the quantity of liquid in relation to the solid is higher (the feed flow rate is maintained), and, consequently, as one increases the switching time interval at one point the maximum capacity of adsorption of the solid is reached.

It is important to note that the desorbent consumption increases with the decrease of the switching time interval and, therefore, to choose the optimum point of operation the capacity of the separation units located downstream in the process have to be taken into consideration.

## 5.4 CONCLUDING REMARKS

Over the years, the Parex process has been chosen as the main technology for the industrial separation of *p*-xylene from its isomers and ethylbenzene. It has proven its advantages towards the crystallization processes in terms of recovery, and its worldwide application has proven the reliability of its design, in particular, of its 24-port rotary valve. This process allows a *p*-xylene purity per pass higher than 99.7 wt% (market demand) with consistently high recoveries over the years of operation.

The correct operation of the seven-zone SMB allows the elimination of the separation constraints that the unit's dead volume may create in a rotary-valve design approach for switching periodically the ports.

For each feed flow rate, an optimum switching time of the rotary valve (controlled by the Operator) allows the maximization of the recovery for a chosen $p$-xylene purity target.

## NOMENCLATURE

| | |
|---|---|
| $A$ | Rate of simulated circulation of the selective adsorbent pore volume (m³/h) |
| $C$ | Bulk liquid concentration (mol/m³) |
| $DC$ | Desorbent consumption (kg of $p$-diethylbenzene/kg of $p$-xylene obtained in the extract) |
| $EOP$ | Extract outlet position |
| $f_F^a$ | $Q_F^a/Q_F$ |
| $f_F^{project}$ | $Q_F/Q_F^{project}$ |
| $f_H^{line}$ | Fraction of the bed line washed by the primary flush stream |
| $f_X^{line}$ | Fraction of the bed line washed by the secondary flush (or recycle) stream |
| $FIP$ | Feed inlet position |
| $H_i$ | Primary flush-in stream |
| $H_o$ | Primary flush-out stream |
| $LI$ | Flow rate of zone $I$ (m³/h) |
| $LII$ | Flow rate of zone $II$ (m³/h) |
| $LIIA$ | Flow rate of zone $IIA$ (m³/h) |
| $LIIB$ | Flow rate of zone $IIB$ (m³/h) |
| $LIII$ | Flow rate of zone $III$ (m³/h) |
| $LIIIA$ | Flow rate of zone $IIIA$ (m³/h) |
| $LIV$ | Flow rate of zone $IV$ (m³/h) |
| $L_2$ | $LII = L_2 + W$ |
| $L_3$ | $LIII = L_3 + W$ |
| $L_4$ | $LIV = L_4 + W$ |
| $Prod$ | Productivity (kg of $p$-xylene obtained in the extract/h m³ of adsorbent) |
| $PUX$ | Purity (%) |
| $Q$ | Flow rate (m³/h) |
| $REX$ | Recovery (%) |
| $SMB$ | Simulated moving bed |
| $t^*$ | Switching time (s) |
| $t_c$ | Cycle time (s) |
| $TMB$ | True moving bed |
| $V$ | Volume (m³) |
| $V_s$ | Selective volume (m³) |
| $V_w$ | Nonselective volume (m³) |
| $X$ | Recycle or secondary flush stream |
| $W$ | Rate of simulated circulation of the nonselective volume (m³/h) |
| $p$-$x$ | $p$-Xylene |
| $m$-$x$ | $m$-Xylene |
| $o$-$x$ | $o$-Xylene |
| eb | Ethylbenzene |
| $p$-deb | $p$-Diethylbenzene |
| n/a | Nonaromatics |
| tol | Toluene |

## Greek Letters

| | |
|---|---|
| $\varepsilon_\mu$ | Particle microporosity $(V_\mu/V_{ads})$ |
| $\gamma$ | Ratio between liquid and solid velocities |

## Subscripts and Superscripts

| | |
|---|---|
| *a* | Aromatic fraction |
| *ads* | Adsorbent |
| *D* | Desorbent stream |
| *E* | Extract stream |
| *F* | Feed stream |
| *H* | Primary flush stream |
| *line* | Bed line |
| *R* | Raffinate stream |
| *X* | Recycle or secondary flush stream |

## REFERENCES

Anderson, G., 1992. Process for Separating *para*-Xylene from a C$_8$ and C$_9$ Aromatic Mixture. US Patent 5171922, UOP.

Ash, G., Barth, K., Hotier, G., Mank, L., Renard, P., 1994. ELUXYL: a new paraxylene separation process. Rev. Inst. Fr. Pét. 49 (5), 541–549.

Ash, G.A., Dao, N.Q., Gloyn, A.J., Haritatos, N.J., Hodgen, P.I., MacPherson, S.R., Morrison, S.G., Nacamuli, G.J., Spindler, P.M., Thom, B.J., Weber, E.P., Wolpert, R., 1999. Process for Converting Hydrocarbon Feed to High Purity Benzene and High Purity Paraxylene. US Patent 6004452, Chevron Chemical Company LLC.

Bárcia, P.S., Guimarães, D., Mendes, P.A.P., Silva, J.A.C., Guillerm, V., Chevreau, H., Serre, C., Rodrigues, A.E., 2011. Reverse shape selectivity in the adsorption of hexane and xylene isomers in MOF UiO-66. Microporous Mesoporous Mater. 139, 67–73.

Bergeot, G., Laroche, C., Leflaive, P., Leinekugel-le-Coq, D., Wolff, L., 2009. Reactive Simulated Mobile Bed for Producing Paraxylene. WO 2009/130402 A1, IFP.

Brookes, T., 2012. New technology developments in the petrochemical industry—refinery integration with petrochemicals to achieve higher value uplift. In: UOP LLC, Egypt Petrochemicals Conference, September 27, Cairo, Egypt.

Broughton, D.B., 1965. Fluid Distributing Means for Packed Chambers. US Patent 3214247, Universal Oil Products Company.

Broughton, D.B., 1982. Separation Process. US 4313015, UOP Inc.

Broughton, D.B., Bieser, H.J., Persak, R.A., 1975. Sixty years of sorbex operations. In: UOP Process Division Technology Conference, UOP Process Division.

Broughton, D.B., Gerhold, C.G., 1961. Continuous Sorption Process Employing Fixed Bed of Sorbent and Moving Inlets and Outlets. US Patent 2985589, Universal Oil Products Company.

Broughton, D.B., Neuzil, R.W., Pharis, J.M., Brearley, C.S., 1970. The parex process for recovering paraxylene. Chem. Eng. Prog. 66 (9), 70–75.

Cannella, W.J., 2007. Xylenes and Ethylbenzene, Kirk-Othmer Encyclopedia of Chemical Technology, electronic ed. John Wiley & Sons, Inc.

Carson, D.B., Purse, F.V., 1962. Rotary Valve. US Patent 3040777, Universal Oil Products Company.

Carson, D.B., 1974. Flow Distribution Apparatus. US Patent 3789989, Universal Oil Products Company.

Cavalcante, C.L., Azevedo, D.C.S., Souza, I.G., Silva, A.C.M., Alsina, O.L.S., Lima, V.E., Araujo, A.S., 2000. Sorption and diffusion of *p*-xylene and *o*-xylene in aluminophosphate molecular sieve AlPO4-11. Adsorption 6, 53–59.

Cheng, L.S., Hurst, J., 2012. Binderless Zeolitic Adsorbents, Methods for Producing Binderless Zeolitic adsorbents, and Processes for Adsorptive Separation of *para*-Xylene from Mixed Xylenes Using the Binderless Zeolitic Adsorbents. US Patent 8283274 B2, UOP LLC.

Cheng, L.S., Johnson, J.A., 2009. Adsorbents with Improved Mass Transfer Properties and Their Use in the Adsorptive Separation of *para*-Xylene. US Patent 2009/0326311, UOP LLC.

Chiang, A.S.T., Lee, C.-K., Chang, Z.-H., 1991. Adsorption and diffusion of aromatics in AlPO4-5. Zeolites 11, 380–386.

Commissaris, S.E., 2004. UOP parex process. In: Meyers, Robert A. (Ed.), Handbook of Petroleum Refining Processes, third ed. McGraw-Hill.

Desideri, R.J., Hirsig, A.R., Dresser, T., Edison, R.R., Peterman, L.G., Truitt, R.E., 1974. Pure *p*-xylene by single stage. Hydrocarb. Process. 53, 81–83.

Fabri, J., Ulrich, G., Simo, T.A., 2012. Xylenes, Ullmann's Encyclopedia of Industrial Chemistry, electronic ed. Wiley-VCH Verlag GmbH & Co. KGaA, Weinheim.

Frey, S.J., 2007. Product Recovery from Simulated Moving Bed Adsorption. US Patent 7208651, UOP LLC.

Gu, Z.-Y., Jiang, D.-Q., Wang, H.-F., Cui, X.-Y., Yan, X.-P., 2009. Adsorption and separation of xylene isomers and ethylbenzene on two Zn–terephthalate metal–organic frameworks. J. Phys. Chem. C 114, 311–316.

Guo, G.-Q., Chen, H., Long, Y.-C., 2000. Separation of *p*-xylene from C$_8$ aromatics on binder-free hydrophobic adsorbent of MFI zeolite. I. Studies on static equilibrium. Microporous Mesoporous Mater. 39, 149–161.

Hotier, G., Il, K.S., 2009. Method for Combined Production of *para*-Xylene and Benzene with Improved Productivity. US Patent 2009/0069612 A1.

Hotier, G., 2009. Simulated Moving Bed Separation Process and Device. US Patent 7473368, Institut Français du Petrole.

Hotier, G., Leflaive, P., Louret, S., Aügier, F., 2009. Process and Device for Simulated Moving Bed Separation with a Reduced Number of Valves and Lines. US 7582208, Institut Français du Pétrole.

Hotier, G., Methivier, A., Pucci, A., 1999. Process for Separating *para*-Xylene, Comprising an Adsorption Step with Injection of Water and a Crystallization Step. US Patent 5948950, Institut Français du Petrole.

Hurst, J.E., Cheng, L.S., Broach, R.W., 2012. *para*-Xylene Separation with Aluminisilicate X-type Zeolite Compositions with Low LTA-type Zeolite. US Patent 2012/0264994 A1, UOP LLC.

Johnson, J.A., 2004. UOP sorbex family of technologies. In: Meyers, Robert A. (Ed.), Handbook of Petroleum Refining Processes, third ed. McGraw-Hill.

Kulprathipanja, S., 1996. Adsorptive Separation of *para*-Xylene with High Boiling Desorbents. US Patent 5495061, UOP.

Kulprathipanja, S., Willis, R., Kuechl, D., Priegnitz, J., Hurst, J., Commissaris, S., Cheng, L., 2009. Binderless Adsorbents Comprising Nano-Size Zeolite X and Their Use in the Adsorptive Separation of *para*-Xylene. US Patent 2009/0326308, UOP LLC.

Laurich, S., 1969. *p*-Xylene Process. US Patent 3467724, Chevron Research Co.

Lee, J.-S., Shin, N.-C., 2009. Simulated Moving Bed Adsorptive Separation Process Comprising Preparation of Feed by Using Single Adsorption Chamber and Device Used Therein. US Patent 7635795, Samsung Total Petrochemicals Co., Ltd.

Lee, J.-S., Shin, N.-C., 2010. Simulated Moving Bed Adsorptive Separation Process Using a Plurality of Adsorption Chambers in Parallel and Crystallizer and Device Used Therein. US Patent 7649124, Samsung Total Petrochemicals Co., Ltd.

Lee, J.-S., Shin, N.-C., 2011. Method for Separating Aromatic Compounds Using Simulated Moving Bed Operation. US Patent 8013202 B2, Samsung Total Petrochemicals Co., Ltd.

Leflaive, F., Wolff, L., Methivier, A., Hotier, G., 2004. Coproduction Process for *para*-Xylene and *ortho*-Xylene Comprising Two Separation Steps. US Patent 6828470, Institut Français du Petrole.

Leflaive, F., Methivier, A., Hotier, G., 2005. Process for Co-producing *para*-Xylene and *meta*-Xylene, Comprising Two Separation Steps. US Patent 6838588, Institut Français du Petrole.

Leflaive, F., Baudot, A., Rodeschini, H., Frising, T., 2009. Process for Separation of C$_8$ Aromatic Compounds with Limited Recycling. US Patent 2009/0149686 A1.

Leflaive, P., Leinekugel-le-Cocq, D., Hotier, G., Wolff, L., 2011a. Procédé et dispositif de séparation chromatographique a contre-courant simule utilisant deux adsorbeurs en parallèle pour la production optimise de paraxylene, FR Patent 2976501, IFP Energies Nouvelles.

Leflaive, P., Wolff, L., Hotier, G., 2011b. Method for Producing Paraxylene Comprising an Adsorption Step and Two Isomerization Steps. US Patent 7915471 B2, IFP Energies Nouvelles.

Locating and Estimating Air Emissions from Sources of Xylene, 1994. United States Environmental Protection Agency, Office of Air Quality Planning and Standards. www.epa.gov.

Luna, F., Coelho, J., Otoni, J., Guimarães, A., Azevedo, D., Cavalcante, C., 2010. Studies of $C_8$ aromatics adsorption in BaY and mordenite molecular sieves using the headspace technique. Adsorption 16, 525–530.

McKay, D.L., Goard, H.W., 1965. Continuous fractional crystallization. Chem. Eng. Prog. 61, 99–104.

Mikitenko, P., Hotier, G.P., 1999. Production of *para*-Xylene from an Affluent from Paraselective Toluene Disproportionation Using a Crystallization Process Combined with Simulated Moving Bed Adsorption. US Patent 5866740, Institut Français du Petrole.

Millard, M., November 15, 2010. Creating value—the modern UOP *para*-xylene complex. In: Indian Petrochem. UOP LLC.

Minceva, M., Gomes, P.S., Meshko, V., Rodrigues, A.E., 2008. Simulated moving bed reactor for isomerization and separation of p-xylene. Chem. Eng. J. 140, 305–323.

Moreira, M.A., Santos, J.C., Ferreira, A.F.P., Loureiro, J.M., Rodrigues, A.E., 2011. Influence of the eluent in the MIL-53(Al) selectivity for xylene isomers separation. Ind. Eng. Chem. Res. 50, 7688–7695.

Neuzil, R.W., 1971. Aromatic Hydrocarbon Separation by Adsorption. US Patent 3558730, Universal Oil Products Company.

Ockerbloom, N.E., 1971. Xylenes and higher aromatics. 1. Sources, specifications, producers, uses and outlook. Hydrocarb. Process. 50, 112–114.

Oroskar, A.R., Prada, R.E., Johnson, J.A., Anderson, G.C., Zinnen, H.A., 1993. Process for Separating *para*-Xylene from a $C_8$ and $C_9$ Aromatic Mixture. US Patent 5177295, UOP.

Otani, S., Iwamura, T., Sando, K., Kanaoka, M., Matsumura, K., Akita, S., Yamamoto, T., Takeuchi, I., Tasuaki, T., Yoshio, N., Mori, T., 1973. Separation Process of Components of Feed Mixture Utilizing Solid Sorbent. US Patent 3761533, Toray Industries, Japan.

Otani, S., Sato, M., Kanaoka, M., Akita, S., Ogawa, D., Tsuchiya, Y., 1977. Development of p-xylene manufacturing processes utilizing a new continuous adsorption technology. Am. Soc. Mech. Eng. 1, 550–556.

Par-Isom™ Process, 2007. UOP LLC. www.uop.com.

Perego, C., Pollesel, P., 2009. Advances in Aromatics Processing Using Zeolite Catalysts. In: Advances in Nanoporous Materials, vol. 1. Elsevier.

Porter, J.R., 2011. Simulated Countercurrent Adsorptive Separation Process. US Patent 7977526 B2.

Porter, J.R., Pilliod, D.L., 2012. Separation Process. US Patent 8168845 B2, ExxonMobil.

Priegnitz, J.W., Johnson, D.M., Cheng, L.S., Commissaris, S.E., Hurst, J.E., Quick, M.H., Kulprathipanja, S., 2012. Binderless Adsorbents and Their Use in the Adsorptive Separation of *para*-Xylene. US Patent 7820869 B2, UOP LLC.

Silva, M.S.P., Mota, J.P.B., Rodrigues, A.E., 2012a. Fixed-bed adsorption of aromatic $C_8$ isomers: breakthrough experiments, modelling and simulation. Sep. Purif. Technol. 90, 246–256.

Silva, M.S.P., Moreira, M.A., Ferreira, A.F.P., Santos, J.C., Minceva, M., Mota, J.P.B., Rodrigues, A.E., 2012b. Adsorbent evaluation based on experimental breakthrough curves: separation of p-xylene from $C_8$ isomers. Chem. Eng. Technol. 35 (10), 1777–1785.

Silva, M.S.P., 2013. Optimization of the Parex Unit of Matosinhos Refinery for the Separation of p-Xylene (Ph.D. thesis). Faculdade de Engenharia da Univesidade do Porto.

Silva, M.S.P., Rodrigues, A.E., Mota, J.P.B., 2015. Modeling and simulation of an industrial-scale parex process. AIChE J. 61, 1345–1363.

UOP ADS-27™ Adsorbent, 2007. UOP LLC. www.uop.com.

UOP ADS-37™ Adsorbent, 2009. UOP LLC. www.uop.com.

Wantanachaisaeng, P., O'Neil, K., 2007. Capturing Opportunities for *para*-Xylene Production. UOP LLC. www.uop.com.

Werba, G., 2011. Techno-Economic Opportunities Realized through Unique Approach to Aromatics Expansion Study, Technology and More—2011 Fall Newsletter. UOP LLC. www.uop.com.

Williams, B.A., Doyle, R.A., Miller, J.T., 2007. Method of Obtaining *para*-Xylene. US Patent 7271305, BP Corporation North America Inc.

Wittcoff, H.A., Reuben, B.G., Plotkin, J.S., 2004. Industrial Organic Chemicals, second ed. John Wiley & Sons, Inc.

Wolff, L., Leflaive, P., 2004. ELUXYL twin raffinate technology. In: AIChE Annual Meeting, Austin.

Wolff, L., Leflaive, P., Methivier, A., 2009. Process for Co-producing *para*-Xylene and Styrene. US Patent 7592499, Institut Français du Petrole.

Yan, T.Y., 1989. Separation of *p*-xylene and ethylbenzene from $C_8$ aromatics using medium-pore zeolites. Ind. Eng. Chem. Res. 28, 572—576.

# CHAPTER 6

# Multicomponent Separations by Simulated Moving Bed (SMB)-Based Processes

Simulated moving bed (SMB) technology offers many advantages over conventional elution chromatography, namely the possibility of continuous operation and a significant reduction of stationary- and mobile-phase consumption (Nicoud, 2000). Nevertheless, its main disadvantage is the limitation of the separation of binary mixtures or the recovery of only one component from a multicomponent mixture either in the raffinate or extract streams.

Several alternatives have been suggested to perform multicomponent separations using SMB technology such as an SMB with five sections (Massuda et al., 1993b; Navarro et al., 1997; Beste and Arlt, 2002), eight sections (Chiang, 1998), or nine sections (Wooley et al., 1998). However, studies using the equilibrium theory have shown that neither a five-section nor an eight-section SMB is able to separate three-component mixtures into three pure fractions (Beste and Arlt, 2002). The use of a cascade of SMB in series for ternary (Wankat, 2001) and quaternary (Müller-Späth et al., 2008) separations is also a possible alternative, but this would clearly increase the investment costs of the process. Another alternative for ternary separations is the pseudo-SMB Japan Organo (JO) process, which is a two-step process used to separate a multicomponent mixture in three fractions (Mata and Rodrigues, 2001). The SMB concept is also exploited in the multicolumn countercurrent solvent gradient purification (MCSGP) process by combining both simulated countercurrent operation and solvent gradient chromatography (Aumann and Morbidelli, 2007). Besides the SMB-based processes, other multicolumn processes such as sequential multicolumn chromatography (SMCC) have shown good results in dealing with multicomponent complex mixtures (Zobel et al., 2014).

## 6.1 PSEUDO-SMB JO PROCESS

The pseudo-SMB JO process was first applied by the Japan Organo Company using patented technology (Ando et al., 1990; Massuda et al., 1993a). This technique of pseudo-SMB chromatography has been successful in the separation of complex multicomponent mixtures, such as the separation of beet molasses mixtures into fractions rich in raffinose, rich in betaine, and containing sucrose and monosaccharides (Massuda et al., 1993a), and in the production of raffinose from beet molasses (Sayama et al., 1992).

*Simulated Moving Bed Technology*
ISBN 978-0-12-802024-1

The process cycle is divided into two steps. In step 1, the system is equivalent to a series of preparative chromatographic columns (without solid flow). During this step, the feed and eluent streams are introduced and the intermediate component is collected. Step 2 is similar to an SMB, but without feed. The less-adsorbed component is collected in the raffinate, whereas the more-retained component is collected in the extract. This step can be mathematically modeled as an equivalent true moving bed (TMB) system. Figure 6.1(a) and (b) show the traditional TMB system and a schematic diagram of both steps 1 and 2 of the JO process, respectively.

In the TMB model, four sections are separated by the inlets (feed and eluent) and outlets (extract and raffinate) streams. Each section contains $N_c$ chromatographic columns with the length $L_C$. In the pseudo-SMB model of the JO process, four sections are also considered in both steps for convenience. In step 1, the feed stream enters at the beginning of section 3 and flows through sections 3 and 4; the eluent stream enters at the beginning of section 1 together with the outlet stream of section 4 and flows through sections 1 and 2. The intermediate component is collected at the end of section 2. In step 2, the system is equivalent to a TMB without the feed stream.

## 6.1.1 Mathematical Model

Step 1 of the pseudo-SMB model can be mathematically modeled as a series of chromatographic columns. During this step, the circuit is cut between sections 2 and 3 and the intermediate stream ($Q_I$), rich in the intermediate adsorbate, is collected.

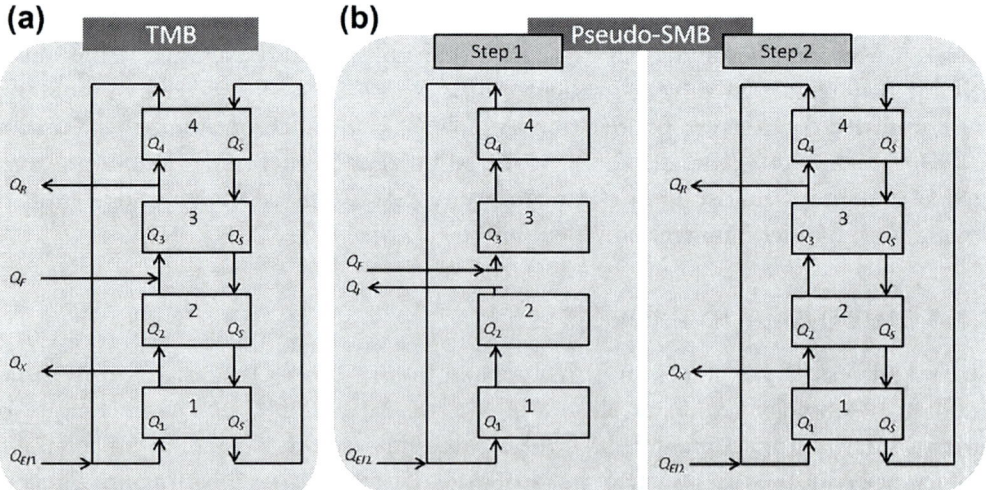

**Figure 6.1** Schematic representation of (a) true moving bed (TMB) and (b) Japan Organo (JO) processes. *(Reprinted from Mata and Rodrigues (2001), with permission from Elsevier.)*

The mathematical model for step 1 comprises the following equations:

*The mass balance for component i in a volume element of the bulk fluid phase of section j is*

$$\frac{\partial C_{ij}}{\partial t} = D_{Lj}\frac{\partial^2 C_{ij}}{\partial z^2} - v_j\frac{\partial c_{ij}}{\partial z} - \frac{(1-\varepsilon)}{\varepsilon}k_p(q_{ij}^* - q_{ij}) \tag{6.1}$$

in which $C_{ij}$ and $q_{ij}$ are the fluid phase and average adsorbed phase concentrations of component $i$ in section $j$, respectively; $z$ is the axial coordinate; $t$ is the time variable; $\varepsilon$ is the bed porosity; $v_j$ is the interstitial fluid velocity in section $j$; $D_{Lj}$ is the axial dispersion coefficient in section $j$; and $k_p$ is the intraparticle mass-transfer coefficient.

*The mass balance for component i in the adsorbed phase of section j is:*

$$\frac{\partial q_{ij}}{\partial t} = k_p(q_{ij}^* - q_{ij}) \tag{6.2}$$

in which $q_{ij}^*$ is the adsorbed-phase concentration in equilibrium with $C_{ij}$.

The initial conditions for step 1 are

$$t = 0 \ (cycle \ 1) \ \rightarrow \ C_{ij} = q_{ij} = 0 \tag{6.3}$$

$$t = t_{S2} \ (cycle \ k) \ \rightarrow \ \begin{matrix} C_{ij} = C_{ij}(cycle \ k-1, \ step \ 2) \\ q_{ij} = q_{ij}(cycle \ k-1, \ step \ 2) \end{matrix} \tag{6.4}$$

and the boundary conditions for each section $j$ are

$$z = 0 \ \rightarrow \ C_{ij,0} = C_{ij} - \frac{D_{Lj}}{v_j}\frac{\partial C_{ij}}{\partial z}\bigg|_{z=0} \tag{6.5}$$

in which $C_{ij,0}$ is the inlet concentration of component $i$ in section $j$.

$$z = L_j \ \rightarrow \ \begin{cases} \textit{Feed node: } \ C_{i3,0} = C_i^F \\ \\ \textit{Eluent node: } \ C_{i1,0} = \frac{Q_4}{Q_1}C_{i4,L_j} \\ \\ \textit{Between sections: } \ C_{ij,L_j} = C_{ij+1,0} \end{cases} \tag{6.6}$$

Step 2 can be modeled by an equivalent TMB model with four sections. During this step the extract stream ($Q_X$), rich in the more-retained component, and the raffinate stream ($Q_R$), rich in the less-retained component, are the outputs of the system, and the eluent stream ($Q_{E2}$) is the only input.

The mathematical model for step 2 is constituted by the following equations:

*The mass balance for a component in a volume element of the bulk fluid phase of section j is*

$$\frac{\partial C_{ij}}{\partial t} = D_{Lj}\frac{\partial^2 C_{ij}}{\partial z^2} - v_j\frac{\partial c_{ij}}{\partial z} - \frac{(1-\varepsilon)}{\varepsilon}k_p(q_{ij}^* - q_{ij}) \tag{6.7}$$

*Moreover, the mass balance for component i in the adsorbed phase of section j is*

$$\frac{\partial q_{ij}}{\partial t} = u_s \frac{\partial q_{ij}}{\partial z} + k_p(q_{ij}^* - q_{ij}) \tag{6.8}$$

The initial conditions for this step are

$$t = 0 \ (cycle \ 1) \rightarrow C_{ij} = C_{ij}(cycle \ 1, \ step \ 1); \ q_{ij} = q_{ij}(cycle \ 1, \ step \ 1) \tag{6.9}$$

$$t = t_{S1} \ (cycle \ k) \rightarrow C_{ij} = C_{ij}(cycle \ k, \ step \ 1); \ q_{ij} = q_{ij}(cycle \ k, \ step \ 1) \tag{6.10}$$

and the boundary conditions are

$$z = 0 \rightarrow C_{ij,0} = C_{ij} - \frac{D_{Lj}}{v_j} \frac{\partial C_{ij}}{\partial z}\bigg|_{z=0} \tag{6.11}$$

and

$$z = L_j \rightarrow \begin{cases} Eluent \ node: \ C_{i1,0} = \dfrac{Q_4}{Q_1} C_{i4,L_j} \\[2ex] Between \ sections, \ raffinate \ and \ extract \ nodes: \ C_{ij,L_j} = C_{ij+1,0} \end{cases} \tag{6.12}$$

## 6.1.2 Determination of Operating Conditions

To obtain the desired separation on a JO process it is necessary to determine the operating conditions, that is, the internal flow rates in each section and the duration of steps 1 and 2. Mata and Rodrigues (2001) suggested for mixtures with linear isotherms a method to determine the operating conditions based on the definition of concentration propagation velocity of the components and on constraints for steps 1 and 2 (Mata and Rodrigues, 2001). In this method, for a given feed flow rate, $Q_F$, and a given $t_{S2}$, a set of conditions must be met to separate all three components of the mixture. The conditions and equations needed to calculate all the variables of the system are shown in Table 6.1.

This methodology was modified by Da Silva and Rodrigues (2008), and its application follows two strategies depending on which is the "harder separation," that is, the separation between the intermediate and the more adsorbed components (Strategy 1) or between the intermediate and the weakly adsorbed components (Strategy 2). Also, new parameters were introduced, namely the number of columns traveled by the intermediate component after the feed port during step 1, $p_{tS1}$, and a parameter that controls the amount of intermediate compound collected during step 1, $p_{QS1}$, and the distance traveled by component $i$ during step 2, $x_{i,j,S2}$, which is related with parameter $p_i$ by the following expressions:

$$x_{A,2/3,S2} = p_A L_c - x_{A,3,S1} = \frac{t_{S2}}{\varepsilon A_c \phi_A}(Q_{2/3} - Q_S K_A) \quad \left(p_A \geq \frac{x_{A,3,S1}}{L_C}\right) \tag{6.13}$$

**Table 6.1** Conditions and Equations for the Determination of Operating Conditions (Adsorption Strength: A < B < C) (Mata et al., 2001)

| | Step 1 | Step 2 |
|---|---|---|
| Variables Assumptions | $t_{S1}$, $Q_{1,S1}$, $Q_{2,S1}$, $Q_{3,S1}$, $Q_{4,S1}$, $Q_F$, $Q_{EI1}$, $Q_I$<br>• $Q_F$ is known<br>• Intermediate retained component injected only in one column<br>• $N_c = 3$ (number of columns per section) | $t_{S2}$, $Q_S$, $Q_{1,S2}$, $Q_{2,S2}$, $Q_{3,S2}$, $Q_{4,S2}$, $Q_X$, $Q_R$, $Q_{EI2}$<br>• $t_{S2}$ is known<br>• Intermediate retained component moves $(x_{B,S1} + L_C)$ with the solid<br>• More retained component moves $(x_{C,S1} + 3L_C)$ with the liquid<br>• $N_c = 3$ (number of columns per section) |
| Equations | $$t_{S1} = \bar{t}_B - \sqrt{\frac{8}{k_p}}\sqrt{(\bar{t}_B - \tau)}$$ $$Q_{EI1} = \frac{\epsilon A_C (2L_C)}{t_{S1}}\left(1 + \frac{1-\epsilon}{\epsilon}K_B\right) - Q_F$$ $$Q_{3,S1} = Q_{4,S1} = Q_F$$ $$Q_{1,S1} = Q_{2,S1} = Q_I$$ $$Q_I = Q_F + Q_{EI1}$$ | $$Q_{3,S2} = \frac{\epsilon A_C}{t_{S2}}\left\{\frac{K_B}{K_B - K_C}\left[x_{C,S2}\left(1 + \frac{1-\epsilon}{\epsilon}K_C\right)\right.\right.$$ $$\left.\left. - x_{B,S2}\left(1 + \frac{1-\epsilon}{\epsilon}K_B\right)\right]\right.$$ $$\left. + x_{B,S2}\left(1 + \frac{1-\epsilon}{\epsilon}K_B\right)\right\}$$ $$Q_S = \frac{\epsilon A_C}{t_{S2}}\left\{\frac{1}{K_B - K_C}\left[x_{C,S2}\left(1 + \frac{1-\epsilon}{\epsilon}K_C\right)\right.\right.$$ $$\left.\left. - x_{B,S2}\left(1 + \frac{1-\epsilon}{\epsilon}K_B\right)\right]\right\}$$ $$Q_{1,S2} = Q_S K_C \beta - (Q_F + Q_{EI1})\frac{t_{S1}}{t_{S2}}$$ $$Q_{2,S2} = Q_{3,S2}$$ $$Q_{4,S2} = \frac{Q_S K_A}{\beta} - Q_F\frac{t_{S1}}{t_{S2}}$$ $$Q_R = Q_{3,S2} - Q_{4,S2}$$ $$Q_X = Q_{1,S2} - Q_{2,S2}$$ $$Q_{EI2} = Q_{1,S2} - Q_{4,S2}$$ |

$$x_{B,2/3,S2} = -(x_{B,3,S1} + p_B L_C) = \frac{t_{S2}}{\varepsilon A_c \phi_B}(Q_{2/3} - Q_S K_B) \quad (0 < p_B < n_{S2}) \qquad (6.14)$$

$$x_{C,2/3,S2} = -(x_{C,3,S1} + p_C L_C) = \frac{t_{S2}}{\varepsilon A_c \phi_C}(Q_{2/3} - Q_S K_C) \quad (p_C > 0) \qquad (6.15)$$

in which,

$$\phi_i = 1 + \frac{1 - \varepsilon}{\varepsilon} K_i \qquad (6.16)$$

Table 6.2 shows the equations used by this methodology in both strategies.

## 6.1.3 Application to a Mixture of Sucrose, Fructose, and Betaine

Mata et al. studied the application of the JO process to a mixture of two sugars, sucrose and fructose, and one nonsugar, betaine (Mata et al., 2001). The experimental response curves for the three components are presented in Figure 6.2. These curves were obtained at 25 °C by injecting 200 μl of monocomponent aqueous solutions in a chromatographic column ($L = 11.8$ cm and $D = 2.6$ cm) packed with the cation-exchange resin Amberlite CR1310 Na$^+$ form, from Rohm & Haas. The samples were analyzed using a Gilson 131 IR detector, connected to the column outlet.

The isotherm parameters and the mass-transfer coefficients were determined considering that all the components of the mixture present linear adsorption behavior. The values of the linear isotherm parameters and the mass-transfer coefficients are presented in Table 6.3.

The pseudo-SMB model can be used to simulate the JO process operation for the separation of the mixture. The characteristics of the JO process considered for purposes of simulation are presented in Table 6.4.

The feed flow rate and the duration of step 2 considered are: $Q_F = 350.0$ ml/min and $t_{S2} = 66$ min. The duration of step 1 is $t_{S1} = 17.8$ min. All the flow rates in the JO process in both steps 1 and 2 are presented in Table 6.5.

Figure 6.3 shows the cyclic steady-state internal liquid phase concentration profiles for both steps 1 and 2. The feed containing the three components with concentration, $C_0 = 100$ g/l, is injected during step 1 with flow rate $Q_F$, at the inlet of section 3. At the same time, the intermediate stream, $Q_I$, rich in fructose, is collected from the end of section 2. During step 2 both $Q_F$ and $Q_I$ are stopped, and a stream rich in sucrose is collected as raffinate, $Q_R$; and a stream rich in betaine is collected as extract, $Q_X$. At the same time, the movement of the solid phase in opposite direction of the fluid phase is assumed. The less-retained component, sucrose, is carried by the liquid phase, and the fructose and betaine, the intermediate and the more-retained component, respectively, are carried by the solid phase in the opposite direction.

**Table 6.2** Operating Conditions of Japan Organo (JO) for a Ternary Separation (Da Silva and Rodrigues, 2008)

**Step 1**

$$t_{S1} = \bar{t}_B - 2\left(\frac{2t_B^2}{Pe} + \frac{2}{k_p}\left(\bar{t}_B - p_{tS1}\frac{\epsilon L_C A_C}{Q_F}\right)\right)^{1/2}$$

$$Q_{1,S1} = p_{QS1}\epsilon\frac{L_C A_C}{t_{S1}}\left(1 + \frac{1-\epsilon}{\epsilon}K_B\right)$$

$$Q_F = Q_{3,S1} = Q_{4,S1}$$

$$Q_{EI1} + Q_F = Q_{1,S1} = Q_{2,S1}$$

**Step 2**

Strategy 1

$$Q_{2/3} = \frac{\epsilon A_C}{t_{S2}}\left(\frac{K_C}{K_C - K_B}\right)\left[\phi_B x_{B,2/3,S2} - \phi_C\frac{K_B}{K_C}x_{C,2/3,S2}\right]$$

$$Q_S = \frac{\epsilon A_C}{t_{S2}}\left(\frac{1}{K_C - K_B}\right)\left[\phi_B x_{B,2/3,S2} - \phi_C x_{C,2/3,S2}\right]$$

$$Q_{1,S2} = \beta_1 Q_S K_C - \frac{t_{S1}}{t_{S2}}(Q_F + Q_{EI1}) \left(\text{with } \beta_1 > 1 \text{ and } \beta_1 > \frac{t_{S1}(Q_F + Q_{EI1})}{t_{S2}Q_S K_C}\right)$$

$$Q_{4,S2} = \beta_4 Q_S K_A - \frac{t_{S1}}{t_{S2}}Q_F \left(\text{with } \frac{t_{S1}Q_F}{t_{S2}Q_S K_A} < \beta_4 < 1\right)$$

Strategy 2

$$Q_{2/3} = \frac{\epsilon A_C}{t_{S2}}\left(\frac{K_B}{K_B - K_A}\right)\left[\phi_A x_{A,2/3,S2} - \phi_B\frac{K_A}{K_B}x_{B,2/3,S2}\right]$$

$$Q_S = \frac{\epsilon A_C}{t_{S2}}\left(\frac{1}{K_B - K_A}\right)\left[\phi_A x_{A,2/3,S2} - \phi_B x_{2/3,S2}\right]$$

The parameter β represents the safety margin.

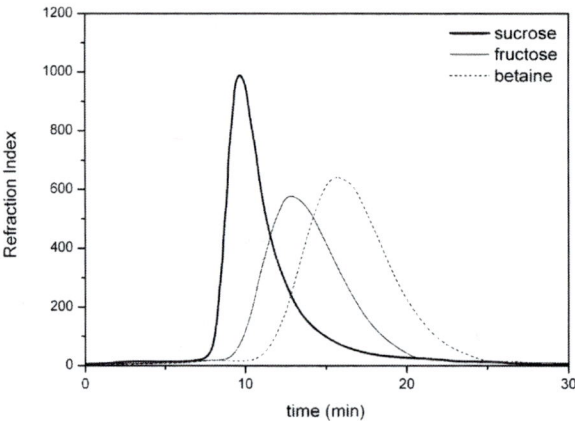

**Figure 6.2** Experimental pulse responses for sucrose, fructose, and betaine. ($C_{SU} = 15$ g/l, $C_{FR} = 15$ g/l, and $C_{BE} = 80$ g/l, respectively).

**Table 6.3** Calculated Values of Adsorption Parameters and Mass-transfer Coefficients for the Ternary Mixture at 25 °C

| Component $i$ | $K_i$ | $k_p$ (s$^{-1}$) |
|---|---|---|
| Sucrose | 0.19 | 0.20 |
| Fructose | 0.44 | 0.13 |
| Betaine | 0.65 | 0.11 |

**Table 6.4** Characteristics of the JO System (Japan Organo Catalog, 1998)

| | |
|---|---|
| Column number (total) | 12 |
| Column length (cm) | 120 |
| Column diameter (cm) | 10.84 |
| Column volume (l) | 11.1 |
| Number of columns per section | 3 |
| Bed porosity, $\varepsilon$ | 0.4 |

**Table 6.5** Flow Rates Used for the Simulation of the JO Process

| Step 1 | Step 2 |
|---|---|
| $Q_1 = 826.4$ ml/min | $Q_1 = 723.5$ ml/min |
| $Q_2 = 826.4$ ml/min | $Q_2 = 454.8$ ml/min |
| $Q_3 = 350.0$ ml/min | $Q_3 = 454.8$ ml/min |
| $Q_4 = 350.0$ ml/min | $Q_4 = 166.4$ ml/min |
| $Q_{EI1} = 476.4$ ml/min | $Q_{EI2} = 557.1$ ml/min |
| $Q_F = 350.0$ ml/min | $Q_X = 268.7$ ml/min |
| $Q_I = 826.4$ ml/min | $Q_R = 288.4$ ml/min |
| | $Q_S = 1413.5$ ml/min |

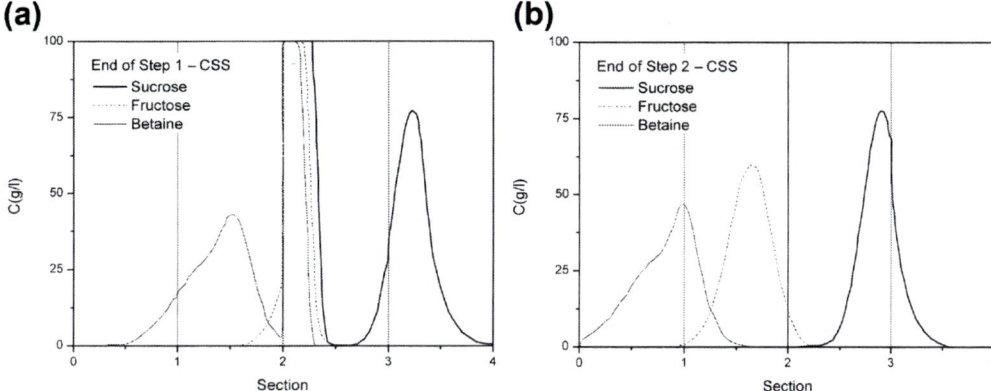

**Figure 6.3** Profiles of sucrose, fructose, and betaine, at the end of: (a) step 1 and (b) step 2.

Figure 6.4 shows the movement of the liquid phase concentration profile of each component during its recovery: (a) the recovery of sucrose occurs from the end of step 1 to the end of step 2; during this time, $t_{S2}$, the sucrose is recovered at the outlet of section 3, in the raffinate stream; (b) fructose is recovered from the end of step 2 to the end of step 1; during this time, $t_{S1}$, fructose is collected at the outlet of section 2, in the intermediate stream; (c) the recovery of betaine is performed from the end of step 1 to the end of step 2; during this time, $t_{S2}$, betaine is recovered from the outlet of section 1, in the extract stream.

## 6.2 MCSGP PROCESS

The MCSGP process was developed at the Swiss Federal Institute of Technology in Zurich by Aumann and Morbidelli (2007) and is used in the chromatographic purification of biomolecules from complex mixtures resulting from upstream biological processes. The MCSGP process is able to split the feed-stream mixture into three fractions in which the target component with intermediate adsorption strength is collected in one fraction, and the weak- and strong-adsorbing impurities are the other two fractions. Conventionally, this kind of separation is performed by batch solvent gradient chromatography in which the column effluent is collected in several consecutive fractions that can be rich in the target component fulfilling the purity requirements or fractions in which a high amount of impurities is present. The yield of the batch chromatography process can be enhanced by recycling to the feed point the fractions rich in the target component but that do not fulfill the purity requirements (Tarafder et al., 2008). However, the use of a continuous process such as MCSGP in large-scale production is much better in terms of yield, solvent consumption, and

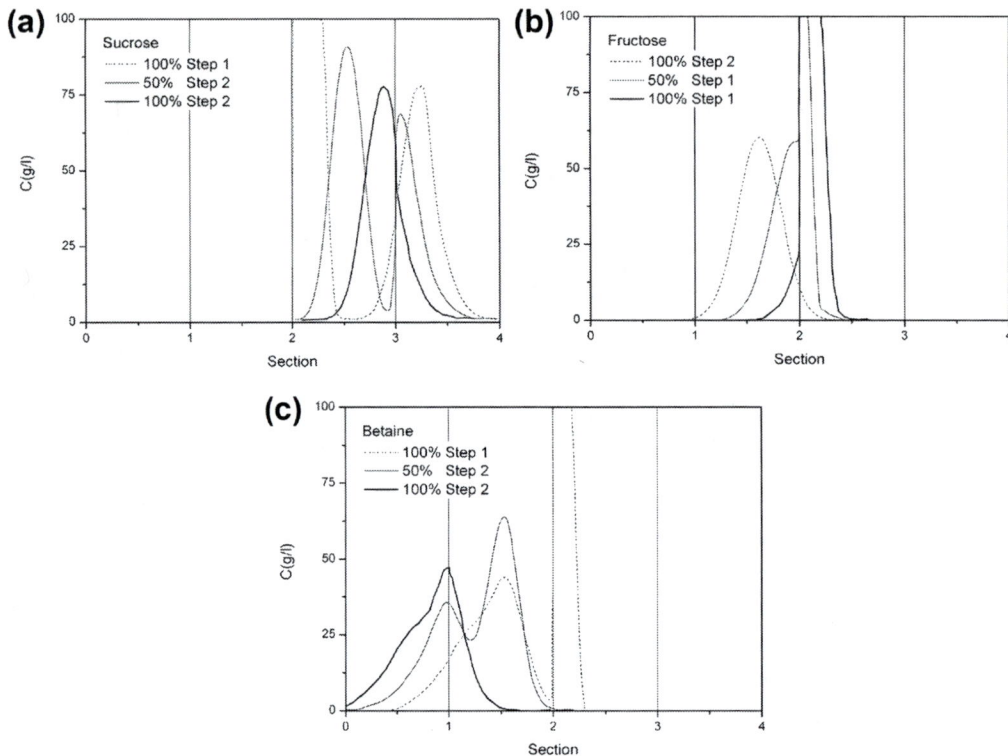

**Figure 6.4** Evolution of concentration profiles during the recovery of (a) sucrose in step 2, (b) fructose in step 1, and (c) betaine in step 2.

productivity relative to the conventional batch solvent gradient chromatography (Aumann and Morbidelli, 2007). The MCSGP process has been applied in different purification problems, such as purification of monoclonal antibody (mAb) variants (Müller–Späth et al., 2008, 2010b) and the purification of peptide in reverse-phase mode (Aumann and Morbidelli, 2007).

Figure 6.5 shows a simplified chromatogram representing a generic problem in the chromatographic solvent gradient purification of biomolecules. The chromatogram can be divided into five fractions indicated by the number in the time axis: (1) strongly adsorbing impurities (S), (2) product contaminated by strongly adsorbing impurities (P + S), (3) product (P), (4) product contaminated by weakly adsorbing impurities (P + W), (5) weakly adsorbing impurities (W).

The original continuous MCSGP process addresses the purification of the target component by performing three tasks: collect fraction 3, drain fractions 1 and 5, and recycle fractions 2 and 4. Figure 6.6 shows a schematic representation of a six-column MSCGP process in which each column is connected to a gradient pump.

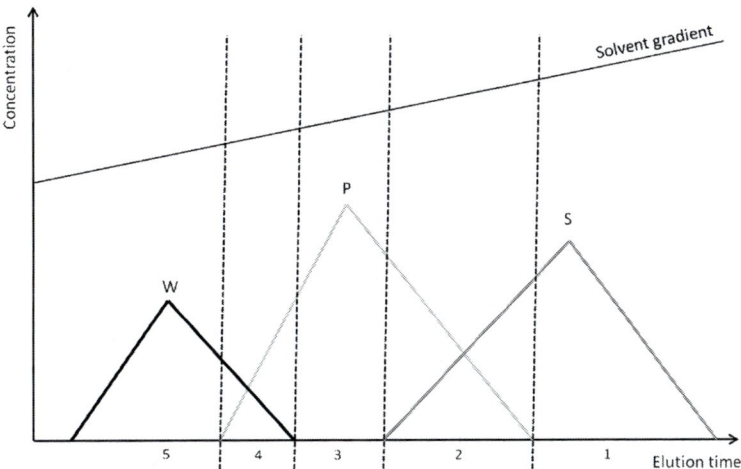

**Figure 6.5** Simplified chromatogram of a purification problem (S: strongly adsorbing impurities, P: product, W: weakly adsorbing impurities).

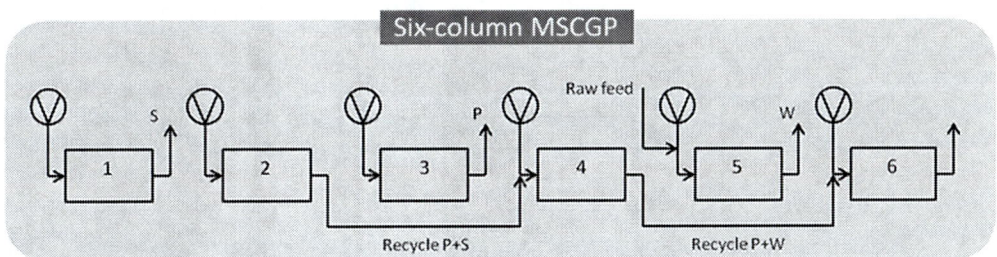

**Figure 6.6** Schematic representation of a six-column multicolumn countercurrent solvent gradient purification (MCSGP) process.

The different tasks necessary to the purification of the target product (P) are integrated within the MCSGP process (Figure 6.6). The fractions of contaminated product (P + S and P + W) are recycled from the outlet of columns 2 and 4 to the inlet of columns 4 and 6, respectively. The fraction with strong-adsorbing impurities is eluted from column 1 and the fraction with weak-adsorbing impurities is eluted from column 5. The target product is collected from the outlet of column 3. After a predetermined interval, all the columns switch position in the liquid-phase direction, similar to the SMB. This periodic switch of the columns position simulates the countercurrent movement between the liquid and the solid, which leads to an improvement of the process stationary-phase efficiency (Pais et al., 2000).

An alternative configuration is a semicontinuous MCSGP process with only three columns, which can be used to reduce the equipment costs (Aumann and Morbidelli, 2008). In this configuration, collection of the S, W, and P fractions and the recycling of the P + S and P + W fractions are performed in two sequential steps (Figure 6.7).

This concept can be directly derived by decoupling the six-column MCSGP process (Figure 6.6) in two parts, the collect/drain part comprising columns 1, 3, and 5, and the recycling part comprising columns 1, 4, and 6. Because no connection exists between these two parts during the six-column MCSGP operation, it is possible to alternate them in time using only three columns resulting in the semicontinuous three-column MCSGP process. Because the two steps are independent, the system can be operated using two different switching times; during time $t_B^*$, the system works in batch mode and during time $t_{CC}^*$, the columns are connected in series and the system works in countercurrent mode.

Other alternative configurations for the MCSGP process have been proposed, such as the four-column MCSGP process (Muller-Späth et al., 2009, 2010a). In addition, the extension of the MCSGP process to the separation of more than three fractions (two target components) has been studied (Krättli et al., 2013).

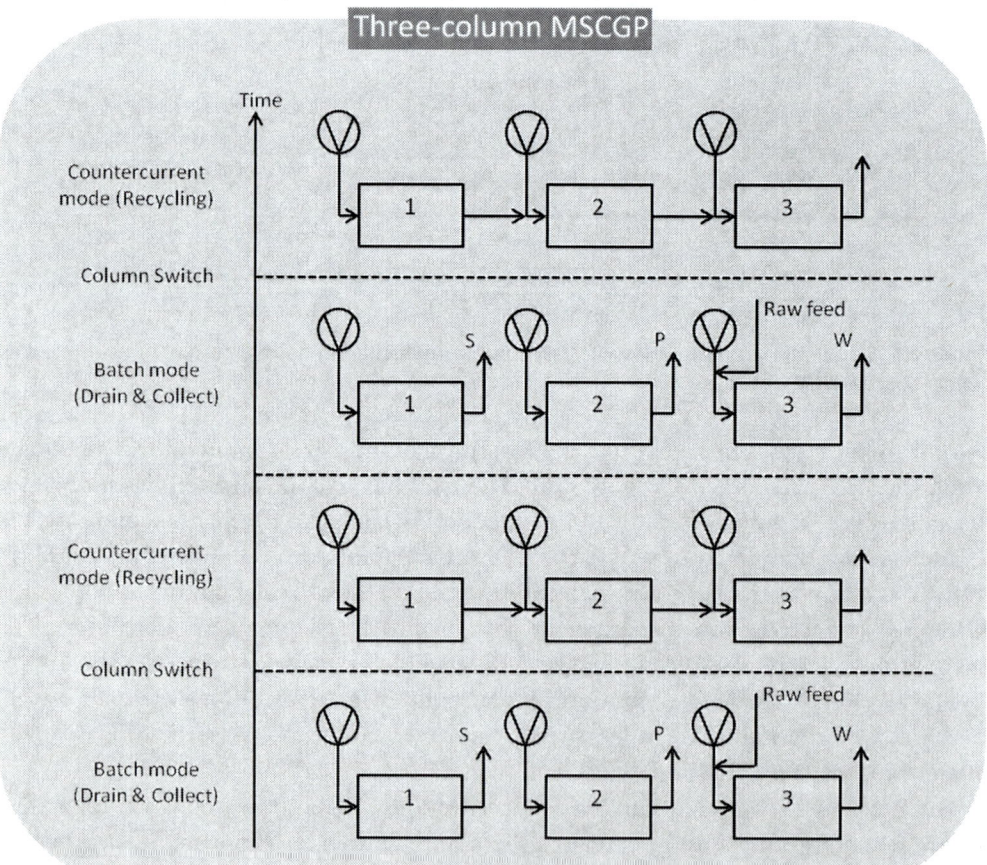

**Figure 6.7** Schematic representation of a three-column MCSGP process operation.

### 6.2.1 Design of the MCSGP Process

For the empirical design of the MCSGP process, any information about adsorption isotherms, mass-transfer limitations, and axial dispersion is unnecessary. The empirical design of the MCSGP is based on the information given by a batch single-column chromatogram (Figure 6.8).

The chromatogram presented in Figure 6.8 shows the solvent gradient elution of calcitonin on a polystyrene reversed-phase resin (Aumann and Morbidelli, 2007). A previous optimized solvent gradient is applied during the entire elution time ($t_F$ to $t_E$). Different tasks are needed to purify the target product (P): loading the feed, collecting the fraction, cleaning in place, and equilibration.

The empirical design procedure begins by transferring the single-column chromatogram (Figure 6.8) to the MCSGP unit (Figure 6.9). As can be noticed in Figure 6.9, each section of the process ($\alpha$, $\beta$, $\gamma$, and $\delta$) is defined based on the time intervals of the single-column chromatogram.

The second step of the design is to determine the flow rates in each column. This step begins by calculating the switching time ($t^*$) through Eqn (6.20), in which the flow rate of column 4 is set equal to the flow rate in the batch column to reproduce the batch single-column chromatogram (Aumann and Morbidelli, 2007). The flow rate in each column is calculated as follows:

$$Q_1 = Q_{Batch}\left(\frac{\varepsilon V}{\varepsilon_{Batch} V_{Batch}}\right)\frac{t_\gamma - t_\beta}{t^*} \tag{6.17}$$

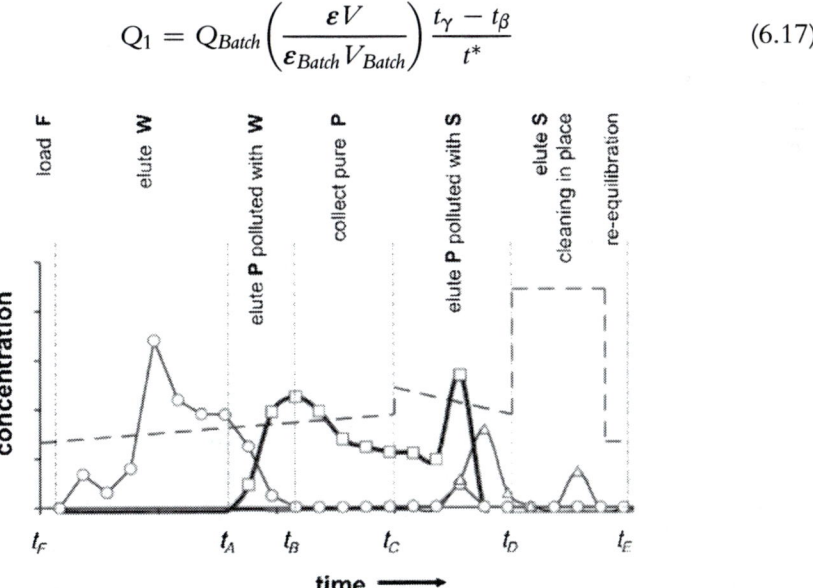

**Figure 6.8** Batch single-column chromatogram. Dashed line: modifier concentration; triangles: strongly adsorbing impurities; squares: product; and circles: weakly adsorbing impurities (Aumann and Morbidelli, 2007). *(Reprinted from Aumann and Morbidelli, 2007, with permission from John Wiley & Sons, Inc.)*

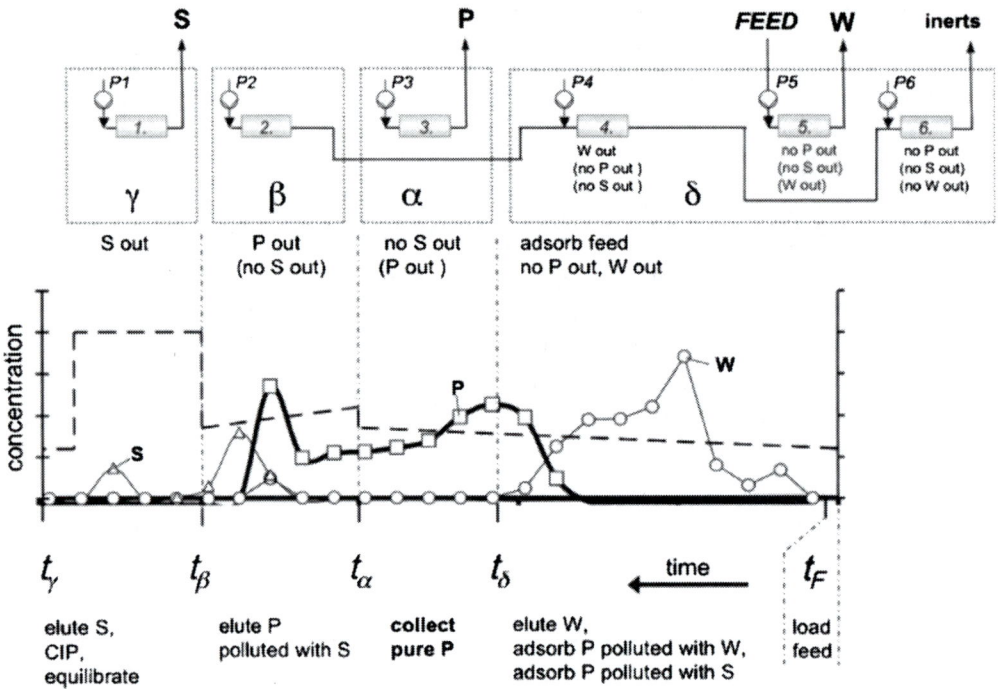

**Figure 6.9** Schematic representation of the empirical design procedure of the MCSGP process with six columns (Aumann and Morbidelli, 2007). *(Reprinted from Aumann and Morbidelli, 2007, with permission from John Wiley & Sons, Inc.)*

$$Q_2 = Q_{Batch} \left( \frac{\varepsilon V}{\varepsilon_{Batch} V_{Batch}} \right) \frac{t_\beta - t_\alpha}{t^*} \qquad (6.18)$$

$$Q_3 = Q_{Batch} \left( \frac{\varepsilon V}{\varepsilon_{Batch} V_{Batch}} \right) \frac{t_\alpha - t_\delta}{t^*} \qquad (6.19)$$

$$Q_4 = Q_{Batch} \left( \frac{\varepsilon V}{\varepsilon_{Batch} V_{Batch}} \right) \frac{t_\delta - t_F}{t^*} \qquad (6.20)$$

The flow rate for column 5 is calculated to ensure that the concentration of the purified product obtained is the same as that of the batch operation during the $t_\delta - t_\alpha$ interval. Therefore, the amount of product introduced through the feed stream has to be the same taken during the interval $t_\delta - t_\alpha$ to obtain a steady-state product concentration profile similar to the batch operation. The flow rate of column 5 can be calculated through the following expression:

$$C_{P,Feed} Q_5 = Q_{Batch} \frac{C_{P,t_\delta} + C_{P,t_\alpha}}{2} \qquad (6.21)$$

**Table 6.6** Modifier Concentrations at the Column Inlets at the Beginning and End of a Switch Time (Aumann and Morbidelli, 2007)

| Column | $C_{mod}$ at $t = 0$ | $C_{mod}$ at $t = t^*$ |
|---|---|---|
| 1 | $C_{batch,mod}$ at $t = t_\beta$ | $C_{batch,mod}$ at $t = t_\gamma$ |
| 2 | $C_{batch,mod}$ at $t = t_\alpha$ | $C_{batch,mod}$ at $t = t_\beta$ |
| 3 | $C_{batch,mod}$ at $t = t_\delta$ | $C_{batch,mod}$ at $t = t_\alpha$ |
| 4 | $C_{batch,mod}$ at $t = t_F$ | $C_{batch,mod}$ at $t = t_\delta$ |
| 5 | $C_{batch,mod}$ at $t = t_F$ | $C_{batch,mod}$ at $t = t_F$ |
| 6 | To be optimized | To be optimized |

After the determination of the flow rates, it is necessary to reproduce also a solvent gradient profile equivalent to the batch operation. The gradient inside the MCSGP process is determined according to Table 6.6, which gives the initial and final modifier concentrations in each column during a switching time interval. The flow rate and modifier concentrations in column 6 are determined by optimizing the separation between the less-adsorbing impurities and the product.

Because the presence of dead volumes inside the MCSGP unit were not considered in the design procedure described above, some deviations can occur between expected and experimental results obtained. Therefore, all the parameters determined by the proposed design procedure should be viewed as a first estimation of the operating conditions, and a subsequent fine-tuning procedure should be applied to achieve the desired performance.

## 6.3 SEQUENTIAL MULTICOLUMN CHROMATOGRAPHY

The purification steps in biopharmaceutical downstream processing often have to deal with a numerous impurities, such as host cell protein and DNA (Roque et al., 2004), that result from the upstream fermentation or cell culture processes used to produce pharmaceutical drugs (e.g., recombinant proteins and mAb). The costs associated with downstream processing are estimated to be 50—80% of the total manufacturing costs (Roque et al., 2004; Müller-Späth et al., 2008).

Chromatography is one of the most used technologies in downstream processing in biopharmaceutical applications (Holzer et al., 2008). The basic concept of batch chromatography is applied by loading the feed into the column (load). In the next step, the non-retained materials are eluted from the column by using a different buffer (wash) followed by another buffer change to remove the retained components from the column (elute). Finally, the column is regenerated and re-equilibrated to prepare for the next feed load (Wiesel et al., 2003). However, when high flow rates are required, the single-column batch chromatography presents low yield because of the decrease in its dynamic binding capacity that results from the incomplete loading of the column stationary phase when the product breakthrough occurs (Mahajan et al., 2012).

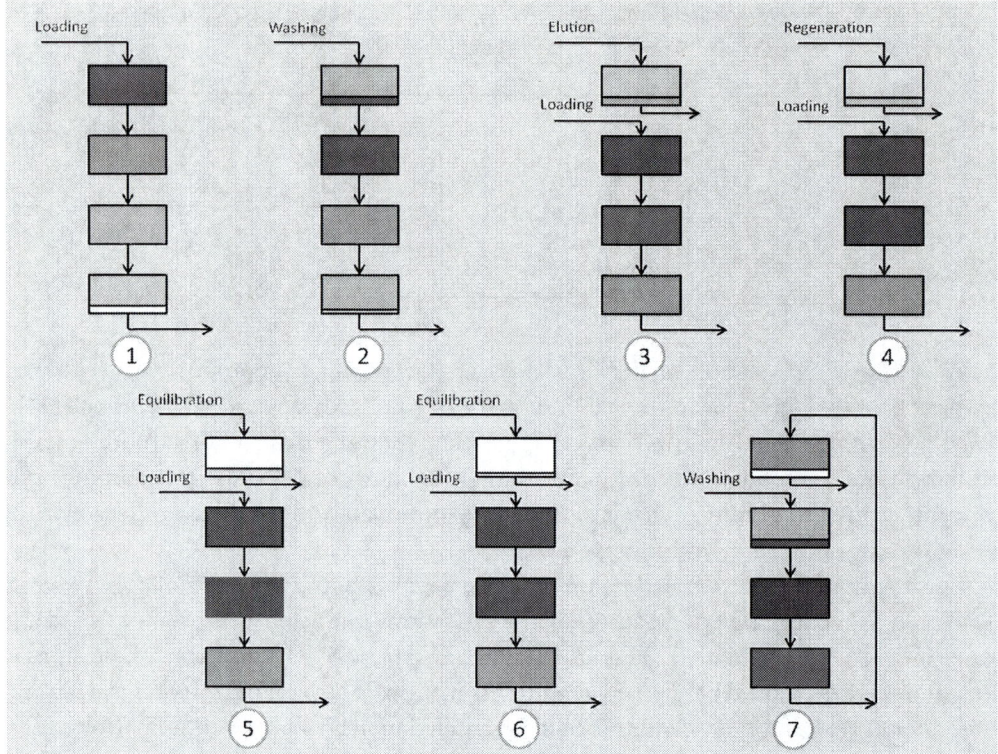

**Figure 6.10** Schematic representation of the SMCC process.

An alternative process that can be used to overcome the limitations of batch chromatography is the SMCC process (Holzer et al., 2008). Figure 6.10 shows an SMCC process setup with four columns disposed in linear sequence. The SMCC process begins by connecting the loading stream to the first column (1). When the first column becomes highly loaded, (2) a wash step begins, pushing the nonretained product from the first column to the second column. This step is important to prevent product loss and it minimizes buffer consumption. After the wash step, the first column is disconnected from the sequence and goes though (3) elution, (4) regeneration, and (5, 6) equilibration steps while the loading stream is transferred to the second column. When the first column becomes fully regenerated and equilibrated, the wash step begins in the high-loaded second column, and (7) the first column is reconnected into the sequence. The same steps are repeated in the rest of the column sequence until the loading stream returns to the first column completing a full cycle.

## 6.3.1 BioSC® Process by Novasep

The BioSC multicolumn chromatographic process developed by Novasep (Figure 6.11) is applied mainly in separation problems associated with biopharmaceutical downstream

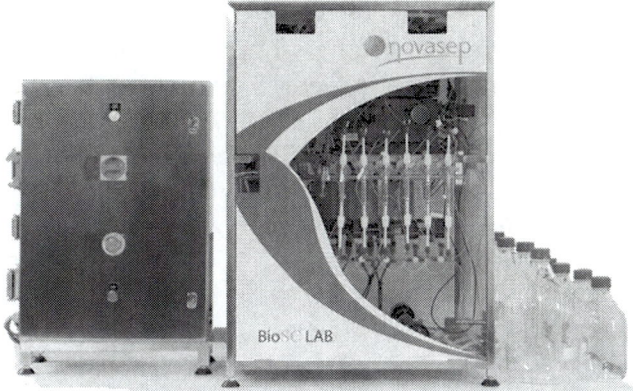

**Figure 6.11** BioSC® process by Novasep (Novasep, 2014).

**Table 6.7** Economic Comparison of Batch and Sequential Multicolumn Chromatography (SMCC) Processes for mAb Capture (Holzer et al., 2008)

| Parameter | Batch mAb Capture | SMCC mAb Capture | Potential Savings |
|---|---|---|---|
| Volume of stationary phase | 455 l | 200 l | >$2,300,000/unit |
| Processing time | 4 h | 2 h | 50% of time |
| Water use | 18,000 l | 5800 l[a] | >$2/$g_{protein}$ ($50,000/batch) |
| Capital expenditure | $1,000,000 | $1,000,000[a] | None |

[a]Capital expenditure costs due to utility downsizing for SMCC are not taken into consideration.

processing. The BioSC process was designed to be a continuous process that is able to more efficiently use the capacity of the stationary phase.

The BioSC is a sequential multicolumn system composed of two to six columns connected in series (Novasep, 2014). The process working principle is the same as that illustrated in Figure 6.10.

It was shown that the use of Novasep's BioSC technology, in the purification of mAb using protein A media, leads to a significant improvement in both productivity and economic savings relatively to the batch chromatography process (Holzer et al., 2008). Table 6.7 shows a comparative economic study between the SMCC and batch operation modes.

## 6.4 CONCLUDING REMARKS

Several SMB-based multicolumn processes can be used to deal with multicomponent purification problems. These processes are designed to take advantage of the countercurrent operation mode that can be simulated by applying the SMB concept, which can lead to improvement of process efficiency.

The improvement of the processes used in the purification of complex multicomponent mixtures is very important in the biopharmaceutical area, in which drugs produced by upstream fermentation processes represent a huge challenge to the downstream purification processes, typically the most expensive part of the manufacturing process. Therefore, the development of continuous chromatographic processes that efficiently use the stationary phase and provide the product with the required purity is an important asset to the improvement of several manufacturing processes.

## NOMENCLATURE

| | |
|---|---|
| $A_C$ | Column cross-section area (m$^2$) |
| $C$ | Liquid phase concentration (mol/m$^3$) |
| $D_L$ | Axial dispersion coefficient (m$^2$/min) |
| $K$ | Linear adsorption isotherm parameter |
| $k_p$ | Intraparticle mass-transfer coefficient (min$^{-1}$) |
| $L_C$ | Column length (m) |
| $N_C$ | Number of columns |
| p | Dimensionless parameter of Eqns (6.13)–(6.15) |
| $q$ | Average adsorbed-phase concentration (mol/m$^3$) |
| Q | Volumetric flow rate (m$^3$/min) |
| $q^*$ | Adsorbed-phase concentration in equilibrium with C (mol/m$^3$) |
| t | Time variable (min) |
| $\bar{t}$ | Retention time (min) |
| $t^*$ | Switching time (min) |
| $u_s$ | Solid-phase velocity (m/min) |
| $v$ | Liquid-phase interstitial velocity (m/min) |
| V | Volume (m$^3$) |
| x | Component displacement on Step 2 of JO (m) |
| z | Axial coordinate (m) |

## Greek Letters

| | |
|---|---|
| $\beta$ | Safety margin |
| $\varepsilon$ | Bed porosity |
| $\tau$ | Residence time (min) |
| $\phi$ | Dimensionless parameter of Eqn 6.16 |

## Subscripts

| | |
|---|---|
| $c$ | Column |
| $El$ | Eluent |
| $F$ | Feed |
| $i$ | Component ($i = A,B,C$) |
| $I$ | Intermediate component |
| $j$ | Section ($j = 1,2,3,4$) |
| $R$ | Raffinate |

| | |
|---|---|
| *s* | Solid |
| *S1* | Step 1 of the JO process |
| *S2* | Step 2 of the JO process |
| *X* | Extract |

## REFERENCES

Ando, M., Tanimura, M., Tamura, M., 1990. Method of Chromatographic Separation. U.S. Patent No. 4.970.002.

Aumann, L., Morbidelli, M., 2007. A continuous multicolumn countercurrent solvent gradient purification (MCSGP) process. Biotechnol. Bioeng. 98, 1043−1055.

Aumann, L., Morbidelli, M., 2008. A semicontinuous 3-column countercurrent solvent gradient purification (MCSGP) process. Biotechnol. Bioeng. 99, 728−733.

Beste, Y.A., Arlt, W., 2002. Side-stream simulated moving-bed chromatography for multicomponent separation. Chem. Eng. Technol. 25, 956−962.

Chiang, A.S.T., 1998. Continuous chromatographic process based on SMB technology. AIChE J. 44, 1930−1932.

Da Silva, E.A.B., Rodrigues, A.E., 2008. Design methodology and performance analysis of a pseudo-simulated moving bed for ternary separation. Sep. Sci. Technol. 43, 533−566.

Holzer, M., Osuna-Sanchez, David, L., 2008. Multicolumn chromatography, a new approach to relieving capacity bottlenecks for downstream processing efficiency. BioProcess Int. 6, 74−82.

Krättli, M., Müller-Späth, T., Morbidelli, M., 2013. Multifraction separation in countercurrent chromatography (MCSGP). Biotechnol. Bioeng. 110, 2436−2444.

Mahajan, E., George, A., Wolk, B., 2012. Improving affinity chromatography resin efficiency using semi-continuous chromatography. J. Chromatogr. A 1227, 154−162.

Massuda, T., Sonobe, T., Matsuda, F., 1993a. Process for Fractional Separation of Multi-component Fluid Mixture. U.S. Patent No. 5.198.120.

Massuda, T., Sonobe, T., Matsuda, F., Horie, M., 1993b. Process for Fractional Separation of Multi-component Fluid Mixture.

Mata, V.G., Pais, L.S., Azevedo, D.C.S., Rodrigues, A.E., 2001. Fraction of multicomponent sugar mixtures using a pseudo simulated moving bed. In: Ribeiro, F.R., Pinto, J.J.C.C. (Eds.), CHEMPOR' 2001-Eighth International Chemical Engineering Conference, Aveiro, Portugal.

Mata, V.G., Rodrigues, A.E., 2001. Separation of ternary mixtures by pseudo-simulated moving bed chromatography. J. Chromatogr. A 939, 23−40.

Müller-Späth, T., Aumann, L., Melter, L., Ströhlein, G., Morbidelli, M., 2008. Chromatographic separation of three monoclonal antibody variants using multicolumn countercurrent solvent gradient purification (MCSGP). Biotechnol. Bioeng. 100, 1166−1177.

Muller-Späth, T., Aumann, L., Morbidelli, M., 2009. Role of cleaning-in-place in the purification of mab supernatants using continuous cation exchange chromatography. Sep. Sci. Technol. 44, 1−26.

Müller-Späth, T., Aumann, L., Ströhlein, G., Kornmann, H., Valax, P., Delegrange, L., Charbaut, E., Baer, G., Lamproye, A., Jöhnck, M., Schulte, M., Morbidelli, M., 2010a. Two step capture and purification of IgG2 using multicolumn countercurrent solvent gradient purification (MCSGP). Biotechnol. Bioeng. 107, 974−984.

Müller-Späth, T., Krättli, M., Aumann, L., Ströhlein, G., Morbidelli, M., 2010b. Increasing the activity of monoclonal antibody therapeutics by continuous chromatography (MCSGP). Biotechnol. Bioeng. 107, 652−662.

Navarro, A., Caruel, H., Rigal, L., Phemius, P., 1997. Continuous chromatographic separation process: simulated moving bed allowing simultaneous withdrawal of three fractions. J. Chromatogr. A 770, 39−50.

Nicoud, R.M., 2000. Simulated Moving Bed Chromatography for Biomolecules, Handbook of Bioseparations. Academic Press, San Diego, CA, pp. 475−509.

Novasep, 2014. http://www.novasep.com/.

Pais, L.S., Loureiro, J.M., Rodrigues, A.E., 2000. Chiral separation by SMB chromatography. Sep. Purif. Technol. 20, 67—77.

Roque, A.C.A., Lowe, C.R., Taipa, M.Â., 2004. Antibodies and genetically engineered related molecules: production and purification. Biotechnol. Prog. 20, 639—654.

Sayama, K., Kamada, T., Oikawa, S., Masuda, T., 1992. Production of raffinose: a new byproduct of the beet sugar industry. Zucherind 177, 893—898.

Tarafder, A., Strohlein, G., Aumann, L., Morbidelli, M., 2008. Role of recycling in improving the performance of chromatographic solvent gradient purifications. J. Chromatogr. A 1183, 87—99.

Wankat, P.C., 2001. Simulated moving bed cascades for ternary separations. Industrial Eng. Chem. Res. 40, 6185—6193.

Wiesel, A., Schmidt-Traub, H., Lenz, J., Strube, J., 2003. Modelling gradient elution of bioactive multicomponent systems in non-linear ion-exchange chromatography. J. Chromatogr. A 1006, 101—120.

Wooley, R., Ma, Z., Wang, N.H.L., 1998. A nine-zone simulating moving bed for the recovery of glucose and xylose from biomass hydrolyzate. Industrial Eng. Chem. Res. 37, 3699—3709.

Zobel, S., Helling, C., Ditz, R., Strube, J., 2014. Design and operation of continuous countercurrent chromatography in Biotechnological production. Industrial Eng. Chem. Res. 53, 9169—9185.

# CHAPTER 7

# Gas-Phase Simulated Moving Bed

## 7.1 INTRODUCTION

The core industrial applications of the simulated moving bed (SMB) technology operate in the liquid phase; nevertheless, SMBs can also be operated in the gas phase in which a desorbent is used to displace the components retained by the adsorbent. Recent developments in this area have been remarkable (Storti et al., 1992; Mazzotti et al., 1996; Juza et al., 1998; Biressi et al., 2000; Cheng and Wilson, 2001a,b; Biressi et al., 2002; Rao et al., 2005; Lamia et al., 2007; Mota and Esteves, 2007; Mota et al., 2007; Kostroski and Wankat, 2008; Granato et al., 2014). One of the more-studied separations with this technology is the olefin/paraffin separation and, more precisely, the propane/propylene separation (Gomes et al., 2009; Lamia et al., 2009; Campo et al., 2014).

Propylene is one of the most important building blocks of the chemical industry (Cheng et al., 2002). This chemical commodity can be polymerized to high-octane gasoline or used in the production of polypropylene, isopropyl alcohol, propylene oxide, and cumene, among others. The low relative volatility of this system makes the propylene/propane separation one of the most energy-consuming separations (Järvelin and Fair, 1993). Over 100 theoretical stages are required by a traditional distillation unit for producing polymer-grade propylene (>99.5%) (Da Silva and Rodrigues, 1999).

Consequently, extensive research on various alternative technologies have been carried out to provide more sustainable solutions: cyclic adsorption processes (Da Silva and Rodrigues, 2001; Rege and Yang, 2002; Grande and Rodrigues, 2005; Grande et al., 2010), membranes (Yang, 2003; Kang et al., 2004; Mendes et al., 2007) and, as mentioned before, the gas-phase SMB.

Regarding the SMB, it is possible to find relevant references in the literature such as the patents by Cheng and Wilson (2001a,b). However, these patents describe the use of pressurization and depressurization steps such as the ones used in pressure-swing adsorption. The application of a "classical" SMB, that makes use of a desorbent, has only been mentioned in the patent of Rodrigues et al. (2008).

Next, the importance of the pair adsorbent/desorbent will be discussed, and afterward, two methods will be described for the design of a gas-phase SMB unit and applied to the propane/propylene separation. The first method is based on the equilibrium theory and the second on the use of mathematical modeling and computer simulation.

*Simulated Moving Bed Technology*
ISBN 978-0-12-802024-1

## 7.1.1 Equilibrium and Kinetics

The correct choice of the pair adsorbent/desorbent is of particular importance not only for achieving a given separation in an SMB unit but also to make it economically feasible. The desorbent should be able to displace the other components from the adsorbent and should itself be easily displaced from the adsorbent. Because the final result of the SMB technology is transforming a difficult separation into two easier separations, and because the two components will be obtained together with the desorbent in the extract and the raffinate, the selection of the desorbent should also consider that the subsequent separation of the pairs of most-retained species/desorbent and less-retained species/desorbent should be easy (Granato et al., 2010).

For the propane/propylene separation, Lamia et al. studied the use of a 13X zeolite using isobutane as desorbent (Lamia et al., 2007). The authors concluded from multi-component breakthrough experiments that isobutane was able to displace both propane and propylene and be displaced by any of these species; in fact, isobutene was more retained than propane and less retained than propylene in 13X zeolite. Additionally, the separation of isobutane from propane and from propylene can be easily done by distillation. For this reason, this adsorbent was selected for the case study.

The properties of the 13X zeolite are presented in Table 7.1, and the equilibrium adsorption parameters for the Toth isotherm of propane, propylene, and isobutane over this zeolite are presented in Table 7.2.

**Table 7.1** Properties of the 13X Zeolite

| | |
|---|---|
| Particle radius (m) | $8.0 \times 10^{-4}$ |
| Crystal diameter (m) | $2.0 \times 10^{-6}$ |
| Pore radius (m) | $1.7 \times 10^{-7}$ |
| Particle density (kg/m$^3$) | 1357 |
| Particle porosity | 0.395 |
| Heat capacity (J/kg K) | 920 |

**Table 7.2** Equilibrium Adsorption Parameters for the Toth Isotherm of Propane, Propylene, and Isobutane over a 13X Zeolite

| | $q_m$ (mol/kg) | $b_\infty$ (kPa$^{-1}$) | $-\Delta H$ (kJ/mol) | $t\ (-)$ |
|---|---|---|---|---|
| Propane | 2.20 | $2.5 \times 10^{-7}$ | 36.88 | 0.892 |
| Propylene | 2.59 | $2.5 \times 10^{-7}$ | 42.45 | 0.658 |
| Isobutane | 1.78 | $2.5 \times 10^{-7}$ | 41.54 | 0.848 |

The Toth isotherm is represented by the following equations:

$$q_i^* = q_{m,i} \frac{b_i C_{p,i} R_g T_p}{\left(1 + \left(\sum_{i=1}^{N} b_i C_{p,i} R_g T_p\right)^t\right)^{1/t}} \tag{7.1}$$

$$b_i = b_{\infty,i} \exp\left(\frac{-\Delta H_i}{R_g T_p}\right) \tag{7.2}$$

in which $b_i$ is the affinity parameter, $q_m$ is the saturation capacity, $t$ is a parameter that represents the solid heterogeneity, $C_{p,i}$ is the concentration of species $i$ in the pores, $T_p$ is the temperature of the particle, and $R_g$ is the ideal gas constant.

## 7.2 SMB DESIGN USING EQUILIBRIUM THEORY METHODOLOGY

The equilibrium theory methodology, defined previously for liquid-phase separations, can also be used to calculate the operating parameters of the gas-phase SMB and to define a separation region (the region in which the separation is complete). As for liquid-phase separations, the separation region for each section $j$ is delimited by the ratios between the interstitial fluid velocity and the solid velocity, represented as the parameter $\gamma_j$, at different conditions of equilibrium.

Figure 7.1 presents a typical separation region of an SMB unit.

### 7.2.1 The Procedure

To obtain the operating parameters of the gas–phase SMB a mass balance has to be done to each section, as illustrated in Figure 7.2 (Gomes et al., 2009).

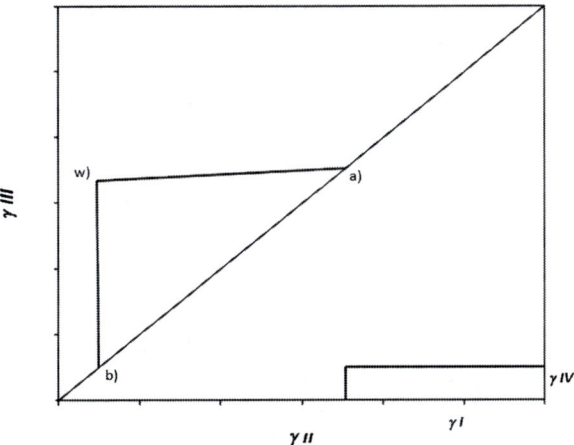

**Figure 7.1** Typical separation region of an SMB unit.

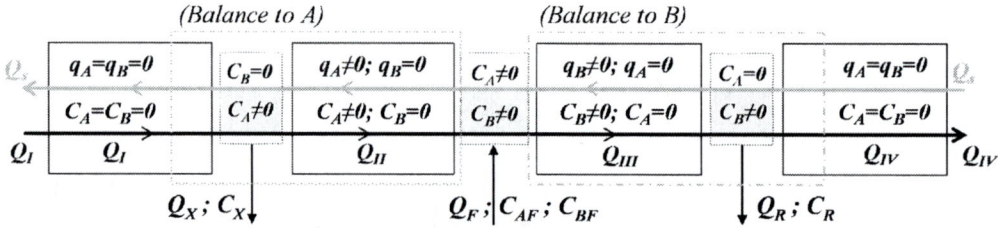

**Figure 7.2** Schematic representation of species balances for each section, according to the equilibrium theory methodology. *(Reprinted from Chemical Engineering Science 64, Gomes, P.S., Lamia, N., Rodrigues, A.E. Design of a gas phase simulated moving bed for propane/propylene separation, pp. 1336–1357, Copyright (2009), with permission from Elsevier.)*

Points (a) and (b) in Figure 7.1 were obtained from the analysis of the regeneration region, comprising sections I and IV.

To achieve a complete separation of A and B from a mixture of these two it is necessary to respect the boundaries of the following equations for both regeneration sections:

$$\frac{1-\varepsilon}{\varepsilon} \cdot \rho_p \cdot \frac{q_{A,I}}{C_{A,I}} < \gamma_I \tag{7.3}$$

$$\gamma_{IV} < \frac{1-\varepsilon}{\varepsilon} \cdot \rho_p \cdot \frac{q_{B,IV}}{C_{B,IV}} \tag{7.4}$$

in which $C_{A,I}$ and $C_{B,IV}$ are the bulk fluid phase concentration of species A and B in sections I and IV, respectively, and $q_{A,I}$ and $q_{B,IV}$ the total average adsorbed phase concentration in mol/kg, $\varepsilon$ the bed porosity, and $\rho_p$ the particle density.

Considering that only desorbent, D, is present in sections I and IV, these equations may be simplified, and the following expressions may be obtained:

$$\gamma_{I,min} = \frac{1-\varepsilon}{\varepsilon} \cdot \rho_p \cdot \lim_{C_{A,I} \to 0} \frac{q_A\left(C_{D,I} = C_{T,I}\right)}{C_{A,I}} \tag{7.5}$$

$$\gamma_{IV,max} = \frac{1-\varepsilon}{\varepsilon} \cdot \rho_p \cdot \lim_{C_{B,IV} \to 0} \frac{q_B\left(C_{D,IV} = C_{T,IV}\right)}{C_{B,IV}} \tag{7.6}$$

Therefore, the coordinates of point (a) in Figure 7.1 are defined by $(\gamma_{I,min}, \gamma_{I,min})$ and of point (b) are defined by $(\gamma_{IV,max}, \gamma_{IV,max})$.

In case of intermediate affinity for the desorbent, it is easy to understand that this last assumption for section IV will underestimate $\gamma_{IV,max}$.

Point (w), the vertex of the triangle, is calculated from the mass balances for each species in sections II and III, that is, in the separation region.

The limits of sections II and III are given by the following equations:

$$\frac{1-\varepsilon}{\varepsilon} \cdot \rho_p \cdot \frac{q_{B,II}}{C_{B,II}} < \gamma_{II} < \frac{1-\varepsilon}{\varepsilon} \cdot \rho_p \cdot \frac{q_{A,II}}{C_{A,II}} \tag{7.7}$$

$$\frac{1-\varepsilon}{\varepsilon}\cdot\rho_p\cdot\frac{q_{B,III}}{C_{B,III}} < \gamma_{III} < \frac{1-\varepsilon}{\varepsilon}\cdot\rho_p\cdot\frac{q_{A,III}}{C_{A,III}} \tag{7.8}$$

The vertex of the triangle is given by the coordinates $(\gamma_{II,min}, \gamma_{III,max})$.

The assumption used in the regeneration region is not valid here. Considering complete separation and instantaneous equilibrium, the more-retained species, A, is expected to be found only in section II, and the less-retained species, B, is expected to be found only in section III. However, in the feed node, both species are present. Therefore, in the separation region, the three species must be taken into account in the adsorption term, as shown in the following expressions:

$$\gamma_{II,min} = \frac{1-\varepsilon}{\varepsilon}\cdot\rho_p\cdot\frac{q_B(C_{A,II}, C_{B,II}, C_{D,II})}{C_{B,II}} \tag{7.9a}$$

$$\gamma_{III,max} = \frac{1-\varepsilon}{\varepsilon}\cdot\rho_p\cdot\frac{q_A(C_{A,III}, C_{B,III}, C_{D,III})}{C_{A,III}} \tag{7.9b}$$

The following assumptions were considered:

$$C_{A,II} = C_{A,F}; \quad C_{B,III} = C_{B,F} \tag{7.10a}$$

$$C_{B,II} = \alpha\cdot C_{A,II}; \quad C_{A,III} = \beta\cdot C_{B,III} \tag{7.10b}$$

$$C_{D,II} = C_{T,F} - (1+\alpha)\cdot C_{A,II} \tag{7.10c}$$

$$C_{D,III} = C_{T,F} - (1+\beta)\cdot C_{B,III} \tag{7.10d}$$

in which the subscript "F" stands for "Feed." The importance of the parameters $\alpha$ and $\beta$ is not less than the importance of the remaining parameters. These two parameters will define the pollution of the less-retained component in section II and of the more-retained component in section III. A valid approximation is considering for the pollution of the less-retained component in section II, $\alpha = \frac{C_{B,F}}{C_{A,F}}$, and for the pollution of the more-retained component in section III, $\beta = \frac{C_{A,F}}{C_{B,F}}$. These approximations will be more accurate as the higher is the feed flow rate, near the optimal operation point.

After defining the boundaries to achieve a complete separation, we can choose a point of operation $(\gamma_{II}, \gamma_{III})$.

By performing a mass balance to sections II and III, it is possible to calculate $\gamma_F$:

$$\gamma_F = \gamma_{III} - \gamma_{II} \tag{7.11}$$

At a given feed flow rate it is possible to calculate the solid flow rate, $Q_s$:

$$Q_s = \frac{Q_F}{\gamma_F}\cdot\frac{1-\varepsilon}{\varepsilon} \tag{7.12}$$

Consequently, the point of operation in terms of flow rates is:

$$Q_{II} = \gamma_{II}\cdot Q_s\cdot\frac{\varepsilon}{1-\varepsilon} \tag{7.13}$$

$$Q_{III} = \gamma_{III} \cdot Q_s \cdot \frac{\varepsilon}{1 - \varepsilon} \qquad (7.14)$$

The minimum desorbent amount required may be obtained from the following equation:

$$Q_{D,min} = Q_{I,min} - Q_{IV,max} \qquad (7.15)$$

in which $Q_{IV,max} = Q_{recycle}$.

It is now possible to calculate the extract, $X$, raffinate, $R$, and recycle flow rates from the following expressions:

$$Q_X = Q_{I,min} - Q_{II} \qquad (7.16)$$

$$Q_R = Q_{III} - Q_{IV,max} \qquad (7.17)$$

$$Q_{recycle} = Q_{I,min} - Q_{D,min} \qquad (7.18)$$

The total volume of adsorbent required can be obtained for a given product recovery, $RE$, and productivity, $PR$:

$$V_{ads} = \frac{RE_{A,X} \cdot Q_F \cdot C_{A,F}}{PR_{A,X}} \qquad (7.19)$$

The volume of each column is given by:

$$V_{column} = \frac{V_{ads}}{(1 - \varepsilon)N_{columns}} \qquad (7.20)$$

The pressure drop may be evaluated by the Ergun equation:

$$\frac{\Delta P}{L_{column} \cdot N_I} \geq 150 \frac{\mu \cdot (1 - \varepsilon)^2}{\varepsilon^3 \cdot d_p^2} u_{0,I} + 1.75 \frac{(1 - \varepsilon) \cdot \rho}{\varepsilon^3 \cdot d_p} |u_{0,I}| u_{0,I} \qquad (7.21)$$

## 7.2.2 Test Case: Propane/Propylene Separation

Following the procedure described previously, it is possible to obtain the separation regions for the propane/propylene separation using 13X as adsorbent and isobutane as desorbent for a feed concentration of 75% propylene and 25% propane at 373 K. The separation regions, calculated at different pressures, are presented in Figure 7.3.

Observing Figure 7.3, it is possible to conclude that when increasing the operating pressure the area of the separation region decreases. This is due to the decrease of the selectivity of the adsorbent with increasing pressure.

Considering a feed flow rate of 1.35 m³/s at 150 kPa and at 373 K, it is possible to calculate the solid flow rate (approximately 0.032 m³/s) and calculate the separation

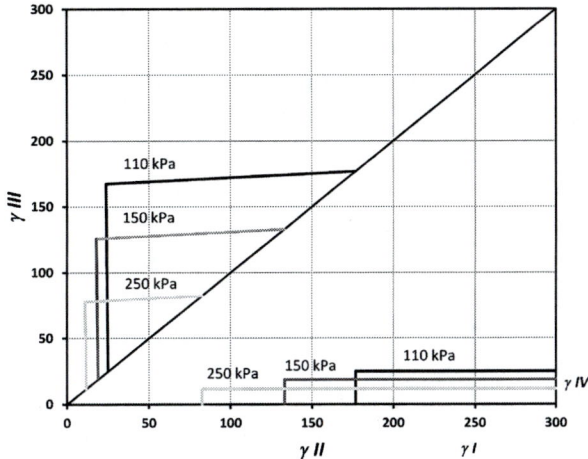

**Figure 7.3** Equilibrium-based separation regions at 373 K and at different pressures.

region in terms of volumetric flow rates. Figure 7.4 presents the separation regions, calculated at 373 K and at different pressures, in terms of volumetric flow rates.

After conducting a literature review for this separation it was verified that the higher purity values (99.3%) were obtained using cyclic adsorption processes by Da Silva and Rodrigues (2001) with a productivity of 0.82 mol/(kg h), whereas the higher productivity values of 2.74 mol/(kg h) were obtained by Ramachandran and Dao (1994) for a propylene purity of 89.9%. Because the gas–phase SMB should perform better to be competitive, let us take as reference the higher productivity value and apply a factor of

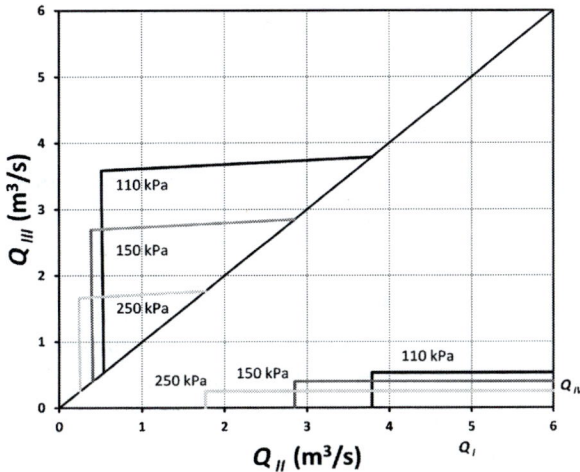

**Figure 7.4** Equilibrium-based separation regions at 373 K and at different pressures, in terms of volumetric flow rates.

150% to it. Assuming a recovery of 100%, it is possible to calculate the total volume of adsorbent required and the total column volume (32 m$^3$, approximately).

Considering that the unit should operate at a pressure above 100 kPa, the 150 kPa operation leads to the constraint $\Delta P \leq 50$ kPa. Taking a 10% tolerance, we can set the pressure drop constraint to 45 kPa. Figure 7.4 shows that the highest velocity (higher-pressure drop) occurs in section I. Selecting a flow rate for section I, $Q_I$, 10% higher (*for example*) than the minimum value (approximately 3.13 m$^3$/s) and defining a ratio of column length/diameter of the column of 0.5 after defining the configuration (e.g., 2-2-2-2), it is possible to calculate the length and diameter of each column ($L_c = 1.18$ m and $d_c = 2.09$ m). Finally, we can choose an operation point in Figure 7.4 (e.g., $Q_{II} = 0.5$ m$^3$/s, for 373 K) and, with the feed flow rate mentioned before, we get $Q_{III} = 1.85$ m$^3$/s. Choosing a flow rate for section IV, for example, 10% below the maximum value (0.36 m$^3$/s), it is possible to obtain all the remaining operating parameters: $Q_X = 2.65$ m$^3$/s, $Q_R = 1.49$ m$^3$/s, $Q_D = 2.79$ m$^3$/s, $Q_{Rec} = 0.36$ m$^3$/s, and $t_s = 76.5$ s.

## 7.3 SMB DESIGN USING MATHEMATICAL MODELS

The SMB unit may be designed with the help of mathematical models. At this stage, that is, for design purposes, modeling the SMB by means of the TMB analogy is the best strategy as it will provide faster simulations, which is also useful for implementing optimization routines. The steady-state results obtained following this strategy can be used to predict the unit's performance parameters.

In this methodology, the separation region is built (and searched) by performing simulations at different operating conditions, that is, different values of $\gamma_F$ and $\gamma_{II}$. The time required to perform each simulation is, thus, of critical importance and is proportional to the complexity of the mathematical model. A detailed model of the particle diffusion and/or film mass transfer will result in a more time-consuming simulation, whereas using an approximation of the intraparticle mass-transfer rate with the Linear Driving Force (LDF) approach will result in a faster simulation. This second is, in fact, used by the majority of researchers in the field.

### 7.3.1 The Mathematical Model

The mathematical model described next is based on the following assumptions:
- the fluid motion is modeled by a plug flow with axial dispersion;
- the solid–phase motion is assumed to be a piston flow;
- variable fluid velocity along the column;
- two particle levels of mass transfer are possible, represented by a *bi*-LDF approximation;

- the thermal effects are considered by means of a heterogeneous energy balance distinguishing temperatures in the fluid, the solid, and at the wall of the column;
- the pressure-drop effects are considered by means of a continuous formulation of the Ergun equation.

The mass, momentum, and energy balance equations of the mathematical model can be found in Table 7.3; further details can be found elsewhere (Da Silva et al., 1999; Da Silva and Rodrigues, 2001; Ribeiro et al., 2008; Gomes et al., 2009).

**Table 7.3** Mass, Momentum, and Energy Balances of the Mathematical Model

**Mass Balances**

Gas phase

$$\frac{\partial}{\partial z}\left(\varepsilon D_{ax} C_T \frac{\partial y_{g,i,j}}{\partial z}\right) - \frac{\partial}{\partial z}(u_0 C_{g,i,j}) - \varepsilon \frac{\partial C_{g,i}}{\partial t} - \frac{(1-\varepsilon)a_p K_f}{1+Bi_{i,j}}\left(C_{g,i,j} - \overline{C_{p,i,j}}\right) = 0$$

Solid phase—macropore

$$\frac{\partial \overline{C_{p,i,j}}}{\partial t} = \frac{15 D_{p,i,j} Bi_{i,j}}{R_p^2(1+Bi_{i,j})}\left(C_{g,i,j} - \overline{C_{p,i,j}}\right) - \frac{\rho_p}{\varepsilon_p}\frac{\partial \overline{q_{i,j}}}{\partial t}$$

Solid phase—micropore

$$\frac{\partial \overline{q_{i,j}}}{\partial t} = \frac{15 D_{c,i,j}}{r_c^2}\left(\overline{q_{i,j}^*} - \overline{q_{i,j}}\right)$$

**Momentum Balance**

$$-\frac{\partial P_j}{\partial z} = \frac{150\mu(1-\varepsilon)^2}{\varepsilon^3 d_p^2} u_{0,j} + \frac{1.75(1-\varepsilon)\rho}{\varepsilon^3 d_p}|u_{0,j}|u_{0,j}$$

**Energy Balances**

Gas phase

$$\frac{\partial}{\partial z}\left(\lambda \frac{\partial T_{g,j}}{\partial z}\right) - u_0 C_{g,T,j} C_p \frac{\partial T_{g,j}}{\partial z} + \varepsilon R_g T_{g,j}\frac{\partial C_{g,T,j}}{\partial t}$$

$$-(1-\varepsilon)a_p h_f(T_{g,j} - T_{p,j}) - \frac{4h_w}{d_{wi}}(T_{g,j} - T_{w,j}) - \varepsilon C_{g,T,j} C_V \frac{\partial T_{g,j}}{\partial t} = 0$$

Solid phase

$$(1-\varepsilon)\left[\varepsilon_p \sum_{i=1}^{n}\overline{C_{p,i,j}} C_{v,i,j} + \rho_p \sum_{i=1}^{n}\overline{q_i} C_{v,ads,i,j} + \rho_p \widehat{C}_{ps}\right]\frac{\partial T_{p,j}}{\partial t}$$

$$= (1-\varepsilon)\varepsilon_p R_g T_{p,j}\frac{\partial \overline{C_{p,T,j}}}{\partial t} + \rho_b \sum_{i=1}^{n}(-\Delta H_{ads})_i \frac{\partial \overline{q_{i,j}}}{\partial t} + (1-\varepsilon)a_p h_f(T_{g,j} - T_{p,j})$$

Column wall

$$\rho_w \widehat{C_{p,w}}\frac{\partial T_{w,j}}{\partial t} = \alpha_w h_w(T_{g,j} - T_{w,j}) - \alpha_{w\ell} U(T_{w,j} - T_\infty)$$

**Table 7.4** Balances to the Nodes Used in the Simulation of SMB

**Desorbent Node**

$$u_{0\text{inlet},I}C_{\text{inlet},i,I} = u_{0,D}C_{D,i} + \left(u_{0,IV}C_{g,i,IV}\right)\big|_{z=L_{IV}}$$

$$u_{0\text{inlet},I}C_{\text{inlet},T,I} = u_{0,D}C_{D,T} + \left(u_{0,IV}C_{g,T,IV}\right)\big|_{z=L_{IV}}$$

$$u_{0\text{inlet},I}C_{\text{inlet},T,I}C_p T_{\text{inlet},I} = u_{0,D}C_{D,T}C_p T_D + \left(u_{0,IV}C_{g,T,IV}C_p T_{g,IV}\right)\big|_{z=L_{IV}}$$

**Extract Node**

$$C_{\text{inlet},i,II} = C_{g,i,I}\big|_{z=L_I}$$

$$u_{0\text{inlet},II} = u_{0,I}\big|_{z=L_I} - u_{0,E}$$

$$T_{\text{inlet},II} = T_{g,I}\big|_{z=L_I}$$

**Feed Node**

$$u_{0\text{inlet},III}C_{\text{inlet},i,III} = u_{0,F}C_{F,i} + \left(u_{0,II}C_{g,i,II}\right)\big|_{z=L_{II}}$$

$$u_{0\text{inlet},III}C_{\text{inlet},T,III} = u_{0,F}C_{F,T} + \left(u_{0,II}C_{g,T,II}\right)\big|_{z=L_{II}}$$

$$u_{0\text{inlet},III}C_{\text{inlet},T,III}C_p T_{\text{inlet},III} = u_{0,F}C_{F,T}C_p T_F + \left(u_{0,II}C_{g,T,II}C_p T_{g,II}\right)\big|_{z=L_{II}}$$

**Raffinate Node**

$$C_{\text{inlet},i,IV} = C_{g,i,III}\big|_{z=L_{III}}$$

$$u_{0\text{inlet},IV} = u_{0,III}\big|_{z=L_{III}} - u_{0,R}$$

$$T_{\text{inlet},IV} = T_{g,III}\big|_{z=L_{III}}$$

The balances to the nodes are presented in Table 7.4.

The columns were considered saturated with desorbent at $t = 0$. The Danckwerts boundary conditions were considered in the model.

The purity of the extract ($PU_X$), and raffinate ($PU_X$), are defined using the variables related to TMB, by:

$$PU_X = \frac{C_{C_3H_6}^X}{C_{C_3H_6}^X + C_{C_3H_8}^X} \tag{7.22}$$

$$PU_R = \frac{C_{C_3H_8}^R}{C_{C_3H_6}^R + C_{C_3H_8}^R} \tag{7.23}$$

Similarly, for the recovery of the more-retained species in the Extract stream ($RE_X$) and the less one in the raffinate port ($RE_R$):

$$RE_X = \frac{Q_X C_{C_3H_6}^X}{Q_F C_{C_3H_6}^F} \tag{7.24}$$

$$RE_R = \frac{Q_R C_{C_3H_8}^R}{Q_F C_{C_3H_8}^F} \tag{7.25}$$

The productivity is expressed in terms of propylene recovered in the Extract stream ($PR_X$), or propane in the Raffinate port ($PR_R$):

$$PR_X = \frac{Q_X C_{C_3H_6}^X}{V_{ads}} = \frac{RE_X Q_F C_{C_3H_6}^F}{V_{ads}} \tag{7.26}$$

$$PR_R = \frac{Q_R C_{C_3H_8}^R}{V_{ads}} = \frac{RE_R Q_F C_{C_3H_8}^F}{V_{ads}} \tag{7.27}$$

The model described previously (Tables 7.3 and 7.4) results in a system of partial differential equations), ordinary differential equations, and algebraic equations, that can be solved by means of numerical solution giving some first simulations based on the estimated operating parameters. The complete mathematical model was implemented in gPROMS environment v.3.5.1 (Process System Enterprise, 2014); the problem was numerically solved using orthogonal collocation on a finite elements method, with 250 discretization intervals and third-order polynomials. The methodology used to calculate the separation region has been described elsewhere by Minceva (2004).

## 7.3.2 Test Case: The Propane/Propylene Separation

A gas-phase SMB unit was simulated with the operating conditions and column geometry obtained from the design using the equilibrium theory methodology. These parameters are summarized in Table 7.5, and the bulk concentration profiles obtained at the steady state are presented in Figure 7.5. As mentioned previously, the feed is at 373 K, and its composition is 75% propylene and 25% propane.

As can be seen in Figure 7.5, the concentration profiles of the less-retained species in sections II and IV and of the more-retained component in sections I and III, do not occupy all section length, that is, no propane is polluting the end of section IV nor the beginning of section II, therefore, room is still available for improving the performance of the unit by utilizing all the adsorption bed. In addition, no propylene is present

**Table 7.5** Operating Conditions and Column Geometry for the First Simulation

| Columns Geometry | Operating Conditions |
|---|---|
| $L_c = 1.18$ m | $T_{T,D} = 373$ K  $P_{T,D} = 150$ kPa |
| $d_c = 2.09$ m | $t_s = 76.5$ s |
| $V_c = 4.05$ m$^3$ | $Q_F = 1.35$ m$^3$/s |
| $n_j = [2\ 2\ 2\ 2]$ | $Q_X = 2.65$ m$^3$/s |
| | $Q_R = 1.49$ m$^3$/s |
| | $Q_D = 2.79$ m$^3$/s |
| | $Q_{IV} = 0.36$ m$^3$/s |

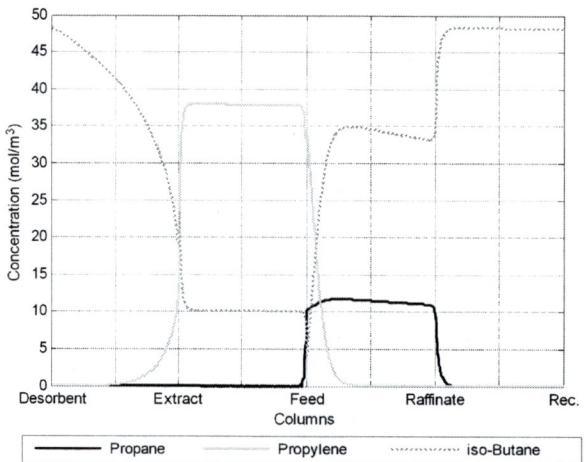

**Figure 7.5** Bulk concentration profiles obtained at steady state. *(Reprinted from Chemical Engineering Science 64, Gomes, P.S., Lamia, N., Rodrigues, A.E. Design of a gas phase simulated moving bed for propane/propylene separation, pp. 1336–1357, Copyright (2009), with permission from Elsevier.)*

**Table 7.6** Performance Parameters Obtained from Simulation

| |
| --- |
| $PU_X = 100\%$ |
| $RE_X = 100\%$ |
| $PR_X = 6.63\ \mathrm{mol/(kg_{ads}\ h)} = 377.9\ \mathrm{kg/(m^3_{ads}\ h)}$ |
| $PU_R = 100\%$ |
| $RE_R = 100\%$ |
| $PR_R = 2.21\ \mathrm{mol/(kg_{ads}\ h)} = 132.0\ \mathrm{kg/(m^3_{ads}\ h)}$ |

in the beginning of section I nor in the end of section III; therefore, it is possible to obtain the 100% performance parameters values presented in Table 7.6.

It is also possible to verify that the value for the operating condition in section IV ($\gamma_{IV}$), was overestimated, giving now more space to increase the recycle flow rate than the possible space to decrease the flow rate in section I. The gas-phase temperature, interstitial velocity, and total pressure profiles at the steady state are shown in Figure 7.6.

Observing Figures 7.5 and 7.6, it is possible to realize that when a more-retained species is displaced by a less-adsorbed one, the temperature decreases and vice versa; this fact is due to the adsorption phenomena itself (exothermic), accounted in the model by the adsorption enthalpies values on the energy balance. However, and based on the solid and fluid specific heats, this temperature variation is observed next to the left of the adsorption or desorption zone, that is, the temperature waves are following the solid movement direction and not the fluid one. The total bulk concentration variation along the unit is similar to the total pressure variation, because the temperature influence is almost imperceptible (varying within a maximum range of 6 K, Figure 7.6(a)), when compared with the pressure variations.

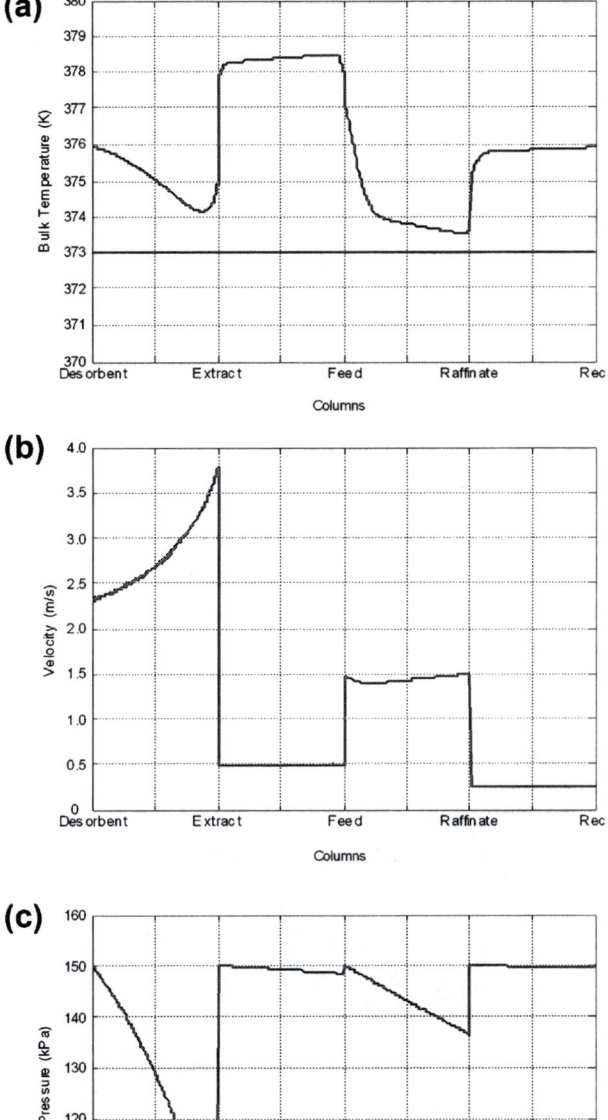

**Figure 7.6** Profiles of (a) gas-phase bulk temperature, (b) interstitial velocity, and (c) total pressure obtained by simulation, at the steady state. *(Reprinted from Chemical Engineering Science 64, Gomes, P.S., Lamia, N., Rodrigues, A.E. Design of a gas phase simulated moving bed for propane/propylene separation, pp. 1336–1357, Copyright (2009), with permission from Elsevier.)*

It is possible to observe in Figure 7.6(c) that the previously pressure drop constraint defined for section I ($\Delta P \leq 50$ kPa) was violated. One should keep in mind that this constraint was defined for constant velocity. The results presented in Figure 7.6(c), even though being based on the same operating parameters, now account for several detailed phenomena, such as the variation of velocity. As can be seen in Figure 7.6(b), the velocity exhibits a large increase near the extract node, which results in an increased pressure drop (see Figure 7.6(c)).

The performance parameters presented in Table 7.6 show that the equilibrium-based estimation is a good approach for designing the gas-phase SMB.

Let us redraw the separation region triangle shown in Figure 7.4, but now account for all the mass-transfer effects, as well as the temperature variations as described by the detailed model. Figure 7.7 shows the complete separation region for column geometry presented in Table 7.5, with $T_{F,D} = 373$ K; $P_0 = 150$ kPa; $t_s = 76.5$ s; $Q_{IV} = 0.27$ m$^3$/s, and $Q_I = 3.76$ m$^3$/s, obtained by consecutive simulations with different $(Q_{II}, Q_{III})$ pairs, (step of 0.075 m$^3$/s).

The $Q_{IV}$ and $Q_I$ flow rates were set with a higher tolerance factor (25%) than the 10% used for the equilibrium assumptions. This fact is related to the influence of these flow rates in the separation region shape, as studied before by Azevedo and Rodrigues (1999), and presented as the Separation Volume Theory.

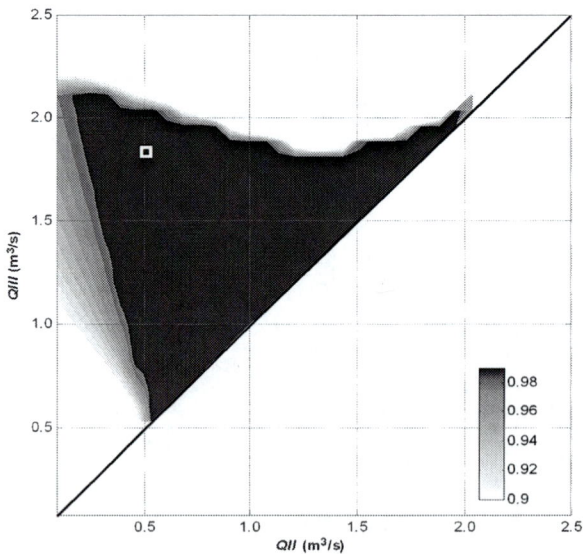

**Figure 7.7** Separation region in terms of volumetric flowrate for column geometry presented in Table 7.5, lower purity values in the raffinate or extract outlets over 90% (0.9)—gray to over 99% (0.99)—black; white square the operating point mentioned in Table 7.5. *(Reprinted from Chemical Engineering Science 64, Gomes, P.S., Lamia, N., Rodrigues, A.E. Design of a gas phase simulated moving bed for propane/propylene separation, pp. 1336–1357, Copyright (2009), with permission from Elsevier.)*

Comparing Figure 7.7 with Figure 7.4 it is possible to verify that this complete separation region fits inside the predicted one (Figure 7.4 for 150 kPa). However, some small differences are noted, namely the vertex point (higher productivity value, maximum $Q_{III}$ for minimum $Q_{II}$), related with the approximations used for the $\alpha$ and $\beta$ parameters in the design procedure based on the adsorption equilibrium, as well as for the consideration of mass-transfer resistances. Nevertheless, the equilibrium-based methodology provides a good estimation, placing the corresponding operating point near to the upper triangle limit.

## 7.4 CONCLUDING REMARKS

In this chapter, the gas-phase simulated moving bed was presented. Two design strategies were described—by means of the equilibrium theory or making use of mathematical models—and the propane/propylene separation using a 13X adsorbent and isobutane as desorbent was used as a test case.

It was demonstrated that the design using the equilibrium theory not only is more simple to use but can also provide very good results. The separation region was obtained through this design methodology, and a point in the region was simulated showing that the separation was possible at that point and complete.

The mathematical model of a gas–phase SMB unit comprising mass and energy balances with two particle levels of mass transfer (macropores and micropores) was presented. The separation region was obtained through simulation and shown not to differ significantly from the region obtained with the equilibrium theory. The design through mathematical modeling can be more precise the more complex that the model is. However, when increasing this complexity the computational time required obtaining the separation region also increases. Nevertheless, the calculation can be expedited by combining the two methodologies, namely, determining the location of the separation zone using the equilibrium theory method, and afterward refining the separation zone through simulation.

It was possible to conclude that the design using the equilibrium theory is ideal for a quick screening of the operating conditions and of the adsorbent/desorbent pair. For performing an economical evaluation of the separation through this technology, the design using mathematical models is advised.

## NOMENCLATURE

| | |
|---|---|
| $a_p$ | Particle specific area $(m^{-1})$ |
| $b_i$ | Parameter of the Toth isotherm $(Pa^{-1})$ |
| $Bi_i$ | Mass Biot number of component $i$, $\left( Bi_{i,j} = \dfrac{a_p k_f R_p^2}{\varepsilon_p 15 D_{p,i}} \right)$ |
| $C_{g,i}$ | Bulk fluid-phase concentration of species $i$ $(mol/m^3)$ |

| $C_{g,T}$ | Total gas-phase concentration in section $j$ (mol/m$^3$) |
|---|---|
| $C_p$ | Gas mixture molar specific heat at constant pressure (J/(mol K)) |
| $C_{p,i}$ | Concentration of component $i$ in the macropores (mol/m$^3$) |
| $\overline{C_{p,i}}$ | Average concentration of component $i$ in the macropores over the pellet (mol/m$^3$) |
| $\widehat{C_{ps}}$ | Particle specific heat at constant pressure (per mass unit) (J/(kg K)) |
| $\overline{C_{p,T}}$ | Average total concentration in the macropores over the pellet (mol/m$^3$) |
| $\widehat{C_{pw}}$ | Wall specific heat at constant pressure (per mass unit) (J/(kg K)) |
| $C_v$ | Gas mixture molar specific heat at constant volume (J/(mol K)) |
| $C_{v,ads}$ | Molar specific heat in the adsorbed phase at constant volume (J/(mol K)) |
| $d_p$ | Adsorbent particle diameter (m) |
| $d_{wi}$ | Internal bed diameter (m) |
| $D_{ax}$ | Axial dispersion coefficient (m$^2$/s) |
| $D_{c,i}$ | Micropore diffusivity of component $i$ (m$^2$/s) |
| $D_{p,i}$ | Macropore diffusivity of component $i$ (m$^2$/s) |
| $h_f$ | Film heat transfer coefficient between the gas and particle (J/(s m$^2$ K)) |
| $h_w$ | Film heat transfer coefficient between the gas and wall (J/(s m$^2$ K)) |
| $k_f$ | Film mass transfer coefficient (m/s) |
| $L$ | Column length (m) |
| $nc$ | Number of components |
| $N$ | Number of columns |
| $P$ | Pressure (Pa) |
| $q_i$ | Average adsorbed phase concentration in the crystal (mol/kg) |
| $\overline{q_i}$ | Average adsorbed phase concentration of component $i$ over the pellet (mol/kg) |
| $q_i^*$ | Adsorbed concentration of component $i$ at the crystal surface in equilibrium with $C_{p,i}$ (mol/kg) |
| $\overline{q_i^*}$ | Adsorbed concentration of component $i$ at the crystal surface averaged over the pellet (mol/kg) |
| $q_m$ | Parameter of the Toth isotherm (mol/kg) |
| $Q$ | Volumetric flow rate (m$^3$/s) |
| $r_c$ | Crystal radius (m) |
| $R_g$ | Ideal gas constant (J/mol K) |
| $R_p$ | Adsorbent particle radius (m) |
| $t$ | Time (s) |
| $t$ | Parameter of the Toth isotherm |
| $T_g$ | Bulk phase temperature (K) |
| $T_p$ | Solid temperature (K) |
| $T_w$ | Wall temperature (K) |
| $T_\infty$ | Ambient temperature (K) |
| $u_0$ | Superficial velocity (m/s) |
| $U$ | Overall heat transfer coefficient (J/(s m$^2$ K)) |
| $V$ | Volume (m$^3$) |
| $y_{g,i}$ | Gas-phase molar fraction of component $i$ |
| $z$ | Axial position (m) |

## Greek Letters

| $\alpha_w$ | Ratio of the internal surface area to the volume of the column wall (m$^{-1}$) |
|---|---|
| $\alpha_{w\ell}$ | Ratio of the log mean surface to the volume of column wall (m$^{-1}$) |
| $\Delta H_{ads}$ | Heat of adsorption (J/mol) |

| | |
|---|---|
| $\varepsilon$ | Bed porosity |
| $\varepsilon_p$ | Particle porosity |
| $\gamma$ | Ratio between fluid and solid interstitial velocities |
| $\lambda$ | Heat axial dispersion coefficient ($J/(s\ m\ K)$) |
| $\mu$ | Bulk gas mixture viscosity ($kg/(m\ s)$) |
| $\rho$ | Bulk gas mixture density ($kg/m^3$) |
| $\rho_b$ | Bed density ($kg/m^3$) |
| $\rho_p$ | Particle density ($kg/m^3$) |
| $\rho_w$ | Wall density ($kg/m^3$) |

## Subscripts

| | |
|---|---|
| *ads* | Adsorbent |
| *D* | Desorbent |
| *F* | Feed |
| *i* | Component ($i = A,B$) |
| *j* | Section ($j = I,II,III,IV$) |
| *R* | Raffinate |
| *s* | Solid |
| *T* | Total |
| *X* | Extract |

## REFERENCES

Azevedo, D.C.S., Rodrigues, A.E., 1999. Design of a simulated moving bed in the presence of mass-transfer resistances. AIChE J. 45, 956–966.

Biressi, G., Quattrini, F., Juza, M., Mazzotti, M., Schurig, V., Morbidelli, M., 2000. Gas chromatographic simulated moving bed separation of the enantiomers of the inhalation anesthetic enflurane. Chem. Eng. Sci. 55, 4537–4547.

Biressi, G., Rajendran, A., Mazzotti, M., Morbidelli, M., 2002. The GC-SMB separation of the enantiomers of isoflurane. Sep. Sci. Technol. 37, 2529–2543.

Campo, M.C., Baptista, M.C., Ribeiro, A.M., Ferreira, A., Santos, J.C., Lutz, C., Loureiro, J.M., Rodrigues, A.E., 2014. Gas phase SMB for propane/propylene separation using enhanced 13X zeolite beads. Adsorpt. J. Int. Adsorpt. Soc. 20, 61–75.

Cheng, L.S., Padin, J., Rege, S.U., Wilson, S.T., Yang, R.T., June 18, 2002. Process for purifying propylene. US Patent 6,406,521.

Cheng, L.S., Wilson, S.T., September 25, 2001a. Process for separating propylene from propane. US Patent 6,293,999.

Cheng, L.S., Wilson, S.T., October 2, 2001b. Vacuum swing adsorption process for separating propylene from propane. US Patent 6,296,688.

Da Silva, F.A., Rodrigues, A.E., 1999. Adsorption equilibria and kinetics for propylene and propane over 13X and 4A zeolite pellets. Ind. Eng. Chem. Res. 38, 2051–2057.

Da Silva, F.A., Rodrigues, A.E., 2001. Vacuum swing adsorption for propylene/propane separation with 4A zeolite. Ind. Eng. Chem. Res. 40, 5758–5774.

Da Silva, F.A., Silva, J.A., Rodrigues, A.E., 1999. A general package for the simulation of cyclic adsorption processes. Adsorpt. J. Int. Adsorpt. Soc. 5, 229–244.

Gomes, P.S., Lamia, N., Rodrigues, A.E., 2009. Design of a gas phase simulated moving bed for propane/propylene separation. Chem. Eng. Sci. 64, 1336–1357.

Granato, M.A., Martins, V.D., Santos, J.C., Jorge, M., Rodrigues, A.E., 2014. From molecules to processes: molecular simulations applied to the design of simulated moving bed for ethane/ethylene separation. Can. J. Chem. Eng. 92, 148–155.

Granato, M.A., Vlugt, T.J.H., Rodrigues, A.E., 2010. Potential desorbents for propane/propylene separation by gas phase simulated moving bed: a molecular simulation study. Ind. Eng. Chem. Res. 49, 5826–5833.

Grande, C.A., Poplow, F., Rodrigues, A.E., 2010. Vacuum pressure swing adsorption to produce polymer-grade propylene. Sep. Sci. Technol. 45, 1252–1259.

Grande, C.A., Rodrigues, A.E., 2005. Propane/propylene separation by pressure swing adsorption using zeolite 4A. Ind. Eng. Chem. Res. 44, 8815–8829.

Järvelin, H., Fair, J.R., 1993. Adsorptive separation of propylene propane mixtures. Ind. Eng. Chem. Res. 32, 2201–2207.

Juza, M., Di Giovanni, O., Biressi, G., Schurig, V., Mazzotti, M., Morbidelli, M., 1998. Continuous enantiomer separation of the volatile inhalation anesthetic enflurane with a gas chromatographic simulated moving bed unit. J. Chromatogr. A 813, 333–347.

Kang, S.W., Kim, J.H., Oh, K.S., Won, J., Char, K., Kim, H.S., Kang, Y.S., 2004. Highly stabilized silver polymer electrolytes and their application to facilitated olefin transport membranes. J. Membr. Sci. 236, 163–169.

Kostroski, K.P., Wankat, P.C., 2008. Separation of dilute binary gases by simulated-moving bed with pressure-swing assist: SMB/PSA processes. Ind. Eng. Chem. Res. 47, 3138–3149.

Lamia, N., Granato, M.A., Gomes, P.S.A., Grande, C.A., Wolff, L., Leflaive, P., Leinekugel-le-Cocq, D., Rodrigues, A.E., 2009. Propane/propylene separation by simulated moving bed II. Measurement and prediction of binary adsorption equilibria of propane, propylene, isobutane, and 1-butene on 13X zeolite. Sep. Sci. Technol. 44, 1485–1509.

Lamia, N., Wolff, L., Leflaive, P., Gomes, P.S., Grande, C.A., Rodrigues, A.E., 2007. Propane/propylene separation by simulated moving bed I. Adsorption of propane, propylene and isobutane in pellets of 13X zeolite. Sep. Sci. Technol. 42, 2539–2566.

Mazzotti, M., Baciocchi, R., Storti, G., Morbidelli, M., 1996. Vapor-phase SMB adsorptive separation of linear/nonlinear paraffins. Ind. Eng. Chem. Res. 35, 2313–2321.

Mendes, A., Santos, J.C., Taveira, P., 2007. Device to Separate Olefins from Paraffins and to Purify Olefins and Use Thereof. PCT/PT2007/000015, WO 2007/111521 A2.

Minceva, M., 2004. Separation/Isomeration of Xylenes by Simulated Moving Bed Technology (Ph.D. thesis). University of Porto, Porto, Portugal.

Mota, J.P.B., Esteves, I.A.A.C., 2007. Optimal design and experimental assessment of time-variable simulated moving bed for gas separation. Ind. Eng. Chem. Res. 46, 6978–6988.

Mota, J.P.B., Esteves, I.A.A.C., Eusbio, M.F.J., 2007. Synchronous and asynchronous SMB processes for gas separation. AIChE J. 53, 1192–1203.

Process System Enterprise, 2014. gPROMS. www.psenterprise.com/gproms.

Ramachandran, R., Dao, L.H., November 15, 1994. Method of producing unsaturated hydrocarbons and separatin the same from saturated hydrocarbons. US Patent 5,365,011.

Rao, D.P., Sivakumar, S.V., Mandal, S., Kota, S., Ramaprasad, B.S.G., 2005. Novel simulated moving-bed adsorber for the fractionation of gas mixtures. J. Chromatogr. A 1069, 141–151.

Rege, S.U., Yang, R.T., 2002. Propane/propylene separation by pressure swing adsorption: sorbent comparison and multiplicity of cyclic steady states. Chem. Eng. Sci. 57, 1139–1149.

Ribeiro, A.M., Grande, C.A., Lopes, F.V.S., Loureiro, J.M., Rodrigues, A.E., 2008. A parametric study of layered bed PSA for hydrogen purification. Chem. Eng. Sci. 63, 5258–5273.

Rodrigues, A.E., Lamia, N., Grande, C.A., Wolff, L., Leflaive, P., Leinekugel-le-Coq, D., July 2008. Procédé de Séparation du Propylène en Mélange avec du Propane par Adsorption en Lit Mobile Simulé en Phase Gaz ou Liquide utilisant une Zéolithe de type Faujasite 13X comme Solide Adsorbant. FR Patent 2,903,981 A1 (2006) and WO Patent 2008/012410 A1.

Storti, G., Mazzotti, M., Furlan, L.T., Morbidelli, M., Carra, S., 1992. Performance of a 6-port simulated moving-bed pilot-plant for vapor-phase adsorption separations. Sep. Sci. Technol. 27, 1889–1916.

Yang, R.T., 2003. Adsorbents: Fundamentals and Applications. John Wiley & Sons, New Jersey.

# CHAPTER 8

# Simulated Moving Bed Reactor

## 8.1 PROCESS INTENSIFICATION: COUPLING REVERSIBLE REACTIONS WITH ADSORPTION

### 8.1.1 Reactive Separations

Process intensification has gained increasing interest in chemical engineering. It consists of the development of novel devices and/or techniques that, compared to those traditionally used, brings significant improvements in chemical manufacturing and processing, substantially reducing equipment-size/production-capacity ratio, consumption of energy, or waste production, and ultimately leads to cheaper, safer, sustainable technologies (Stankiewicz and Moulijn, 2000).

A well-known example of process intensification is the multifunctional reactor, in which reactors are integrated with at least one more function that usually takes place in other equipment. One of the most important classes of the multifunctional reactors is the so-called reactive separation. As the name indicates, it is defined by the integration of reaction and separation, and is of considerable interest for oxygenated compound production (esters, acetals, ethers, and so on), which involves equilibrium-limited reactions that are favored by the continuous removal of one reaction product from the reaction medium.

In the state of the art, known reactive separations include reactive distillation (RD), reactive extraction, membrane reactor, and chromatographic reactor, among others.

RD is commercially used, and the best example of its application is the methyl acetate synthesis developed and patented by the Eastman Kodak Company (Agreda and Partin, 1984). The entire process occurs in a single column and represents one-fifth of the capital investment and requires one-fifth of the energy of the traditional process (reaction followed by separation by distillation) (Krishna, 2002). Nevertheless, RD presents some disadvantages for systems that exhibit formation of azeotropes and/or other volatility restrictions regarding the boiling points of the reactants and/or products that make this technology unfeasible (Taylor and Krishna, 2000). In this sense, chromatographic reactors, in which the separation is based on different affinities between the solid (adsorbent) and the reactants/products, arise as an attractive solution. Among the chromatographic reactors, one of the most interesting for oxygenated compound production is the simulated moving bed reactor (SMBR).

### 8.1.2 SMBR Concept

The SMBR is an extension of the SMB technology, in which the continuous countercurrent chromatographic separation is combined with chemical reaction. It also comprises

*Simulated Moving Bed Technology*
ISBN 978-0-12-802024-1

fixed-bed columns arranged in series and in a close circuit, but these columns are packed with an adsorbent with catalytic properties or a mixture of adsorbent and catalyst particles. As in the SMB, the conventional SMBR configuration has two inlet streams (feed and desorbent) and two outlet streams (extract and raffinate), and the countercurrent solid movement is simulated by a synchronous shift of these streams by one column in the direction of the fluid, at a regular time interval called the switching time. In the SMBR, instead of feeding the components to be separated, the reactants are fed and in contact with the catalyst/adsorbent are converted into products, which are simultaneously separated and collected in the outlet streams (extract and raffinate) diluted in desorbent (usually one of the reactants is used as desorbent).

For instance, if the feed comprises two reactants (A and B), in which A is used as desorbent, and A and B react to give two products, C and D, the latter being more adsorbed than the former, then a mixture of D and A is obtained in the extract and a mixture of C and A in the raffinate (see Figure 8.1).

The inlet/outlet streams divide the unit into four different sections, each one with a specific role and having a specified number of columns:

- section 1, located between desorbent and extract nodes, in which the solid is regenerated by desorption of the most-adsorbed product (D) using the desorbent (A);
- section 2, located between the extract and feed nodes, and section 3, located between the feed and raffinate nodes, in which the reaction takes place and the products (C and D) are separated as they are formed;
- section 4, located between the raffinate and desorbent nodes, in which the desorbent (A), before being recycled to section 1, is regenerated by adsorption of the less-retained product (C).

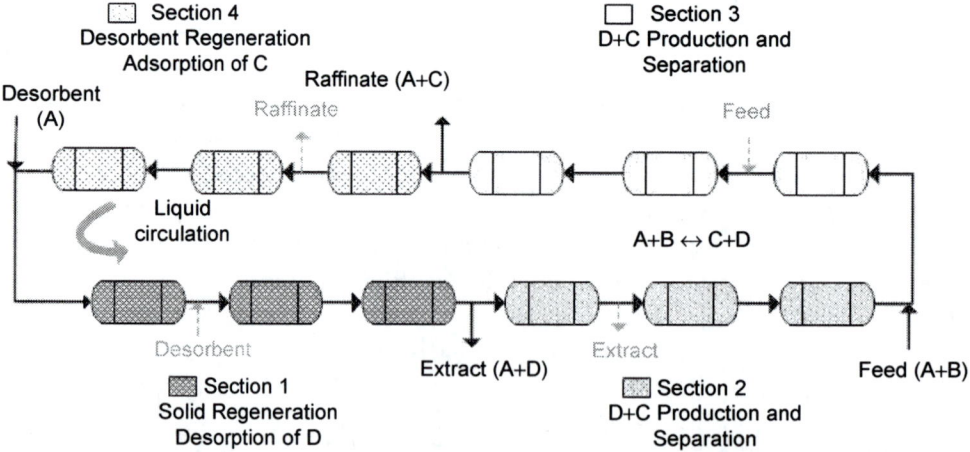

**Figure 8.1** Schematic representation of an SMBR unit with four sections and three columns per section considering a reaction of type A + B ⇔ C + D, in which C is the less-adsorbed product and D is the more-adsorbed one.

## 8.2 SMBR DESIGN METHODOLOGY

Similarly to the SMB, the proper SMBR design involves the correct choice of the operating conditions, particularly the flow rates in each section of the unit and the switching time. The equilibrium theory neglecting axial dispersion and mass-transfer resistance effects can be used to form an initial guess about the SMBR operating conditions.

The constraints in each section in terms of interstitial fluid and solid velocity ratio ($\gamma_j$), considering the reaction $A + B \Leftrightarrow C + D$, in which C is the less-adsorbed product, D is the more-adsorbed one, and A is also used as desorbent, for the true moving bed reactor (TMBR), are given by

$$\gamma_1 > \frac{(1 - \varepsilon_b)}{\varepsilon_b} \left( \varepsilon_p + (1 - \varepsilon_p) \frac{q_{D1}}{\overline{C}_{p,D1}} \right) \tag{8.1}$$

$$\frac{(1 - \varepsilon_b)}{\varepsilon_b} \left( \varepsilon_p + (1 - \varepsilon_p) \frac{q_{C2}}{\overline{C}_{p,C2}} \right) < \gamma_2 < \gamma_3 < \frac{(1 - \varepsilon_b)}{\varepsilon_b} \left( \varepsilon_p + (1 - \varepsilon_p) \frac{q_{D3}}{\overline{C}_{p,D3}} \right) \tag{8.2}$$

$$\gamma_4 < \frac{(1 - \varepsilon_b)}{\varepsilon_b} \left( \varepsilon_p + (1 - \varepsilon_p) \frac{q_{C4}}{\overline{C}_{p,C4}} \right) \tag{8.3}$$

These equations, based on separation conditions, give us the lower limit in sections 1 and 2 and the upper limit in sections 3 and 4, which (after applying the equivalence between TMBR and SMBR: $\gamma_{SMBR} = \gamma_{TMBR} + 1$) can be used as an initial guess for the construction of the SMBR reactive/separation regions. These are feasible regions in the $\gamma_2$–$\gamma_3$ plane that define the proper operating conditions in sections 2 and 3, for predetermined conditions in sections 1 and 4, which ensure complete solid and desorbent regeneration, respectively, to achieve a preset product purity and limiting reactant conversion.

The construction of these regions must then be based on dynamic simulations or in analytical solutions (when possible), taking into account the mass-transfer resistances, and reaction kinetics and equilibrium, because these parameters play important roles in the reactive/separation regions shape and position and consequently in the SMBR process performance (Fricke et al., 1999; Fricke and Schmidt-Traub, 2003; Minceva and Rodrigues, 2005).

For instance, in the case of linear isotherms and considering an irreversible reaction of type $A \Rightarrow C + D$, the size of the feasible region depends on the Damköhler number: smaller Damköhler numbers lead to reactive/separation regions inside the separation

**Figure 8.2** Separation (dashed line), reactive and regeneration regions for linear isotherms and irreversible reaction (A ⇒ C + D), in which D is the more-retained product and C is the less-retained one. The circles indicate the $\gamma_2$, $\gamma_3$ coordinates for which maximum productivity and the $\gamma_1$, $\gamma_4$ coordinates for which minimum desorbent consumption are achieved.

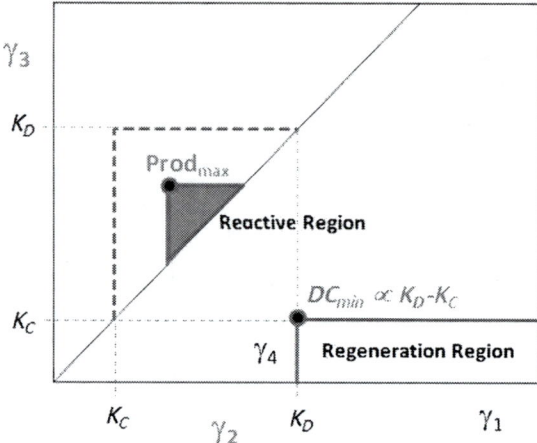

**Figure 8.3** Separation (dashed line), reactive and regeneration regions for linear isotherms and reversible reaction (A ⇔ C + D), in which D is the more-retained product and C is the less-retained one. The circles indicate the $\gamma_2$, $\gamma_3$ coordinates for which maximum productivity and the $\gamma_1$, $\gamma_4$ coordinates for which minimum desorbent consumption are achieved.

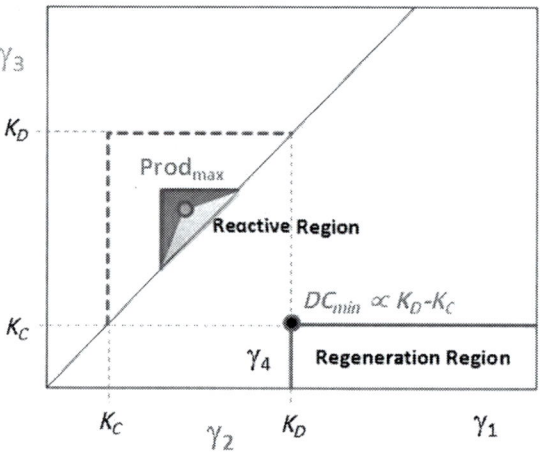

region (Figure 8.2), and in this case, the reaction is the limiting step, whereas for larger Damköhler numbers the reactive/separation region is equal to the region defined by the separation conditions (separation region), and the product separation is the limiting step (Fricke and Schmidt-Traub, 2003; Minceva and Rodrigues, 2005). In the case of linear isotherms and reversible reaction (A ⇔ C + D), not only the reaction kinetics affects the size of the reactive/separation region, but also the reaction equilibrium constant. The reactive/separation region decreases with a decrease in the equilibrium constant (Figure 8.3) (Fricke and Schmidt-Traub, 2003).

## 8.3 SMBR PERFORMANCE PARAMETERS

To evaluate SMBR process feasibility under different operating conditions the following performance parameters are usually used. Once again, a reaction of type A + B ⇔ C + D, in which C is the less-adsorbed product, D is the more-adsorbed one, and A is also used as desorbent, was considered.

*Purity*

The purity of the raffinate (*PUR*) and extract (*PUX*) streams at cyclic steady state (*css*) over a complete cycle is defined as

$$\text{Raffinate Purity: } PUR(\%) = 100 \, \frac{\displaystyle\int_{t}^{t+N_c t^*} C_{R,C} dt}{\displaystyle\int_{t}^{t+N_c t^*} \left( C_{R,B} + C_{R,C} + C_{R,D} \right) dt} \tag{8.4a}$$

$$\text{Extract Purity: } PUX(\%) = 100 \, \frac{\displaystyle\int_{t}^{t+N_c t^*} C_{X,D} dt}{\displaystyle\int_{t}^{t+N_c t^*} \left( C_{X,B} + C_{X,C} + C_{X,D} \right) dt} \tag{8.4b}$$

In the above equations, the numerator indicates the amount of the desired product collected in the outlet stream (extract or raffinate) within a complete cycle, whereas the denominator indicates the sum of products and unconverted reactants withdrawn, with the exception of the reactant used as desorbent. Both extract and raffinate purities are presented in desorbent-free basis.

*Conversion*

The limiting reactant conversion (*X*) is given by the amount of the limiting reactant fed during a cycle discounting the amounts withdrawn at both extract and raffinate streams:

$$X = 1 - \frac{Q_X \displaystyle\int_{t}^{t+N_c t^*} C_{X,B} dt + Q_R \displaystyle\int_{t}^{t+N_c t^*} C_{R,B} dt}{Q_F C_{F,B} N_c t^*} \tag{8.5}$$

*Productivity*

The productivity (*PR*) is defined considering the target product produced and withdrawn in the corresponding outlet stream. In the following expression, the desired product is considered the less-adsorbed species (*C*), and therefore is collected in the raffinate stream.

$$PR\left(\frac{\text{kg}_C}{\text{dayL}_{adsorbent}}\right) = \frac{M_{W,C}Q_R \displaystyle\int_{t}^{t+N_c t^*} C_{R,C}dt}{(1-\varepsilon)V_{unit}N_c t^*} \qquad (8.6)$$

This parameter is an important economic indicator, which specifies how much product is produced per amount of adsorbent/catalyst used in the unit.

*Desorbent Consumption*

The desorbent consumption (*DC*) reflects on the costs involved in the separation of the desorbent from the products, being another important performance parameter. It is calculated from the total amount of desorbent (*A*) fed to the system through the feed and desorbent streams, discounting the amount consumed by reaction.

$$DC\left(L_A/Kg_C\right) = N_c t^* \frac{\left(Q_D C_{D,A} + Q_F\left(C_{F,A} - v_A X C_{F,B}\right)\right)V_{mol,A}}{M_{W,C}Q_R \displaystyle\int_{t}^{t+N_c t^*} C_{R,C}dt} \qquad (8.7)$$

## 8.4 SMBR MATHEMATICAL MODEL

As in the SMB case, two main strategies model an SMBR: (1) the TMBR approach, modeling the TMBR and using the equivalence between a TMBR unit and an SMBR unit (the same as for TMB and SMB units, which was previously presented in Chapter 2, Section 2.1.1.2), and (2) the real SMBR approach, modeling the real SMBR unit by considering the series of the fixed-bed columns and incorporating the periodic ports switch. The equations are similar to the ones described in Chapter 2, Section 2.1.1.1 for TMB/SMB units; the difference is that now we must also take into account the reaction term when writing the mass balance in the particle.

The SMBR model equations, considering axial dispersion flow for the bulk fluid phase, linear driving force (LDF) approximation for the inter- and intraparticle mass-transfer rates, multi-component adsorption equilibrium at the adsorbent phase, liquid velocity variations due to changes in bulk composition (due to reaction and adsorption/desorption), constant bed length and packing porosity, and isothermal operation, are presented as follows.

Bulk fluid mass balance to component *i* in column *k*:

$$\frac{\partial C_{ik}}{\partial t} + \frac{\partial C_{ik}v_k}{\partial z} + \frac{(1-\varepsilon)}{\varepsilon}\frac{3}{r_p}k_{L,ik}\left(C_{ik} - \overline{Cp}_{ik}\right) = D_{ax,k}\frac{\partial}{\partial z}\left(C_T\frac{\partial y_{ik}}{\partial z}\right) \qquad (8.8)$$

in which $C_{ik}$ and $\overline{Cp}_{ik}$ are the bulk and average particle concentrations in the fluid phase of species $i$ in column $k$, respectively, $k_{L,ik}$ is the global mass-transfer coefficient of the component $i$, $\varepsilon$ is the bulk porosity, $y_{ik}$ is the component $i$ molar fraction in the fluid phase in column $k$, $C_T$ is the bulk total concentration, $t$ is the time variable, $z$ is the axial coordinate, $D_{ax,k}$ and $v_k$ are the axial dispersion coefficient and the interstitial velocity in column $k$, respectively, and $r_p$ is the particle radius.

The global mass-transfer coefficient ($k_L$) is defined as

$$\frac{1}{k_L} = \frac{1}{k_e} + \frac{1}{\varepsilon_p k_i} \tag{8.9}$$

in which $k_e$ and $k_i$ are the external and internal mass-transfer coefficients to the liquid phase, respectively. The methods used to obtain these values and the determination of axial dispersion coefficient, $D_{ax}$ are presented in detail in Silva and Rodrigues (2002).

Interstitial fluid velocity variation is calculated using the total mass balance:

$$\frac{dv_k}{dz} = -\frac{(1-\varepsilon)}{\varepsilon} \frac{3}{r_p} \sum_{i=1}^{NC} k_{L,ik} V_{mol,i} \left( C_{ik} - \overline{Cp}_{ik} \right) \tag{8.10}$$

in which $V_{mol,i}$ is the molar volume of component $i$.

Pellet mass balance to component $i$, in column $k$:

$$\varepsilon_p \frac{\partial \overline{Cp}_{ik}}{\partial t} + \left(1 - \varepsilon_p\right) \frac{\partial q_{ik}}{\partial t} = \frac{3}{r_p} k_{L,ik} \left( C_{ik} - \overline{Cp}_{ik} \right) + \sigma_i \rho_p r \left( \overline{Cp}_{ik} \right) \tag{8.11}$$

in which $\overline{q}_{ik}$ is the average adsorbed phase concentration of species $i$ in column $k$ in equilibrium with $\overline{Cp}_{ik}$, $\varepsilon_p$ is the particle porosity, $\sigma_i$ is the stoichiometric coefficient of component $i$, $\rho_p$ the particle density, and $r$ is the chemical reaction rate relative to the average particle concentrations in the fluid phase.

Initial and Danckwerts boundary conditions:

$$t = 0: \quad C_{ik} = \overline{Cp}_{ik} = C_{ik,0} \tag{8.12}$$

$$z = 0: \quad v_j C_{ik} - D_{ax,j} C_T \frac{\partial y_{ik}}{\partial z}\bigg|_{z=0} = v_j C_{ik,F} \tag{8.13a}$$

$$z = L_c: \quad \frac{\partial y_{ik}}{\partial z}\bigg|_{z=L_c} = 0 \tag{8.13b}$$

in which $F$ and $0$ refer to the feed and initial states, respectively.

Mass balances at the nodes of the inlet and outlet lines of the SMBR:

$$\text{Desorbent node: } C_{i(j=4,z=L_c)} = \frac{v_1}{v_4} C_{i(j=1,z=0)} \tag{8.14a}$$

Extract $(j = 2)$ and Raffinate $(j = 4)$ nodes: $C_{i(j-1, z = L_c)} = C_{i(j, z = 0)}$    (8.14b)

Feed node: $C_{i(2, z = L_c)} = \dfrac{\nu_3}{\nu_2} C_{i(3, z = 0)} - \dfrac{\nu_F}{\nu_2} C_i^F$    (8.14c)

in which

$$\nu_1 = \nu_4 + \nu_{Ds} \quad \text{Desorbent } (Ds) \text{ node;} \tag{8.15a}$$

$$\nu_2 = \nu_1 - \nu_X \quad \text{Extract } (X)\text{node;} \tag{8.15b}$$

$$\nu_3 = \nu_2 + \nu_F \quad \text{Feed } (F)\text{node;} \tag{8.15c}$$

$$\nu_4 = \nu_3 - \nu_R \quad \text{Raffinate } (R)\text{node.} \tag{8.15d}$$

## 8.5 ANALYTICAL SOLUTION FOR LINEAR ISOTHERMS AND LINEAR REACTION

### 8.5.1 Linear SMBR Model

An analytical solution for linear SMBR in the presence of mass-transfer resistances was derived (Minceva et al., 2005). The SMBR was modeled using the steady-state equivalent TMBR concept. The developed mathematical model considers plug flow for liquid and solid phases, internal mass transfer described by the LDF model, the irreversible reaction that occurs in the liquid phase (according to A → C + D, in which each species has a linear adsorption isotherm), isothermal operation, constant bed porosity, and all sections with the same length.

The steady-state model equations, for section $j$, are

$$0 = -\nu_j \frac{dC_{i,j}}{dz} - \frac{1 - \varepsilon}{\varepsilon} k_{p,ij} \left( K_i C_{ij} - \bar{q}_{ij} \right) + \sigma_i k_c C_{A,j} \tag{8.16}$$

$$0 = u_s \frac{d\bar{q}_{i,j}}{dz} + k_{p,ij} \left( K_i C_{i,j} - \bar{q}_{i,j} \right) \tag{8.17}$$

in which $C_{i,j}$ is the liquid-phase concentration of component $i$ in section $j$, $\bar{q}_{ij}$ is the average solid-phase concentration of component $i$ in section $j$, $\nu_j$ is the interstitial liquid velocity in section $j$, $u_s$ is the interstitial solid velocity, $K_i$ is the adsorption equilibrium constant of component $i$, $k_{p,i}$ is the mass-transfer coefficient of component $i$, and $k_c$ is the kinetic constant.

Introducing the dimensionless variable for axial position $x = z/L_j$, and the following dimensionless parameters:

fluid/solid interstitial velocity ratio

$$\gamma_j = \nu_j / u_s \tag{8.18}$$

number of mass–transfer units

$$\alpha_{p,i,j} = L_j k_{p,i}/u_s \tag{8.19}$$

Damköhler number

$$Da_j = L_j k_c/u_s, \tag{8.20}$$

the model equations become

$$0 = -\gamma_j \frac{dC_{i,j}}{dx} - v\alpha_{p,i,j}\left(K_i C_{i,j} - \bar{q}_{i,j}\right) + \sigma_i Da_j C_{A,j} \tag{8.21}$$

$$0 = \frac{d\bar{q}_{i,j}}{dx} + \alpha_{p,i,j}\left(K_i C_{i,j} - \bar{q}_{i,j}\right). \tag{8.22}$$

These equations were analytically solved applying Laplace transforms (Minceva et al., 2005) considering the following boundary conditions for each section:

$x = 0$

$$C_{i,j} = C_{i,j}(0) \tag{8.23}$$

$$\bar{q}_{i,j} = \bar{q}_{i,j}(0) \tag{8.24}$$

$x = 1$

$$C_{A,j}(1) = M_{A,j} C_{A,j}(0) + N_{A,j} \bar{q}_{A,j}(0) \tag{8.25}$$

$$\bar{q}_{A,j}(1) = P_{A,j} C_{A,j}(0) + Q_{A,j} \bar{q}_{A,j}(0) \tag{8.26}$$

in which the coefficients $M_{A,j}$, $N_{A,j}$, $P_{A,j}$, and $Q_{A,j}$ are defined as functions of the model parameters ($v$, $K_A$, $\alpha_{p,A,j}$, and $Da_j$), and the operating conditions ($\gamma_j$) (see Table 8.1).

To calculate the liquid and solid concentrations at the beginning and end of all sections, the four sections were connected according to the boundary conditions:

$$\gamma_1 C_{i,1}(0) = \gamma_4 C_{i,4}(1) \tag{8.27}$$

$$\bar{q}_{i,1}(0) = \bar{q}_{i,4}(1) \tag{8.28}$$

$$C_{i,2}(0) = C_{i,1}(1) \tag{8.29}$$

$$\bar{q}_{i,2}(0) = \bar{q}_{i,1}(1) \tag{8.30}$$

$$\gamma_3 C_{i,3}(0) = \gamma_2 C_{i,2}(1) + (\gamma_3 - \gamma_2)C_{i,F} \tag{8.31}$$

$$\bar{q}_{i,3}(0) = \bar{q}_{i,2}(1) \tag{8.32}$$

$$C_{i,4}(0) = C_{i,3}(1) \tag{8.33}$$

$$\bar{q}_{i,4}(0) = \bar{q}_{i,3}(1) \tag{8.34}$$

**Table 8.1** Definition of the Coefficients $m_{A,j}$, $N_{A,j}$, $P_{A,j}$ and $Q_{A,j}$ Used in Eqns (8.25), (8.26), and (8.35) for Reactant A

$$M_{A,j} = [(\psi_{A,j} + \Theta_{A,j} - \alpha_{p,A,j})/(2\Theta_{A,j})]e^{\psi_{A,j}+\Theta_{A,j}} - [(\Psi_{A,j} - \Theta_{A,j} - \alpha_{p,A,j})/(2\Theta_{A,j})]e^{\psi_{A,j}-\Theta_{A,j}}$$

$$N_{A,j} = [(\nu\alpha_{p,A,j})/(2\Theta_{A,j}\gamma_j)](e^{\psi_{A,j}+\Theta_{A,j}} - e^{\psi_{A,j}-\Theta_{A,j}})$$

$$P_{A,j} = [(\alpha_{p,A,j}K_A)/(2\Theta_{A,j})](e^{\Psi_{A,j}-\Theta_{A,j}} - e^{\Psi_{A,j}+\Theta_{A,j}})$$

$$Q_{A,j} = e^{\alpha_{p,A,j}} - [(\nu\alpha_{p,A,j}^2 K_A)/(2\Theta_{A,j}\gamma_j)]\{(2\Theta_{A,j}e^{\alpha_{p,A,j}})/[(\alpha_{p,A,j} - \psi_{A,j})^2 - (\Theta_{A,j})^2] + e^{\psi_{A,j}+\Theta_{A,j}}/(\psi_{A,j}+\Theta_{A,j}-\alpha_{p,A,j}) - e^{\psi_{A,j}-\Theta_{A,j}}/(\psi_{A,j}-\Theta_{A,j}-\alpha_{p,A,j})\}$$

$$\Theta_{A,j} = \sqrt{(\psi_{A,j})^2 - (\sigma_A\alpha_{p,A,j}Dq_j/\gamma_j)}$$

in which $\psi_{A,j} = [\alpha_{p,A,j}(\gamma_j - \nu K_A) + \sigma_A Dq_j]/(2\gamma_j)$

For each section, the system of algebraic equations was obtained through the elimination of the liquid and solid phase concentrations in Eqns (8.27) and (8.34), by using Eqns (8.25) and (8.26), respectively. The system to solve is written in matrix form $\mathbf{P} \cdot \mathbf{X} = \mathbf{Q}$, in which $\mathbf{X}$ is the vector of the variables (liquid and solid concentrations at the beginning of each section):

$$
\begin{bmatrix}
M_{A,1} & -1 & 0 & 0 & N_{A,1} & 0 & 0 & 0 \\
0 & M_{A,2} & -\dfrac{\gamma_3}{\gamma_2} & 0 & 0 & N_{A,2} & 0 & 0 \\
0 & 0 & M_{A,3} & -1 & 0 & 0 & N_{A,3} & 0 \\
-\dfrac{\gamma_1}{\gamma_4} & 0 & 0 & M_{A,4} & 0 & 0 & 0 & N_{A,4} \\
P_{A,1} & 0 & 0 & 0 & Q_{A,1} & -1 & 0 & 0 \\
0 & P_{A,2} & 0 & 0 & 0 & Q_{A,2} & -1 & 0 \\
0 & 0 & P_{A,3} & 0 & 0 & 0 & Q_{A,3} & -1 \\
0 & 0 & 0 & P_{A,4} & -1 & 0 & 0 & Q_{A,4}
\end{bmatrix}
\cdot
\begin{bmatrix}
C_{A,1}(0) \\
C_{A,2}(0) \\
C_{A,3}(0) \\
C_{A,4}(0) \\
\overline{q}_{A,1}(0) \\
\overline{q}_{A,2}(0) \\
\overline{q}_{A,3}(0) \\
\overline{q}_{A,4}(0)
\end{bmatrix}
$$

$$
=
\begin{bmatrix}
0 \\
-C_{A,F}\left(\dfrac{\gamma_3}{\gamma_2} - 1\right) \\
0 \\
0 \\
0 \\
0 \\
0 \\
0
\end{bmatrix}
\tag{8.35}
$$

This solution provides the concentrations of the liquid and solid phases at the beginning of each section, which are needed for the calculation of the concentration profiles of the reactant A given by the following expressions:

$$
C_{A,j}(z) = M^*_{A,j} e^{(\psi_{A,j} + \Theta_{A,j})z} + N^*_{A,j} e^{(\psi_{A,j} - \Theta_{A,j})z}
\tag{8.36}
$$

$$\bar{q}_{A,j}(z) = \bar{q}_{A,j}(0)e^{\alpha_{p,A,j}z} - \left[P_{A,j}^{*}e^{\alpha_{p,A,j}z} + Q_{A,j}^{*}e^{(\psi_{A,j}+\Theta_{A,j})z} + R_{A,j}^{*}e^{(\psi_{A,j}-\Theta_{A,j})z}\right] \quad (8.37)$$

in which the coefficients are

$$M_{A,j}^{*} = \frac{\left(\psi_{A,j} + \Theta_{A,j} - \alpha_{p,A,j}\right)C_{A,j}(0) + \frac{\nu\alpha_{p,A,j}}{\gamma_{j}}\bar{q}_{A,j}(0)}{2\Theta_{A,j}} \quad (8.38)$$

$$N_{A,j}^{*} = \frac{\left(\psi_{A,j} - \Theta_{A,j} - \alpha_{p,A,j}\right)C_{A,j}(0) + \frac{\nu\alpha_{p,A,j}}{\gamma_{j}}\bar{q}_{A,j}(0)}{-2\Theta_{A,j}} \quad (8.39)$$

$$P_{A,j}^{*} = \frac{\frac{\nu\alpha_{p,A,j}^{2}K_{A}}{\gamma_{j}^{2}}\bar{q}_{A,j}(0)}{\left(\psi_{P,j} - \psi_{A,j}\right)^{2} - \Theta_{A,j}^{2}} \quad (8.40)$$

$$Q_{A,j}^{*} = \frac{\alpha_{p,A,j}K_{A}\left(\psi_{A,j} - \alpha_{p,A,j}\right)C_{A,j}(0) + \frac{\nu\alpha_{p,A,j}}{\gamma_{j}}\bar{q}_{A,j}(0)}{2\Theta_{A,j}\left(\psi_{A,j} + \Theta_{A,j} - \psi_{i,j}\right)} \quad (8.41)$$

$$R_{A,j}^{*} = \frac{\alpha_{p,A,j}K_{A}\left(\psi_{A,j} - \Theta_{A,j} - \alpha_{p,A,j}\right)C_{A,j}(0) + \frac{\nu\alpha_{p,A,j}}{\gamma_{j}}\bar{q}_{A,j}(0)}{-2\Theta_{A,j}\left(\psi_{A,j} - \Theta_{A,j} - \psi_{i,j}\right)} \quad (8.42)$$

The concentration profiles of species $A$ were used to calculate the liquid and solid concentrations of the products (B and C) at the end of each section as a function of the concentrations at the beginning of each section:

$$C_{i,j}(1) = M_{i,j}C_{i,j}(0) + N_{i,j}\bar{q}_{i,j}(0) - \Phi_{i,j} \quad (8.43)$$

$$\bar{q}_{i,j}(1) = P_{i,j}C_{i,j}(0) + Q_{i,j}\bar{q}_{i,j}(0) - \Lambda_{i,j} \quad (8.44)$$

$$\Phi_{i,j} = F_{i,j}C_{A,j}(0) + G_{i,j}\bar{q}_{A,j}(0) \quad (8.45)$$

$$\Lambda_{i,j} = R_{i,j}C_{A,j}(0) + S_{i,j}\bar{q}_{A,j}(0) \quad (8.46)$$

The coefficients $M_{i,j}$, $N_{i,j}$, $F_{i,j}$, $G_{i,j}$, $P_{i,j}$, $Q_{i,j}$, $R_{i,j}$, and $S_{i,j}$ are defined in Table 8.2.

As in the case of the reactant A, the liquid and solid concentrations of the products (B and C) at the beginning and end of each section were calculated by connecting all sections according to the boundary conditions (Eqns 8.27–8.34). Afterward, the system of algebraic equations was written in matrix form $\mathbf{P} \cdot \mathbf{X} = \mathbf{Q}$, in which $\mathbf{X}$ is the vector of

**Table 8.2** Definition of the Coefficients $M_{i,j}$, $N_{i,j}$, $F_{i,j}$, $G_{i,j}$, $P_{i,j}$, $Q_{i,j}$, $R_{i,j}$, and $S_{i,j}$ Used in Eqns (8.43)–(8.46) for Products B and C

$$M_{i,j} = [\gamma_j/(\gamma_j - \nu K_i)] + \{1 - [\gamma_j/(\gamma_j - \nu K_i)]\} e^{\psi_{i,j}}$$

$$N_{i,j} = -[\nu/(\gamma_j - \nu K_i)](1 - e^{\psi_{i,j}})$$

$$F_{i,j} = (\sigma_i/\sigma_A)[\gamma_j/(\gamma_j - \nu K_i)] - (\sigma_i Da_j/\gamma_j)\big[((\psi_{i,j} - \alpha_{p,i,j})(\psi_{i,j} - \alpha_{p,A,j})/\psi_{i,j}((\psi_{i,j} - \psi_{A,j})^2 - \Theta^2{}_{A,j}))e^{\psi_{i,j}}$$

$$+ ((\psi_{A,j} + \Theta_{A,j} - \alpha_{p,i,j})(\psi_{A,j} + \Theta_{A,j} - \alpha_{p,A,j})/2\Theta_{A,j}(\psi_{A,j} + \Theta_{A,j})(\psi_{A,j} - \Theta_{A,j} - \psi_{i,j}))e^{\psi_{A,j}+\Theta_{A,j}} - ((\psi_{A,j} - \Theta_{A,j} - \alpha_{p,i,j})$$

$$(\psi_{A,j} - \Theta_{A,j} - \alpha_{p,A,j})/2\Theta_{A,j}(\psi_{A,j} - \Theta_{A,j})(\psi_{i,j} - \Theta_{A,j} - \psi_{A,j}))e^{\psi_{A,j}-\Theta_{A,j}}\big]$$

$$G_{i,j} = -(\sigma_i/\sigma_A)(\nu/\alpha_{p,i,j})(\gamma_j - \nu K_i) - (\sigma_i \nu \alpha_{p,A,j} Da_j/\gamma_j^2)\big[(e^{\psi_{i,j}}/\psi_{i,j}((\psi_{i,j} - \psi_{A,j})^2 - \Theta^2{}_{A,j}))$$

$$+ (e^{\psi_{A,j}+\Theta_{A,j}}/2\Theta_{A,j}(\psi_{A,j} + \Theta_{A,j})(\psi_{A,j} + \Theta_{A,j})) - (e^{\psi_{A,j}-\Theta_{A,j}}/2\Theta_{A,j}(\psi_{A,j} - \Theta_{A,j})(\psi_{A,j} - \psi_{i,j})(\psi_{A,j} - \Theta_{A,j}))\big]$$

$$P_{i,j} = -[(\gamma_j K_i)/(\gamma_j - \nu K_i)](1 - e^{\psi_{i,j}})$$

$$Q_{i,j} = [(\nu K_i)/(\gamma_j - \nu K_i)] + [1 + (\gamma_j e^{\psi_{i,j}})/(\nu K_i)]$$

$$R_{i,j} = (\sigma_i/\sigma_A)[(\gamma_j K_i)/(\gamma_j - \nu K_i)] + [(\sigma_i \alpha_{p,i,j} Da_j K_i)/\gamma_j]\big\langle\{\{(\psi_{i,j} - \alpha_{p,A,j})e^{\psi_{i,j}}/[\psi_{i,j}((\psi_{i,j} - \psi_{A,j})^2 - \Theta^2{}_{A,j})]\}\}$$

$$+ \{(\psi_{A,j} + \Theta_{A,j} - \alpha_{p,A,j})e^{\psi_{A,j}+\Theta_{A,j}}/[2\Theta_{A,j}(\psi_{A,j} + \Theta_{A,j})(\psi_{A,j} - \Theta_{A,j} - \psi_{i,j})]\}$$

$$- \{(\psi_{A,j} - \alpha_{p,A,j})e^{\psi_{A,j}-\Theta_{A,j}}/[2\Theta_{A,j}(\psi_{A,j} - \Theta_{A,j})(\psi_{A,j} - \Theta_{A,j} - \psi_{i,j})]\}\big\rangle$$

$$S_{i,j} = -(\sigma_i/\sigma_A)[(\nu K_i)/(\gamma_j - \nu K_i)] + [(\sigma_i \nu \alpha_{p,A,j} \alpha_{p,i,j} K_i Da_j)/\gamma_j^2]\big\langle\{\{(e^{\psi_{i,j}}/[\psi_{i,j}((\psi_{i,j} - \psi_{A,j})^2 - \Theta^2{}_{A,j})]\}$$

$$+ \{(e^{\psi_{A,j}+\Theta_{A,j}}/[2\Theta_{A,j}(\psi_{A,j} + \Theta_{A,j})(\psi_{A,j} - \Theta_{A,j} - \psi_{i,j})]\} - \{e^{\psi_{A,j}-\Theta_{A,j}}/[2\Theta_{A,j}(\psi_{A,j} - \Theta_{A,j} - \psi_{i,j})(\psi_{A,j} - \Theta_{A,j})]\}\big\rangle$$

$$\psi_{i,j} = [\alpha_{p,i,j}(\gamma_j - \nu K_i)]/\gamma_j$$

the variables (liquid and solid concentrations of B and C at the beginning of each section):

$$
\begin{bmatrix}
M_{i,1} & -1 & 0 & 0 & N_{i,1} & 0 & 0 & 0 \\
0 & M_{i,2} & -\dfrac{\gamma_3}{\gamma_2} & 0 & 0 & N_{i,2} & 0 & 0 \\
0 & 0 & M_{i,3} & -1 & 0 & 0 & N_{i,3} & 0 \\
-\dfrac{\gamma_1}{\gamma_4} & 0 & 0 & M_{i,4} & 0 & 0 & 0 & N_{i,4} \\
P_{i,1} & 0 & 0 & 0 & Q_{i,1} & -1 & 0 & 0 \\
0 & P_{i,2} & 0 & 0 & 0 & Q_{i,2} & -1 & 0 \\
0 & 0 & P_{i,3} & 0 & 0 & 0 & Q_{i,3} & -1 \\
0 & 0 & 0 & P_{i,4} & -1 & 0 & 0 & Q_{i,4}
\end{bmatrix}
\cdot
\begin{bmatrix}
C_{i,1}(0) \\
C_{i,2}(0) \\
C_{i,3}(0) \\
C_{i,4}(0) \\
\bar{q}_{i,1}(0) \\
\bar{q}_{i,2}(0) \\
\bar{q}_{i,3}(0) \\
\bar{q}_{i,4}(0)
\end{bmatrix}
=
\begin{bmatrix}
\Phi_{i,1} \\
\Phi_{i,2} \\
\Phi_{i,3} \\
\Phi_{i,4} \\
\Lambda_{i,1} \\
\Lambda_{i,2} \\
\Lambda_{i,3} \\
\Lambda_{i,4}
\end{bmatrix}
\tag{8.47}
$$

The solution gives the concentrations of the products (B and C) in the liquid and solid phases at the beginning of each section, later used in the calculation of the respective concentration profiles:

$$
C_{P,j}(z) = J_{P,j}^* + L_{P,j}^* e^{\psi_{P,j}z} + M_{P,j}^* e^{(\psi_{A,j}+\Theta_{A,j})z} + N_{P,j}^* e^{(\psi_{A,j}-\Theta_{A,j})z}
\tag{8.48}
$$

$$
\bar{q}_{P,j}(z) = \bar{q}_{P,j}(0)e^{\alpha_{p,P,j}z} - \Big[ P_{P,j}^* e^{\alpha_{p,P,j}z} + Q_{P,j}^* e^{(\psi_{A,j}+\Theta_{A,j})z} + R_{P,j}^* e^{(\psi_{A,j}-\Theta_{A,j})z} \\
+ S_{P,j}^* e^{\psi_{P,j}z} + T_{P,j}^* \Big]
\tag{8.49}
$$

in which the coefficients are

$$
J_{P,j}^* = \frac{\alpha_{p,P,j}}{\gamma_j \psi_{P,j}} \left\{ \big[ \gamma_j C_{P,j}(0) - v\bar{q}_{P,j}(0) \big] - \frac{\sigma_P D a_j \alpha_{p,A,j}}{\left(\psi_{A,j}^2 - \Theta_{A,j}^2\right)} \big[ \gamma_j C_{A,j}(0) - v\bar{q}_{A,j}(0) \big] \right\}
\tag{8.50}
$$

$$
L_{P,j}^* = \frac{\psi_{P,j} - \alpha_{p,P,j}}{\gamma_j \psi_{P,j}} \left\{ \left[ \gamma_j C_{P,j}(0) + \frac{v\alpha_{p,P,j}}{\psi_{P,j} - \alpha_{p,P,j}} \bar{q}_{P,j}(0) \right] \right.
$$

$$
\left. + \frac{\sigma_P D a_j \left(\psi_{P,j} - \alpha_{p,A,j}\right)}{\left(\left(\psi_{P,j} - \psi_{A,j}\right)^2 - \Theta_{A,j}^2\right)} \left[ \gamma_j C_{A,j}(0) + \frac{v\alpha_{p,A,j}}{\psi_{P,j} - \alpha_{p,A,j}} \bar{q}_{A,j}(0) \right] \right\}
\tag{8.51}
$$

$$M_{P,j}^* = \frac{\sigma_P Da_j}{\gamma_j} \frac{\left(\psi_{A,j} + \Theta_{A,j} - \alpha_{p,A,j}\right)\left(C_{A,j}(0)\left(\psi_{A,j} + \Theta_{A,j} - \alpha_{p,A,j}\right) + \frac{\nu\alpha_{p,A,j}}{\gamma_j}\bar{q}_{A,j}(0)\right)}{2\Theta_{A,j}\left(\psi_{A,j} + \Theta_{A,j} - \psi_{P,j}\right)\left(\psi_{A,j} + \Theta_{A,j}\right)}$$

$$(8.52)$$

$$N_{P,j}^* = -\frac{\sigma_P Da_j}{\gamma_j} \frac{\left(\psi_{A,j} - \Theta_{A,j} - \alpha_{p,A,j}\right)\left(C_{A,j}(0)\left(\psi_{A,j} - \Theta_{A,j} - \alpha_{p,A,j}\right) + \frac{\nu\alpha_{p,A,j}}{\gamma_j}\bar{q}_{A,j}(0)\right)}{2\Theta_{A,j}\left(\psi_{A,j} - \Theta_{A,j} - \psi_{P,j}\right)\left(\psi_{A,j} - \Theta_{A,j}\right)}$$

$$(8.53)$$

$$P_{P,j}^* = \frac{\nu\alpha_{p,P,j}K_P}{\gamma_j\left(\alpha_{p,P,j} - \psi_{P,j}\right)}\bar{q}_{P,j}(0) \tag{8.54}$$

$$Q_{P,j}^* = \frac{\sigma_P Da_j\alpha_{p,P,j}K_P}{\gamma_j} \frac{(\psi_{A,j} + \Theta_{A,j} - \alpha_{p,A,j})C_{A,j}(0) + \frac{\nu\alpha_{p,A,j}}{\gamma_j}\bar{q}_{A,j}(0)}{2\Theta_{A,j}\left(\psi_{A,j} + \Theta_{A,j} - \psi_{P,j}\right)\left(\psi_{A,j} + \Theta_{A,j}\right)} \tag{8.55}$$

$$R_{P,j}^* = -\frac{\sigma_P Da_j\alpha_{p,P,j}K_P}{\gamma_j} \frac{\left(\psi_{A,j} - \Theta_{A,j} - \alpha_{p,A,j}\right)C_{A,j}(0) + \frac{\nu\alpha_{p,A,j}}{\gamma_j}\bar{q}_{A,j}(0)}{2\Theta_{A,j}\left(\psi_{A,j} - \Theta_{A,j} - \psi_{P,j}\right)\left(\psi_{A,j} - \Theta_{A,j}\right)} \tag{8.56}$$

$$S_{P,j}^* = \frac{\alpha_{p,P,j}K_P}{\gamma_j\psi_{P,j}}\left\{\left[\gamma_j C_{P,j}(0) - \frac{\nu\alpha_{p,P,j}}{\alpha_{p,P,j} - \psi_{P,j}}\bar{q}_{P,j}(0)\right]\right.$$

$$\left. - \frac{\sigma_P Da_j\left(\alpha_{p,A,j} - \psi_{P,j}\right)}{\gamma_j\left(\left(\psi_{P,j} - \psi_{A,j}\right)^2 - \Theta_{A,j}^2\right)}\left[\gamma_j C_{A,j}(0) - \frac{\nu\alpha_{p,A,j}}{\alpha_{p,A,j} - \psi_{P,j}}\bar{q}_{A,j}(0)\right]\right\} \quad (8.57)$$

$$T_{P,j}^* = -\frac{\alpha_{p,P,j}K_P}{\gamma_j\psi_{P,j}}\left\{\left[\gamma_j C_{P,j}(0) - \nu\bar{q}_{P,j}(0)\right]\right.$$

$$\left. - \frac{\sigma_P Da_j\alpha_{p,A,j}}{\gamma_j\left(\psi_{A,j}^2 - \Theta_{A,j}^2\right)}\left[\gamma_j C_{A,j}(0) - \nu\bar{q}_{A,j}(0)\right]\right\} \quad (8.58)$$

The liquid-phase concentration in the extract $(C_{i,X} = C_{i,2}(0))$ and raffinate nodes $(C_{i,R} = C_{i,4}(0))$ can be applied to evaluate the SMBR performance (productivity, purity, and desorbent consumption).

## 8.5.2 Analytical versus Numerical Solution

The SMBR internal concentration profiles and performances were calculated by the analytical solution of the linear SMBR in the presence of mass transfer for the reaction of type $A \rightarrow B + C$ and compared with those obtained numerically for both real SMBR and equivalent TMBR. The model parameters and the operating conditions considered are presented in Table 8.3.

The steady-state TMBR model and the equivalent transient real SMBR model were solved using gPROMS-general PROcess Modeling System version: 2.3.1 (www. psentreprise.com). The axial domain was discretized using third-order orthogonal collocation method in finite elements over 20 elements per column.

Figure 8.4 shows the comparison of the steady-state concentration profiles calculated analytically and numerically, in which it can be observed that the equivalent TMBR concentration profiles calculated numerically and analytically are identical, which validates the proposed analytical solution. The CSS SMBR concentration profiles are presented at the middle of the switching time and a slight difference between those profiles and the ones given by the equivalent TMBR is observed around the feed and raffinate ports. This difference leads to a small difference in the performance parameters calculated by the TMBR model/analytical solution and by the real SMBR model, as can be observed in Table 8.4. For the SMBR, the performances were calculated by two ways: (1) using the concentrations at the middle of the switching time; and (2) using the average concentration over a switching time period, after reaching the CSS.

The developed analytical solution provides a simple and fast methodology to determine the performance of a linear SMBR without the need to calculate the internal concentration profiles. This is an important solution that can be easily applied to construct

**Table 8.3** SMBR Model Parameters and Operating Conditions

| SMBR Geometry | | Model Parameters | | Operating Conditions | |
|---|---|---|---|---|---|
| 12 columns | $L_c = 29$ cm | $\varepsilon$ | 0.4 | $t^*$ | 3.4 min |
| | $d_c = 2.6$ cm | $K_A$ | 0.205 | $C_{A,F}$ | 80 g/dm$^3$ |
| Configuration | 3-3-3-3 | $K_B$ | 0.795 | $\gamma_1$ | 0.880 |
| | | $K_C$ | 0.405 | $\gamma_2$ | 0.493 |
| | | $k_c$ | 0.61/min | $\gamma_3$ | 0.659 |
| | | $K_{p,i}$ | 2.667/min | $\gamma_4$ | 0.190 |
| | | $\sigma_A$ | $-1$ | $Q_1^{SMBR}$ | 34.05 cm$^3$/min |
| | | $\sigma_{B/C}$ | 1 | | |

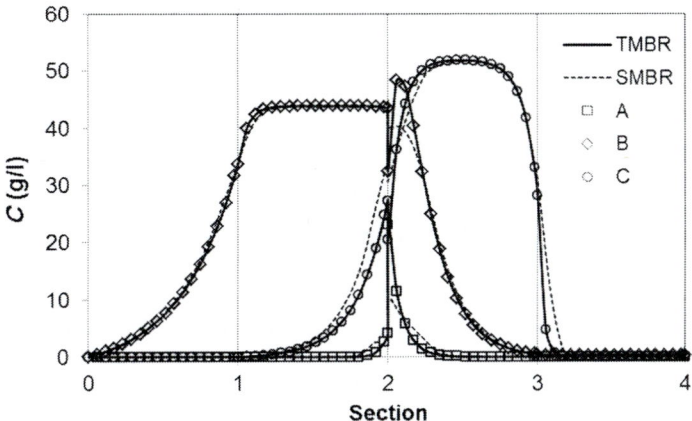

**Figure 8.4** Concentration profiles calculated by the steady-state equivalent TMBR numerical model (full lines), real SMBR numerical model (dashed lines) and by the proposed analytical solution (symbols).

**Table 8.4** Comparison of the SMBR Performance Determined by the TMBR Analytical and Numerical Solutions, and by the Real SMBR Numerical Solution

| Performance | Analytical TMBR | Numerical TMBR | Numerical SMBR (At Half of the Switching Time) | Numerical SMBR (Average over the Switching Time) |
|---|---|---|---|---|
| Extract purity (%) | 99.88 | 99.88 | 99.88 | 99.77 |
| Raffinate purity (%) | 98.33 | 98.33 | 98.42 | 97.83 |
| Conversion (%) | 100 | 100 | 100 | 100 |

the reactive–separation regions for different SMBR systems; it allows a fast assessment about the effect of adsorption equilibrium, reaction rate, mass transfer, and SMBR unit configuration on the SMBR performance for a reaction of type $A \rightarrow B + C$ (Minceva et al., 2005).

## 8.6 DEVELOPMENT OF SMBR FOR THE SYNTHESIS OF OXYGENATES

To develop an SMBR process for the production of oxygenates involving equilibrium-limited reactions, the acquisition of basic data as reaction kinetics and equilibrium and multicomponent adsorption isotherms is fundamental. These data allow the development of a mathematical model to describe the SMBR behavior and to define suitable operating conditions, which are then used to perform SMBR experiments that allow validating the mathematical model previously developed. After validation, this model is used to further

**Table 8.5** Summary of the Steps Performed to Implement an SMBR-Based Process for Oxygenates Production

**1. Batch reactor**

*1.1. Equilibrium experiments*
- Thermodynamic equilibrium constant
- Standard properties of reaction
- Standard formation properties of species

*1.2 Kinetic experiments*
- Mass transfer: pore diffusion limitations
- Reaction rate law
- Batch reactor mathematical model

**2. Fixed bed**

*2.1 Tracer experiments*
- Bed porosity
- Axial dispersion

*2.2 Binary adsorption experiments*
- Equilibrium adsorption isotherms
- Mass-transfer limitations in absence of reaction
- Fixed-bed column mathematical model

*2.3 Reaction experiments*
- Mass-transfer limitations
- Fixed-bed reactor model validation

**3. SMBR**

*3.1 Equilibrium theory: SMB*
- Regeneration region: sections 1 and 4
- Separation region: sections 2 and 3

*3.2 Modeling and simulation*
- Reactive–separation region: sections 2 and 3
- Internal concentration profiles at cyclic steady state
- Extract and raffinate concentration histories

*3.3 Operation*
- SMBR experiments
- Model validation and SMBR optimization

study and to optimize the SMBR process. A summary of the procedure usually undertaken at LSRE is presented in Table 8.5.

In the following subsections, the most important steps are highlighted using the production of ethyl lactate as the case study.

## 8.6.1 Case Study: Ethyl Lactate

Ethyl lactate can be produced from common biorefinery building blocks (ethanol and lactic acid) and can be used as green solvent, contributing positively to the environment because it could replace a range of environment-damaging halogenated and toxic petroleum-derived solvents (Pereira et al., 2011).

The synthesis of ethyl lactate involves a liquid-phase reversible reaction between ethanol and lactic acid, in acid medium, according to Scheme 8.1.

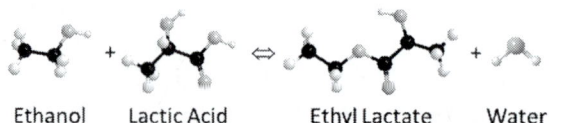

Ethanol    Lactic Acid       Ethyl Lactate       Water

**Scheme 8.1** Reaction illustrated for ethanol + lactic acid ↔ ethyl lactate + water.

As mentioned above (Section 8.1.2), the SMBR requires the use of a solid with catalytic and adsorptive properties or a mixture of solids, one adsorbent plus one catalyst. For the ethyl lactate process development, a commercial acid ion–exchange resin, the Amberlyst-15 wet (A15) was selected based on its efficiency to catalyze esterification reactions (Pöpken et al., 2000; Sanz et al., 2002) and simultaneously its ability to preferentially adsorb water (by-product) (Mazzotti et al., 1997; Sanz et al., 2002; Silva and Rodrigues, 2002). Therefore, the following studies were conducted using this solid as catalyst and water-selective adsorbent.

### 8.6.1.1 Kinetic Studies in a Batch Reactor
The ethyl lactate synthesis was studied in a batch reactor using A15 as catalyst (Pereira et al., 2008b).

#### 8.6.1.1.1 Thermodynamic Equilibrium Constant
The thermodynamic equilibrium constant ($K_{eq}$) was determined in the temperature range 50–90 °C. For each temperature, the equilibrium constant was calculated according to Eqn (8.59) using the equilibrium composition ($x_i$) measured experimentally and the activity coefficients ($\gamma_i^*$) computed by the UNIversal QUasi-Chemical (UNIQUAC) model.

$$K_{eq} = \frac{a_{EL}a_W}{a_{Eth}a_{La}} = \frac{x_{EL}x_W}{x_{Eth}x_{La}} \frac{\gamma_{EL}^* \gamma_W^*}{\gamma_{Eth}^* \gamma_{La}^*} \tag{8.59}$$

The equilibrium constant dependence on temperature was estimated by fitting the experimental values of $\ln K_{eq}$ versus $1/T$ to the Van't Hoff equation:

$$\ln K_{eq} = \frac{\Delta S^\circ}{R} - \frac{\Delta H^\circ}{RT} \tag{8.60}$$

which gives

$$\ln K_{eq} = 2.9625 - 515.13/T(K) \tag{8.61}$$

From the slope of $\ln K_{eq}$ versus $\frac{1}{T}$, it was, therefore, possible to find an average reaction enthalpy of 4.28 kJ/mol.

#### 8.6.1.1.2 Kinetic Experiments
When performing a kinetic study, in order to evaluate the true kinetics, the experiments must be conducted in the absence of mass-transfer limitations. Therefore, a kinetic study should always start by evaluating external and internal mass-transfer resistances.

*Preliminary studies*

The influence of the external mass-transfer resistance, in a batch reactor is quantified by performing experiments at different stirring speed and keeping constant the remaining conditions, whereas the effect of the internal mass-transfer resistance is assessed by per-forming experiments with different sizes of catalyst particles (once again keeping constant the remaining operational conditions). In both cases, the experiments should be conduct-ed at the highest temperature at which higher influence of the mass-transfer resistances is expected.

These experiments were performed for the ethyl lactate (EL) synthesis, and the results are shown in Figures 8.5 and 8.6, from which it can be concluded that the absence of mass-transfer resistances is guaranteed with a stirring speed above 600 rpm and using the unsieved resin (average diameter of 0.685 mm). Therefore, the remaining kinetic experiments were performed using a stirring speed of 600 rpm and the unsieved resin.

*Kinetic studies*

In the absence of mass-transfer limitations, kinetic experiments were carried out to determine the reaction rate and to study the effect of the different operating conditions: catalyst mass (Figure 8.7), initial molar ratio of reactants (Figure 8.8), and temperature (Figure 8.9).

**Figure 8.5** Lactic acid conversion history at different stirrer speeds for a molar ratio of ethanol to lactic acid of 1.82 at 324.18 K, using 2.4 wt.% of A15 as catalyst with particle diameter $0.5 < dp < 0.6$ mm.

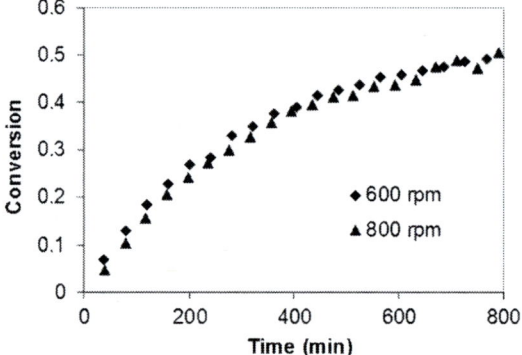

**Figure 8.6** Effect of A15 particle size on the lactic acid conversion history for a molar ratio of ethanol to lactic acid of 1.82 at 324.18 K and using 2.4 wt.% of catalyst.

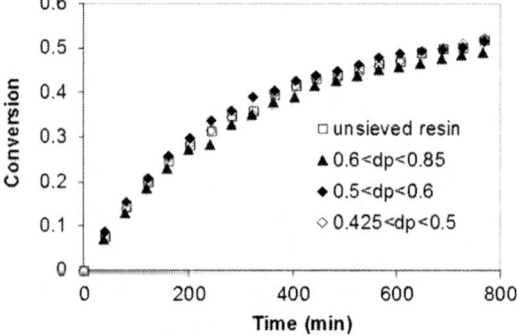

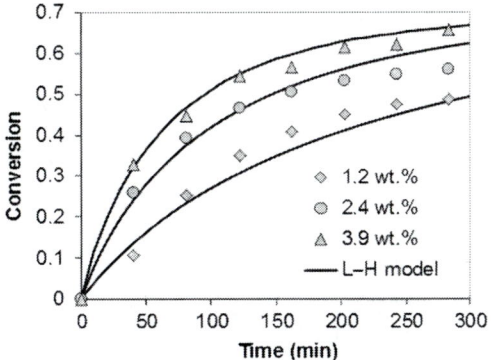

**Figure 8.7** Effect of A15 loading on the lactic acid conversion history for a molar ratio of ethanol to lactic acid of 1.82 at 353.49 K. *(Reprinted with permission from Industrial & Engineering Chemistry Research 47, Pereira, C.S.M., Pinho, S.P., Silva, V.M.T.M., Rodrigues, A.E., 1453–1463. Copyright (2008) American Chemical Society.)*

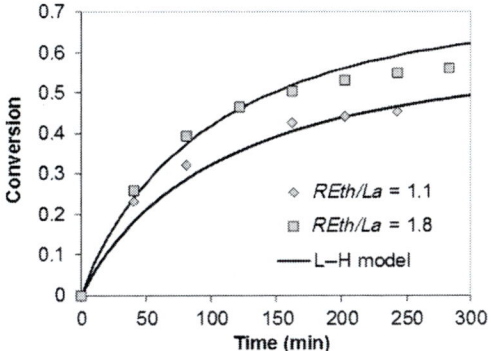

**Figure 8.8** Effect of initial molar ratio of ethanol to lactic acid ($R_{Eth/La}$) on the lactic acid conversion history at 353.40 K and using 2.4 wt.% of A15 catalyst. *(Reprinted with permission from Industrial & Engineering Chemistry Research 47, Pereira, C.S.M., Pinho, S.P., Silva, V.M.T.M., Rodrigues, A.E., 1453–1463. Copyright (2008) American Chemical Society.)*

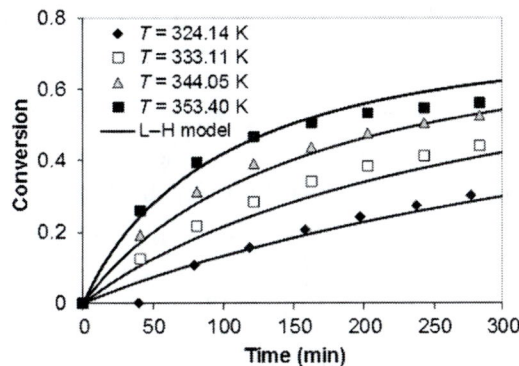

**Figure 8.9** Effect of the reaction temperature on the lactic acid conversion history for a molar ratio of ethanol to lactic acid of 1.82 and using 2.4 wt.% of A15 catalyst. *(Reprinted with permission from Industrial & Engineering Chemistry Research 47, Pereira, C.S.M., Pinho, S.P., Silva, V.M.T.M., Rodrigues, A.E., 1453–1463. Copyright (2008) American Chemical Society.)*

The experimental results, as observed in Figures 8.7—8.9, are well described by the simplified Langmuir—Hinshelwood (L—H) model written in terms of activities of components due to the nonideality of the reaction mixture:

$$r = k_c \frac{a_{Eth}a_{La} - \frac{a_{EL}a_W}{K_{eq}}}{\left(1 + K_{s,Eth}a_{Eth} + K_{s,W}a_W\right)^2} \tag{8.62}$$

with

$$k_c(\text{mol/g min}) = 2.70 \times 10^7 \ exp\left(\frac{-49980}{RT(K)}\right) \quad \text{(activation energy of 49.98 kJ/mol)} \tag{8.63}$$

$$K_{s,W} = 15.19 \ exp\left(\frac{12.01}{T(K)}\right) \tag{8.64}$$

$$K_{Eth} = 1.22 \ exp\left(\frac{359.63}{T(K)}\right) \tag{8.65}$$

The L—H model (Eqn (8.62)) considers the adsorption of both reactants (ethanol and lactic acid), the surface reaction between them and the desorption of both products (ethyl lactate and water). Moreover, it was assumed that the most polar molecules, water and ethanol, have the strongest adsorption strength on the A15 surface, being therefore neglected the adsorption of lactic acid and ethyl lactate.

The model parameters (Eqns (8.63—8.65)) were determined by minimizing the difference between experimental and theoretical conversion ($X$) calculated from

$$\frac{dX}{dt} = \frac{|\nu_l|w_{cat}r}{n_{l,0}} \tag{8.66}$$

with initial condition: $t = 0$; $X = 0$.

In Eqn (8.66), $n_{l,0}$ and $\nu_l$ are, respectively, the initial moles number and the stoichiometric coefficient of the limiting reactant, $t$ is the time, $w_{cat}$ is the mass of catalyst, and $r$ is the reaction rate expressed in moles i/mass of catalyst/min.

### 8.6.1.2 Fixed Bed: Adsorption/Reaction Studies

This study involved two main types of experiments: (1) adsorption experiments—to determine the adsorption parameters over A15 resin and (2) reactive adsorption experiments—to validate both reaction and adsorption data by using a fixed-bed reactor (FBR) mathematical model (Pereira et al., 2009a).

#### 8.6.1.2.1 Adsorption Experiments

To determine the multicomponent equilibrium isotherm for each component involved in the ethyl lactate synthesis, dynamic binary adsorption experiments with nonreactive pairs were conducted using a fixed-bed column packed with A15 resin.

The adsorption experiments comprised nonreactive pairs to have information just on the adsorptive equilibrium, because, as mentioned before, the A15 has a dual role; it acts as catalyst and as water-selective adsorbent. Considering a mixture comprising A and B, the methodology was as follows: the resin was saturated with component $A$, and then the feed concentration of component $B$ was changed stepwise. The experimental data were then used to calculate the number of moles adsorbed/desorbed of each component for all the experiments (Eqns (8.67) and (8.68)).

$$n_{exp}^{ads} = Q \int_0^\infty [C_F - C_{out}(t)] dt \qquad (8.67)$$

$$n_{exp}^{des} = Q \int_0^\infty [C_{out}(t) - C_F] dt \qquad (8.68)$$

The multicomponent Langmuir adsorption isotherm was considered:

$$\overline{q}_i = \frac{Q_{Ads,i} K_i \overline{C}_{p,i}}{1 + \sum_{j=1}^{NC} K_j \overline{C}_{p,j}} \qquad (8.69)$$

in which $Q_{Ads,i}$ is the monolayer capacity and $K_i$ is the equilibrium constant for component $i$.

To reduce the number of adsorption parameters from 8 (one molar monolayer capacity and one equilibrium constant for each component) to 5, for each temperature, a novel approach was implemented that assumes a constant monolayer capacity in terms of volume for all species; it was considered that $Q_{Ads,i} = Q_V/V_{mol,i}$, in which $Q_V$ is the volumetric monolayer capacity and $V_{mol,i}$ is the molar volume of species $i$.

Then, the adsorption parameters were optimized by minimizing the difference between experimental and theoretical values calculated by Eqns (8.70) and (8.71), according to the objective function, Eqn (8.72):

$$n_{theo}^{ads} = \left( [\varepsilon + (1-\varepsilon)\varepsilon_p](C_F - C_0) + (1-\varepsilon)(1-\varepsilon_p)[q(C_F) - q(C_0)] \right) V \qquad (8.70)$$

$$n_{theo}^{des} = \left( [\varepsilon + (1-\varepsilon)\varepsilon_p](C_0 - C_F) + (1-\varepsilon)(1-\varepsilon_p)[q(C_0) - q(C_F)] \right) V \qquad (8.71)$$

$$fob = \sum_{k=1}^{NE} \left[ \left( n_{exp}^{ads} - n_{theo}^{ads} \right)^2 + \left( n_{exp}^{des} - n_{theo}^{des} \right)^2 \right] \qquad (8.72)$$

The optimized Langmuir adsorption parameters over A15 as well as the molar volume of each component are presented in Table 8.6.

**Table 8.6** Adsorption Equilibrium Parameters for Langmuir Isotherms over Amberlyst 15-Wet at 50 °C

| Component | $Q_V$ (ml/l$_{wet\ solid}$) | $Q_{Ads}$ (mol/l$_{wet\ solid}$) | $K$ (l/mol) | $V_{mol}$ (ml/mol) |
|---|---|---|---|---|
| **Ethanol** | 383.5 | 6.30 | 3.068 | 60.87 |
| **Lactic acid** | | 4.94 | 4.085 | 77.56 |
| **Ethyl lactate** | | 3.23 | 1.815 | 118.44 |
| **Water** | | 20.58 | 7.055 | 18.63 |

The possible nonreactive binary mixtures to perform the breakthrough experiments are ethanol/water, ethyl lactate/ethanol, and lactic acid/water. The experimental and simulated breakthrough curves for each one of these pairs at 50 °C are shown in Figures 8.10–8.12. The simulated curves were obtained using the equations of the SMB mathematical model (Section 8.4) applied to one fixed-bed column and considering the multicomponent Langmuir adsorption isotherms with the parameters of Table 8.6.

All adsorption experimental results are very well described by the model, with exception of the experiment in which pure water displaces an 86 wt.% lactic acid solution (Figure 8.12). This deviation might be due to viscous fingering phenomenon, because the lowest viscous fluid (water: 0.55 cP at 323.15 K) is displacing the highest viscous fluid (85% lactic acid solution: 9.40 cP at 323.15 K), inducing an unstable interface between them. The difference in their viscosities is of about 9 cP, and in this case fingering effects are significant (Catchpoole et al., 2006), which are not taken into account in the global mass-transfer coefficient. The effect of viscous fingering can be described by increasing the axial dispersion (Mallmann et al., 1998). Therefore, to verify this assumption, the axial dispersion coefficient was increased; it was verified that the theoretical curve using an axial dispersion coefficient increased by a factor of 7 (thin line in Figure 8.12) fits the experimental data very well.

From the adsorption data, it can be concluded that the most-adsorbed component in A15 is water followed by ethanol, and lactic acid being the less-adsorbed ethyl lactate. From the SMBR process performance point of view, this is a good result because the reactants should have intermediate selectivity relative to the products and the products should have high selectivity between them (water/lactic acid selectivity—7.2; water/ethanol selectivity—7.5; lactic acid/ethyl lactate—3.4; ethanol/ethyl lactate—3.3; and water/ethyl lactate selectivity—24.8).

### 8.6.1.2.2 Reactive Adsorption Experiments

Two different reaction experiments were performed using the fixed-bed column packed with A15 to test the model and the simulation tools under different conditions. One (run 1), in which a mixture of ethanol and lactic acid was fed to the fixed-bed reactor previously saturated with ethanol, and the other (run 2), in which just lactic acid solution (86 wt.%

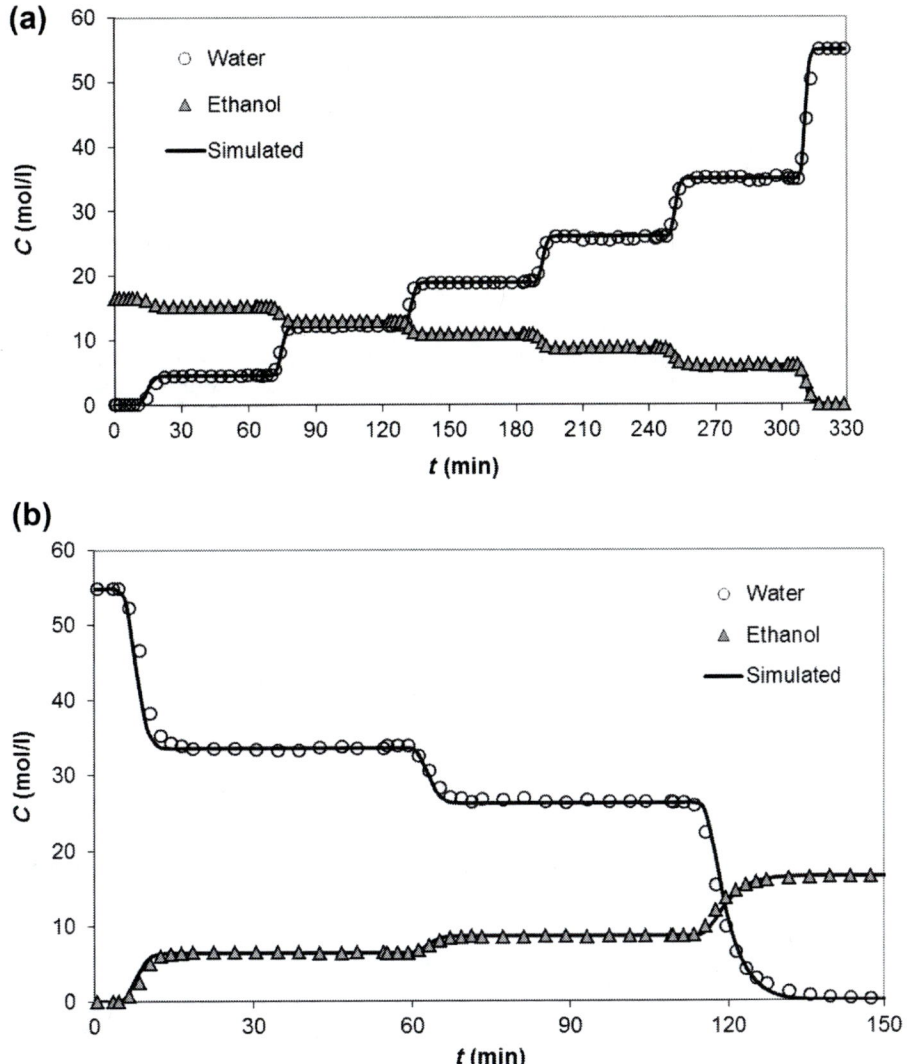

**Figure 8.10** Breakthrough experiments: outlet concentration of ethanol and water as a function of time ($Q = 5$ ml/min; $T = 323.15$ K). (a) water displacing ethanol; (b) ethanol displacing water.

in water) was fed to the reactor also saturated with ethanol. In Figure 8.13, the concentration histories of ethanol, lactic acid, ethyl lactate, and water, at the fixed-bed column outlet, are shown, for both runs.

In run 1 (Figure 8.13(a)), as lactic acid solution enters the column it is adsorbed and reacts with ethanol to produce ethyl lactate and water in the stoichiometric amount. Ethyl lactate is the first eluted component, because it has less affinity with A15 resin

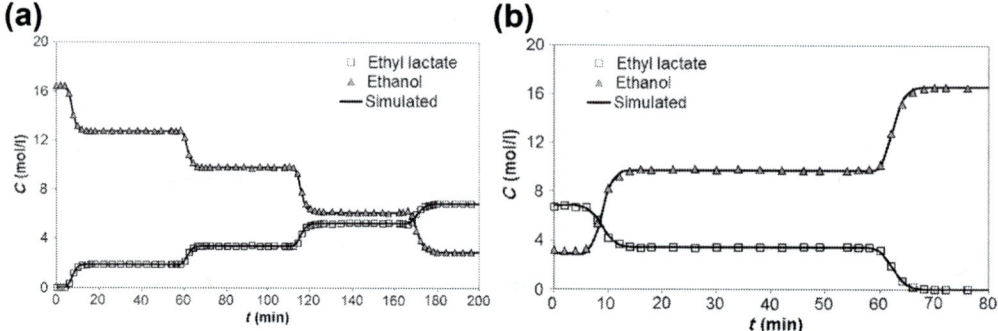

**Figure 8.11** Breakthrough experiments: outlet concentration of ethanol and ethyl lactate as a function of time ($Q = 5$ ml/min; $T = 323.15$ K). (a) ethyl lactate displacing ethanol; (b) ethanol displacing ethyl lactate.

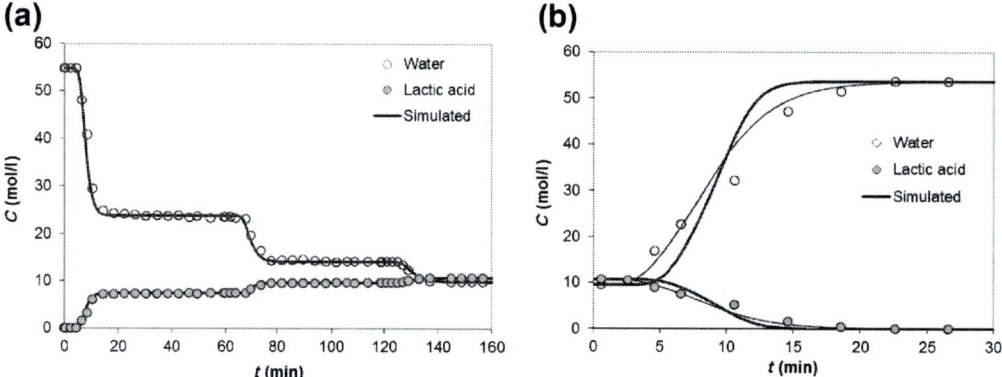

**Figure 8.12** Breakthrough experiments: outlet concentration of water and lactic acid as a function of time ($Q = 5$ ml/min; $T = 323.15$ K). (a) Lactic acid displacing water; (b) water displacing lactic acid.

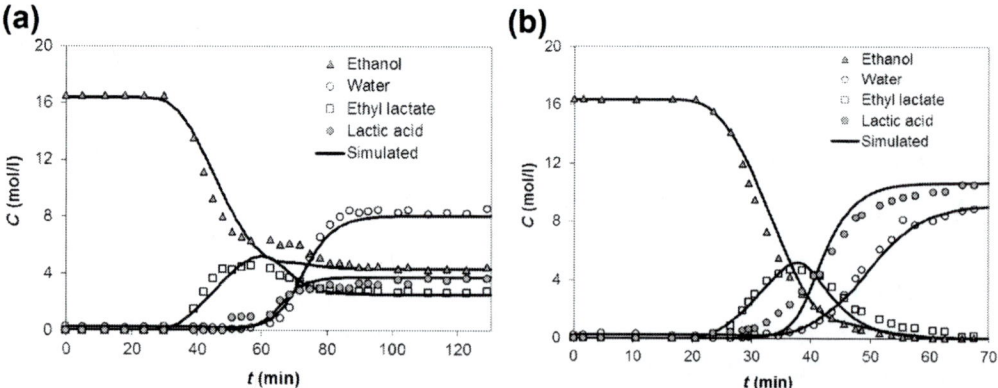

**Figure 8.13** Concentration histories at the outlet of the fixed-bed adsorptive reactor for production steps. (a) Run 1: column initially saturated with ethanol and then fed with a mixture of ethanol and lactic acid solution ($Q = 5$ ml/min; $T = 323.15$ K, $C_{La,F} = 5.88$ mol/l, $C_{Eth,F} = 6.60$ mol/l and $C_{W,F} = 5.87$ mol/l. (b) Run 2: column initially saturated with ethanol and then fed with lactic acid solution (86 wt.% in water); ($Q = 1.3$ ml/min; $T = 323.15$ K).

than water. This process continues until equilibrium is reached; the resin is completely saturated with ethanol and lactic acid solution, and the selective separation of ethyl lactate and water is no longer possible. The local compositions remain constant and steady state is achieved being the outlet stream composed of a reactive mixture at the equilibrium composition. Similarly, in run 2 (Figure 8.13(b)) lactic acid and water are adsorbed as they enter the column, displacing the initially adsorbed ethanol. Simultaneously, lactic acid reacts with ethanol in the solid phase until its complete depletion and the formed ethyl lactate is adsorbed, forming a dispersive front with ethanol once this is more adsorbed than ethyl lactate. The lactic acid solution adsorption continues until the saturation of the A15 resin, displacing the formed ethyl lactate. Finally, lactic acid and water exit the column, water being the last eluted component due to the higher affinity of A15 toward this compound.

Due to the simultaneous reaction and separation steps, in run 1, it was possible to obtain an ethyl lactate maximum concentration (at about 57 min) 56% higher than the equilibrium concentration, showing the potential of sorption–enhanced reaction technologies, as the FBR and the SMBR, for ethyl lactate synthesis.

For both experiments (runs 1 and 2), after steady state was achieved, the column was regenerated using ethanol to displace all the adsorbed species. The regeneration steps are presented in Figure 8.14, in which it can be observed that, as expected, ethyl lactate and lactic acid are quickly desorbed, whereas a large amount of ethanol is required to completely desorb water.

In both steps, production (Figure 8.13) and regeneration (Figure 8.14), the fixed–bed reactor mathematical model using the previously determined kinetic and adsorption data predicts well the experimental results. Therefore, this model can be extended to describe the dynamic behavior of the SMBR for the EL production.

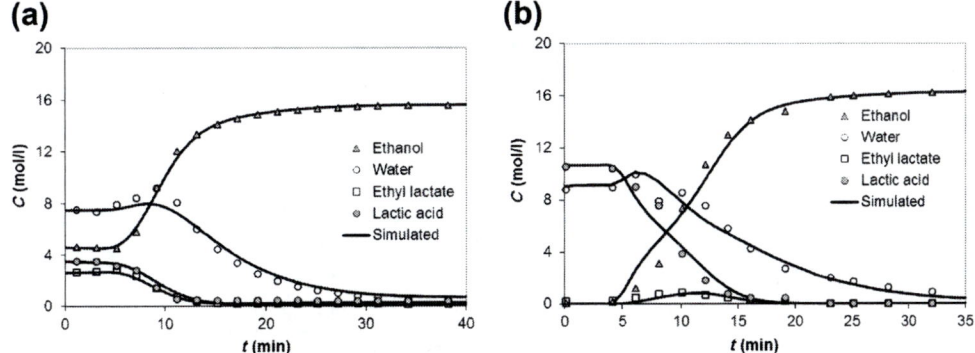

**Figure 8.14** Concentration histories at the outlet of the fixed-bed adsorptive reactor for regeneration steps with ethanol ($Q = 4.3$ml/min, $T = 323.15$ K). (a) Run 1. (b) Run 2.

### 8.6.1.3 Ethyl Lactate Production by SMBR

After the acquisition of the basic data (reaction kinetics and equilibrium and multicomponent adsorption isotherms) and using the methodology described in Section 8.2, which involves first the equilibrium theory and then the SMBR mathematical model, it was possible to define suitable SMBR operating conditions for the efficient production of EL. The SMBR experimental demonstration was performed to prove the SMBR suitability for the EL synthesis and to validate and (if necessary) refine the SMBR mathematical model.

#### 8.6.1.3.1 SMBR Experimental Validation

The technology demonstration of the SMBR process for the EL synthesis was implemented by using the Licosep 12-26 SMB pilot unit available at Laboratory of Separation and Reaction Engineering (LSRE) (Figure 1.13(a)), in which 12 columns Superformance SP 230 × 26 (length × internal diameter, mm), by Götec Labortechnik (Mühltal, Germany), packed with the A15 resin were connected. The operating temperature was 50 °C. The flow rates of feed and raffinate streams were $Q_F = 1.80$ ml/min and $Q_R = 8.80$ ml/min, whereas the desorbent and recycle flow rates were $Q_D = 27.40$ ml/min and $Q_{Rec} = 24.70$ ml/min. Lactic acid solution (85% in water) was introduced from the feed port and ethanol (used as reactant and desorbent) was fed through the desorbent stream. Ethyl lactate (the less-retained species) was collected in the raffinate stream and water (the most-adsorbed species) was obtained in the extract stream, both streams diluted in ethanol (desorbent). Two different switching times (2.9 and 3.0 min) and two different configurations (3-3-3-3 and 3-3-4-2) were tested (Pereira et al., 2009b). The experimentally obtained performance parameters as well as the values obtained by simulation, for two of the performed experiments, are presented in Table 8.7. In Figure 8.15, the corresponding CSS concentration profiles obtained experimentally at the middle of the switching time are shown and compared with the profiles predicted by the SMBR mathematical model.

**Table 8.7** SMBR Operating Conditions and Performance Parameters Obtained Experimentally (Bold) and by Simulation (Inside Brackets)

|  | Experiment 1 | Experiment 2 |
|---|---|---|
| $t^*$ (min) | 2.9 | 3 |
| Configuration | 3-3-4-2 | 3-3-4-2 |
| PUR (%) | **73.54** (79.80) | **75.17** (81.32) |
| PUX (%) | **95.22** (99.83) | **97.76** (99.85) |
| $X$ (%) | **99.91** (97.87) | **99.96** (98.00) |
| PR ($kg_{EL}\ L_{resin}^{-1}\ day^{-1}$) | **3.31** (3.68) | **3.36** (3.68) |
| DC ($L_{Eth}/kg_{EL}$) | **12.99** (11.70) | **12.77** (11.68) |

Adapted from Pereira et al. (2009b).

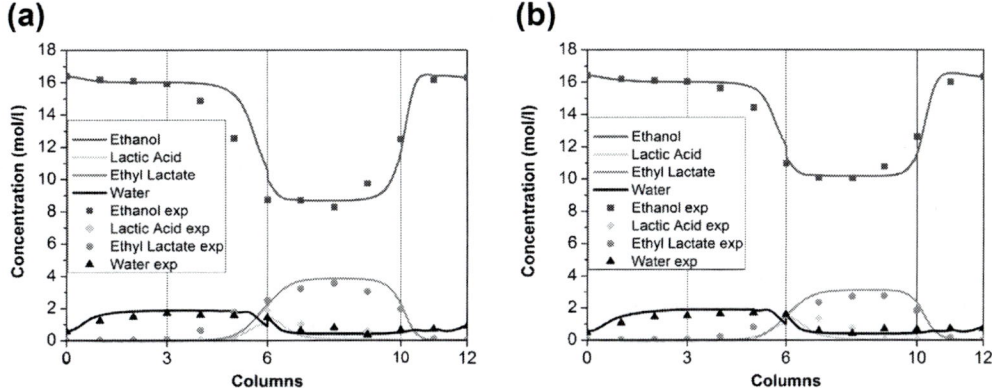

**Figure 8.15** Experimental and simulated concentration profiles in SMBR unit at the middle of switching time at cyclic steady state (13th cycle): (a) Experiment 1; (b) Experiment 2. *(Reprinted from Chemical Engineering Science 64, Pereira, C.S.M., Zabka, M., Silva, V.M.T.M., Rodrigues, A.E. A novel process for the ethyl lactate synthesis in a simulated moving bed reactor (SMBR), 3301–3310, Copyright (2009), with permission from Elsevier.)*

In these experiments, the ethyl lactate purity (in the raffinate stream) is low, which is due to the fact that both experiments were performed under conditions of incomplete resin regeneration in section 1 caused by the SMB Licosep unit limitations, in which the maximum allowable desorbent flow rate is 30 ml/min and, so, the used flow rate in section 1 was below the lower limit given by the equilibrium theory (see Section 8.2). Consequently, in these conditions, the adsorbed water was transported by the resin to sections 4 and 3, decreasing both ethyl lactate productivity (due to ethyl lactate hydrolyses producing ethanol and lactic acid) and ethyl lactate concentration in the raffinate stream. However, the lactic acid conversion is around 100%. As can be observed in Figure 8.15, lactic acid is almost completely consumed in sections 2 and 3, in which the esterification reaction takes place.

For experiment 2, the experimental composition of extract and raffinate streams collected for a whole cycle at the third, fifth, seventh, ninth, and 11th cycles, and the extract and raffinate concentration histories predicted by the SMBR model are shown in Figure 8.16.

As it can be observed, the SMBR mathematical model (described in Section 8.4) using the previously determined reaction rate expression (Eqn (8.62) with parameters given by Eqns (8.61) and (8.63–8.65)) and the multicomponent Langmuir adsorption isotherms (Eqn (8.69) with the parameters of Table 8.6) predicts reasonably well the experimental performance parameters (Table 8.7), the concentration profiles, (Figure 8.15) and the extract/raffinate histories (Figure 8.16).

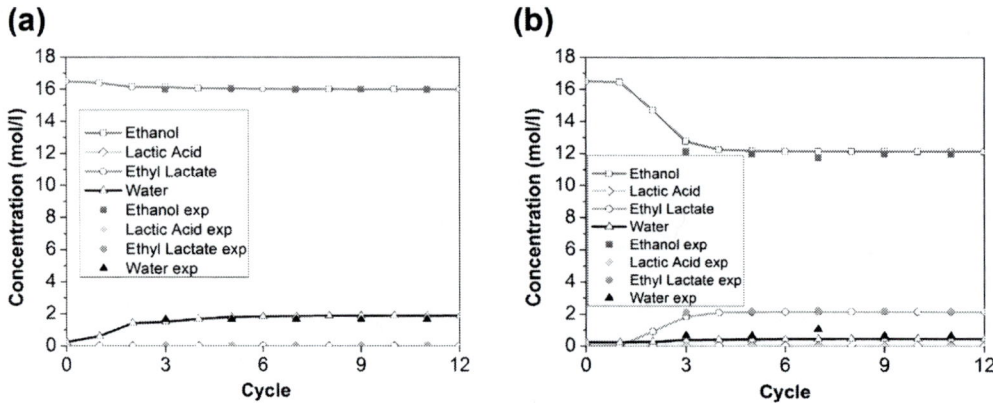

**Figure 8.16** Experimental and theoretical average concentration of all species (ethanol, lactic acid, ethyl lactate, and water) for the conditions of the experiment 2 in the extract (a) and raffinate (b) streams. *(Reprinted from Chemical Engineering Science 64, Pereira, C.S.M., Zabka, M., Silva, V.M.T.M., Rodrigues, A.E. A novel process for the ethyl lactate synthesis in a simulated moving bed reactor (SMBR), 3301–3310, Copyright (2009), with permission from Elsevier.)*

### 8.6.1.3.2 SMBR Simulated Results

To understand the behavior of the SMBR technology for the ethyl lactate production, the reactive/separation regions and/or the optimal operating points were determined for different operating conditions. The effects of feed composition, columns arrangement (SMBR configuration), and switching time were evaluated, keeping constant the remaining operating conditions.

The reference conditions used were: $T = 50\,°C$, $Q_D = 58.0$ ml/min, $Q_{Rec} = 27.0$ ml/min, feed concentration: lactic acid solution (85 wt.% in water) (evaluated in subsection Effect of Feed Composition), configuration = 3-3-4-2 (evaluated in section Effect of SMBR Configuration), and $t^* = 2.7$ min (evaluated in section Effect of the Switching Time).

The reactive/separation regions were determined from the average concentrations over a cycle obtained by the SMBR mathematical model at CSS setting a criterion of 95% for the conversion of lactic acid, and for the purities of extract and raffinate streams. The CSS SMBR model was successively solved for several values of $\gamma_2$ and $\gamma_3$, keeping the values of $\gamma_1$ (4.698) and $\gamma_4$ (1.492). The reactive/separation region is located within the region between the diagonal $\gamma_2 = \gamma_3$, the horizontal line $\gamma_3 = 1.492$ and $\gamma_3$ axis. The $\gamma_3$ value must be higher than $\gamma_2$, because the diagonal $\gamma_2 = \gamma_3$ corresponds to zero feed flow rate. The algorithm used in the construction of the reactive/separation region began by setting a feed flow rate of 0.01 ml/min and the value of $\gamma_2$ equal to 1.492. Then, the feed flow rate was kept constant and the $\gamma_2$ values were gradually increased. The value of $\gamma_3$ was calculated from the mass balance in the feed node for each value of $\gamma_2$. For each set of $\gamma_2$ and $\gamma_3$, the conversion and the purities of extract and raffinate streams were calculated and the values that satisfy the criterion of 95% were selected to build the reaction/separation region. After this set of simulations, the feed flow rate was increased and the same procedure was repeated. The simulations procedure ends when the maximum value of feed flow rate that gives the required product purities and conversion is achieved (vertex of the reaction/separation region which corresponds to the optimal operating point). Above that feed flow rate value the requirements cannot be fulfilled for any $\gamma_2$, $\gamma_3$ pair.

*Effect of Reaction Kinetics and Equilibrium*

The reactive/separation region considering the conditions of the reference case was determined and compared with the separation region calculated considering the separation of ethyl lactate and water with feed composition: $C_W = 12.17$ mol/l and $C_{EL} = 6.47$ mol/l, which corresponds to 100% conversion of lactic acid solution. These regions are shown in Figure 8.17, in which can be clearly observed that under these conditions the SMBR process for the ethyl lactate production is limited by the reaction. This result was expected because the lactic acid esterification reaction is slow and limited by the thermodynamic equilibrium, and, as mentioned in Section 8.2, these parameters strongly affect the SMBR performance (see Figure 8.3).

The reaction kinetics as well as the equilibrium constant (endothermic reaction) can be improved by increasing the system temperature, but this change will negatively affect the separation of the products: if ethyl lactate adsorption is more affected than water adsorption, the productivity will be enhanced, whereas the desorbent consumption will increase. In the opposite case, the desorbent consumption will decrease, but the productivity will be negatively compromised. Therefore, all these effects on the process performance must be taken into account and the final decision on the best conditions to be used should be based on an economical assessment.

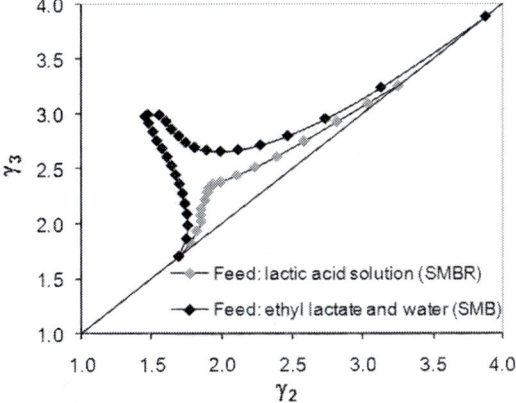

**Figure 8.17** Reactive/separation and separation region ($\gamma_1 = 4.698$; $\gamma_4 = 1.492$; 95% purity). *(Reprinted from Chemical Engineering Science 64, Pereira, C.S.M., Zabka, M., Silva, V.M.T.M., Rodrigues, A.E. A novel process for the ethyl lactate synthesis in a simulated moving bed reactor (SMBR), 3301–3310, Copyright (2009), with permission from Elsevier.)*

Other options are to use new materials with better characteristics for both reaction and adsorption as carbon-based catalysts, or to use a mixture of two solids (catalyst/adsorbent) that increases the catalytic properties without losing efficiency on the separation of the products.

*Effect of Feed Composition*

The reactive/separation regions and the optimal operating points for different feed compositions (varied by adding pure ethanol to the lactic acid solution) are shown in Figures 8.18 and 8.19, respectively.

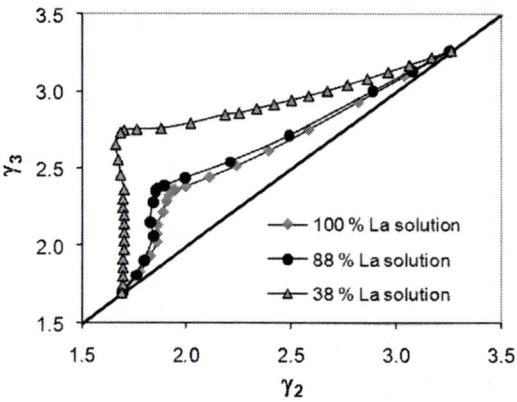

**Figure 8.18** Reactive/separation region for different molar fractions of lactic acid solution in ethanol ($\gamma_1 = 4.698$; $\gamma_4 = 1.492$; 95% purity). *(Reprinted from Chemical Engineering Science 64, Pereira, C.S.M., Zabka, M., Silva, V.M.T.M., Rodrigues, A.E. A novel process for the ethyl lactate synthesis in a simulated moving bed reactor (SMBR), 3301–3310, Copyright (2009), with permission from Elsevier.)*

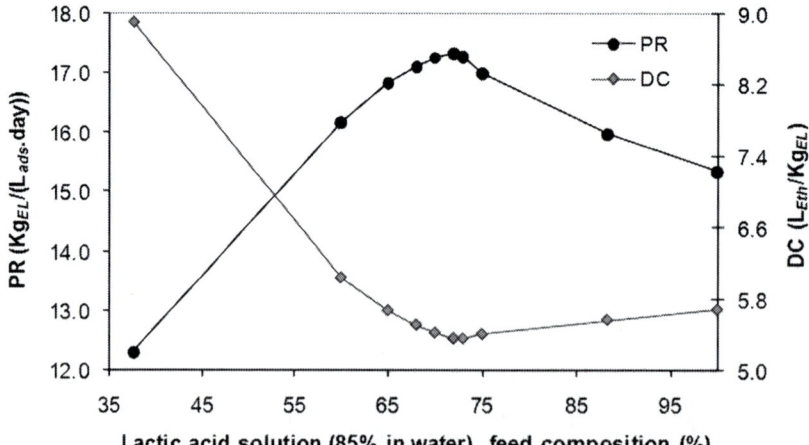

**Figure 8.19** SMBR performance for the optimal operating points as a function of the lactic acid solution molar fraction in the feed ($\gamma_1 = 4.698$; $\gamma_4 = 1.492$; 95% purity). *(Reprinted from Chemical Engineering Science 64, Pereira, C.S.M., Zabka, M., Silva, V.M.T.M., Rodrigues, A.E. A novel process for the ethyl lactate synthesis in a simulated moving bed reactor (SMBR), 3301–3310, Copyright (2009), with permission from Elsevier.)*

From this study, it was possible to infer that for $\gamma_1 = 4.698$, $\gamma_4 = 1.492$, and SMBR configuration 3-3-4-2, the feed molar composition of 72% of lactic acid leads to the highest productivity (17.33 kg$_{EL}$/(L$_{ads}$ day)) and lowest desorbent consumption (5.36 L$_{Eth}$/kg$_{EL}$).

Two different scenarios were noticed: (1) till the optimum composition (72% lactic acid molar fraction in the feed), lactic acid is the limiting reactant and the increase of its concentration improves both productivity and desorbent consumption; (2) above this optimum value, ethanol becomes the limiting reactant due to a lack of this species in the reaction sections (sections 2 and 3), lactic acid is not fully converted into the products, which leads to the contamination of preferably the raffinate stream (due to the smaller resin selectivity between lactic acid and ethyl lactate (3.4) than the one between water and lactic acid (7.2)) and compromises the SMBR performance by decreasing the productivity and increasing the desorbent consumption.

*Effect of SMBR Configuration*

Three different columns arrangement were tested: 3-3-3-3, 3-3-4-2, and 3-3-5-1. For each case, the reactive/separation regions and the optimal performance parameters are presented in Figure 8.20 and Table 8.8, respectively. As can be observed, the influence of these different configurations into the SMBR process performance is negligible. This is caused by the fact that, as discussed in the previous subsection, ethanol is lacking in the reaction zone and so the fed lactic acid is not fully converted, even when increasing section 3, contaminating in this way the raffinate stream. However, when looking to the concentration profiles in Figure 8.21, it is possible to conclude that just one column in section 4 is enough to guarantee the complete regeneration of ethanol before being recycled to section 1.

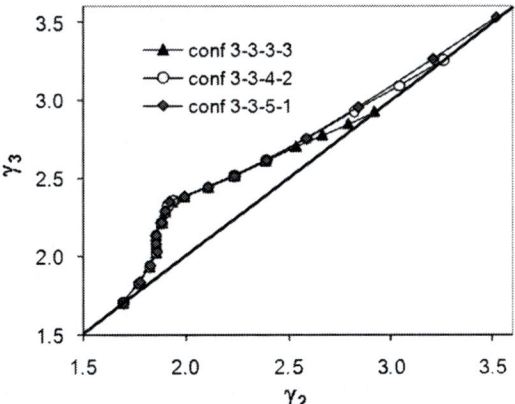

**Figure 8.20** Reactive/separation regions for different columns arrangement ($\gamma_1 = 4.698$; $\gamma_4 = 1.492$; 95% purity). *(Reprinted from Chemical Engineering Science 64, Pereira, C.S.M., Zabka, M., Silva, V.M.T.M., Rodrigues, A.E. A novel process for the ethyl lactate synthesis in a simulated moving bed reactor (SMBR), 3301–3310, Copyright (2009), with permission from Elsevier.)*

**Table 8.8** SMBR Performance Parameters for the Optimal Operation Points for Different SMBR Configurations (Pereira et al., 2009b)

| Columns arrangement | 3-3-3-3 | 3-3-4-2 | 3-3-5-1 |
|---|---|---|---|
| Raffinate productivity ($kg_{EL}/L_{A15}/day$) | 14.97 | 15.33 | 15.36 |
| Desorbent consumption ($L_{Eth}/kg_{EL}$) | 5.83 | 5.68 | 5.67 |

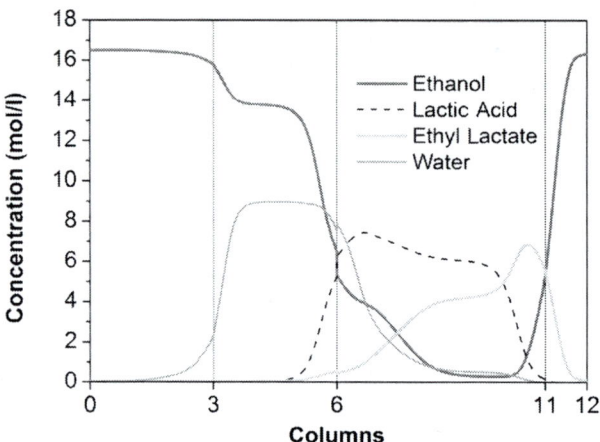

**Figure 8.21** Concentration profiles at the middle of the switching time at cyclic steady state for the optimal operating point of the SMBR configuration 3-3-5-1. *(Reprinted from Chemical Engineering Science 64, Pereira, C.S.M., Zabka, M., Silva, V.M.T.M., Rodrigues, A.E. A novel process for the ethyl lactate synthesis in a simulated moving bed reactor (SMBR), 3301–3310, Copyright (2009), with permission from Elsevier.)*

The lactic acid esterification reaction is very slow, even at 50 °C, but to allow a complete lactic acid conversion it is required not only to increase the reaction zone (sections 2 and 3), but also to feed to the unit a lactic acid concentration that does not lead to a lack of ethanol in the reaction zone.

*Effect of the Switching Time*

The switching time effect on the SMBR performance was evaluated by the analysis of the optimal operating points determined by two ways: (1) varying the switching time and keeping constant the remaining operating conditions (reference case: $T = 50\,°C$, $Q_D = 58.0$ ml/min, $Q_{Rec} = 27.0$ ml/min, feed concentration: lactic acid solution (85 wt.% in water) and configuration = 3-3-4-2), which will affect both $\gamma_1$ and $\gamma_4$; (2) varying the switching time and the desorbent and recycle flow rates to keep constant $\gamma_1$ (4.698) and $\gamma_4$ (1.492) values. The obtained results are shown in Figures 8.22 and 8.23. As can be observed, in the first case (Figure 8.22), the performance of the SMBR improves, as the switching time decreases until 2.1 min; the use of a switching time of 2.1 min instead of 2.8 min increases the productivity in 22% (maximum of 18.06 $kg_{EL}/(L_{ads}$ day)) and decreases the desorbent consumption in 20% (minimum of

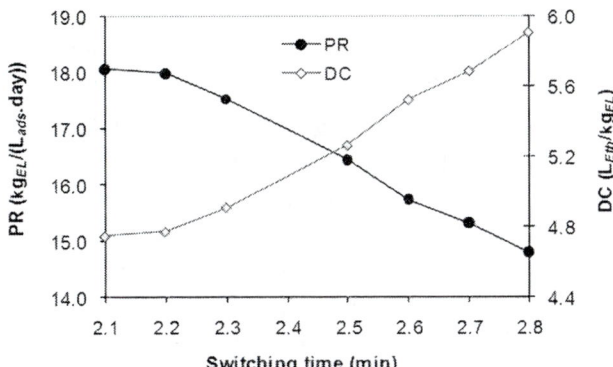

**Figure 8.22** SMBR performance parameters for the optimal operating points as a function of the switching time. Operating conditions: $T = 50\,°C$, $Q_D = 58.0$ ml/min, $Q_{Rec} = 27.0$ ml/min, feed concentration: lactic acid solution (85 wt.% in water) and configuration = 3-3-4-2. *(Reprinted from Chemical Engineering Science 64, Pereira, C.S.M., Zabka, M., Silva, V.M.T.M., Rodrigues, A.E. A novel process for the ethyl lactate synthesis in a simulated moving bed reactor (SMBR), 3301–3310, Copyright (2009), with permission from Elsevier.)*

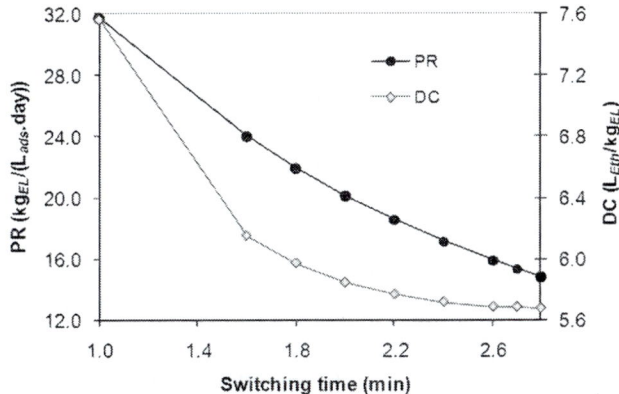

**Figure 8.23** SMBR performance parameters for the optimal operating points as a function of the switching time, keeping constant $\gamma_1$ (4.698) and $\gamma_4$ (1.492) values. *(Reprinted from Chemical Engineering Science 64, Pereira, C.S.M., Zabka, M., Silva, V.M.T.M., Rodrigues, A.E. A novel process for the ethyl lactate synthesis in a simulated moving bed reactor (SMBR), 3301–3310, Copyright (2009), with permission from Elsevier.)*

4.75 $L_{Eth}/kg_{EL}$). Below this optimal switching time value (2.1 min), the minimum purity of 95% is no longer achieved, caused by the violation of $\gamma_1$ restriction, which implies that the solid is not completely regenerated in section 1, and therefore, water is transported with the solid contaminating by this way the raffinate stream, leading to a low purity. In the second case (constant $\gamma_1$ (4.698) and $\gamma_4$ (1.492) values, Figure 8.23), the productivity also increases by decreasing the switching time, but the desorbent consumption is negatively affected because, to keep $\gamma_1$ constant, the desorbent flow rate must be increased when decreasing the switching time. The reduction, in the switching time, from 2.8 min to 1.0 min, leads to an increase in productivity of 114% (31.7 $kg_{EL}$/ ($L_{ads}$ day)), and also to an increase in the desorbent consumption of 33.3% (7.6 $L_{Eth}$/$kg_{EL}$).

Due to the contradictory effects of the switching time into the SMBR performance, the selection of the optimal switching time value requires a deeper study of the whole production plant, including the downstream separation units, to check if the productivity enhancement compensates the penalty on the desorbent consumption.

In summary, the SMBR theoretical assessment allows us to realize the complexity of this type of process, in which the operating parameters all strongly interact in terms of their impact on the performance of the SMBR (extract and raffinate purities, conversion, desorbent consumption, and ethyl lactate productivity). Sometimes the effects caused on the SMBR performance are even contradictory (for instance, the productivity is enhanced but the desorbent consumption is increased), as is the case of the switching time.

## 8.7 OTHER SMBR APPLICATIONS: ACETALS PRODUCTION

The SMBR production of acetals, namely, diethyl acetal (DEE) (Silva and Rodrigues, 2005), dimethyl acetal (DME) (Pereira et al., 2008a), dibutyl acetal (DBE) (Graça et al., 2011), and glycerol ethyl acetal (GEA) (Faria, 2014), was also evaluated in our laboratory; indeed, DEE was the first studied system that uses an SMB-based technology for the simultaneous production and separation of acetal by means of ion-exchange acid resins (Rodrigues and Silva, 2004).

The methodology followed for the SMBR process design was similar to the one described in the previous subsections, but of course taking into account the particularities of each system. Indeed, for the synthesis of glycerol ethyl acetal through the direct acetalization of glycerol with acetaldehyde, an extra step was required: the selection of an appropriate solvent to be used as desorbent. Usually one of the reactants is used as desorbent; however, for the GEA case this is not possible because glycerol is highly viscous (533 cP at 30 °C), which would lead to a large pressure drop, and acetaldehyde oligomerizes in an acidic medium. Therefore, a two-stage methodology was developed and applied for the selection of the most suitable solvent for the production of this acetal by reactive–adsorptive processes, as the SMBR, which involves (Faria et al., 2013): (1) a theoretical rating procedure aiming the identification of common solvents used in similar processes based on an extensive bibliographic research of their properties and applications; (2) experimental tests for a shorter list of potential solvents (selected from the previous step) to assess its miscibility, adsorption, and reaction properties with the species present in the GEA synthesis system. Dimethyl sulfoxide (DMSO) was identified as the solvent with the highest potential benefit to be used as desorbent in the GEA SMBR production process. Then, the basic data required for the implementation of an SMBR-based process, the reaction thermodynamic and kinetic data, and adsorption data were obtained following the previously presented methodologies (as for the ethyl lactate example) but in the presence of the selected solvent (DMSO) (Faria et al., 2013, 2014).

In Table 8.9, a summary of the obtained results (best performance parameters) for DME, DEE, and DBE is presented.

The SMBR outlet streams are always diluted in the desorbent used, and so, to recover the products with the desired purity and to recycle the desorbent back to the reactor, it is required to couple the SMBR with separation units. This integration was evaluated for the production of DEE (Silva et al., 2009). Typically, in the chemical industry, distillation is used to perform separations; however, in the DEE synthesis by SMBR technology, water and DEE are collected in the extract and raffinate streams, respectively, both diluted in ethanol (desorbent), and the recovery of ethanol by distillation from each one of these streams leads to the formation of azeotropes (azeotropes involved in DEE synthesis: ethanol/water, ethanol/DEE, and DEE/water binary azeotropes, and ethanol/water/DEE ternary azeotrope). The use of azeotropic distillation to break azeotropes is expensive, which will be reflected on the overall costs of the process. To overcome this issue, our group proposed an innovative methodology that consists of (1) recycling DEE/ethanol azeotrope formed in the distillation of the raffinate stream back to the SMBR unit through the feed port and (2) using distillation coupled with pervaporation to recover ethanol from the extract stream comprising water/ethanol (see Figure 8.24).

**Table 8.9** Optimal Performance Parameters for the Referred Operating Conditions

|  | Operating Conditions | Productivity ($kg_{acetal}/(L_{A15}$ day)) | Desorbent Consumption ($L_{Desorbent}/kg_{acetal}$) |
|---|---|---|---|
| **DME** | $T = 20\ °C$; Configuration: trun -3-3-3-3; $Q_D = 25$ ml/min; $Q_{Rec} = 25$ ml/min; $t^* = 3.1$ min; $x_{acet,F} = 0.60$; $x_{Meth,,F} = 0.40$. | 5.21 | 7.10 |
| **DEE** | $T = 10\ °C$; Configuration: 3-3-3-3; $Q_D = 50$ ml/min; $Q_{Rec} = 20$ ml/min; $t^* = 3.9$ min; $x_{acet,F} = 0.51$; $x_{Eth,,F} = 0.49$. | 28.82 | 2.34 |
| **DBE** | $T = 25\ °C$; Configuration: 3-3-3-3; $Q_D = 228$ ml/min; $Q_{Rec} = 19$ ml/min; $t^* = 3.5$ min; $x_{acet,F} = 0.30$; $x_{ButF} = 0.50$. | 55.29 | 6.96 |

DME: dimethyl acetal (Pereira et al., 2008a); DEE: diethyl acetal (Silva and Rodrigues, 2005); DBE: dibutyl acetal (Graça et al., 2011).

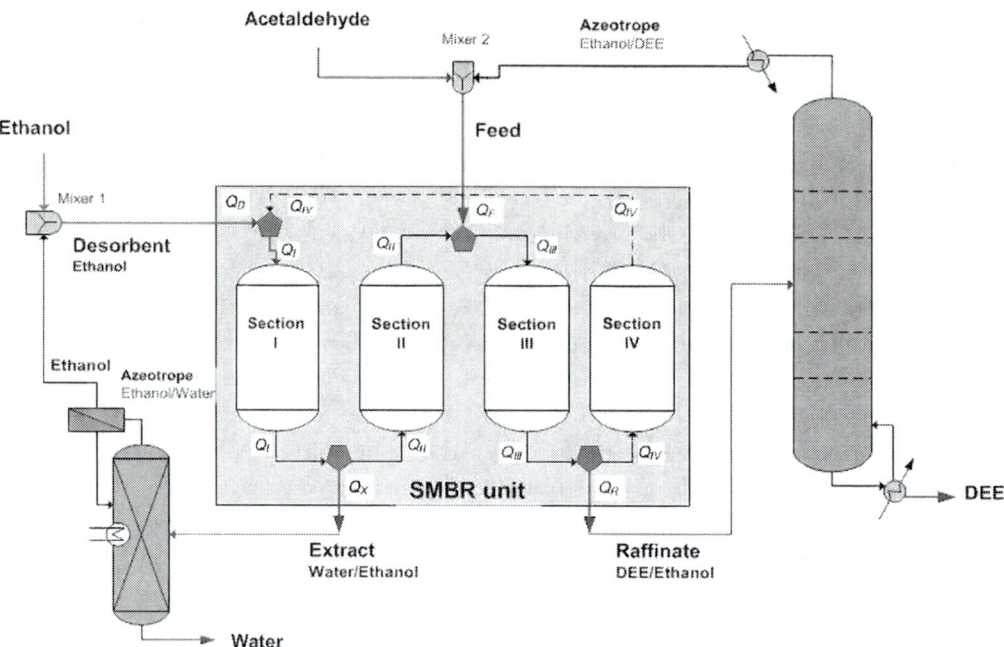

**Figure 8.24** Process scheme for the production of DEE using SMBR technology integrated with distillation and pervaporation units. *(From Chemical Engineering and Technology 35, Rodrigues, A.E., Pereira, C.S.M., Santos, J.C. Chromatographic Reactors, Copyright (2012) WILEY-VCH Verlag GmbH & Co. KGaA, Weinheim, Reprinted by permission of John Wiley & Sons, Inc.)*

Other SMBR applications can be found in the open literature, for example, the production of methyl acetate (Lode et al., 2003), fructose (Zhang et al., 2004; Borges da Silva et al., 2006), and MTBE (Zhang et al., 2001), among others.

Wrapping up, the production of all studied compounds by SMBR exhibits high conversion (far beyond the equilibrium value) and high productivity at moderate temperatures, but also significant desorbent consumption due to the use of a highly selective resin toward water (by-product in all the studied systems) that implies diluted outlet streams and consequently high costs associated with the downstream separation units. This motivated the development of a new hybrid process combining SMBR with membranes, which is presented in the next chapter (Chapter 9).

## 8.8  SMBR CONCLUDING REMARKS

The SMBR is a multifunctional reactor that combines continuous countercurrent chromatography with reaction, which avoids constraints arising from thermodynamic equilibrium and may lead to more efficient as well as more economical processes.

Through a convenient methodology, considering the detailed study of the reaction kinetics and adsorption equilibrium, it is possible to model, design, simulate, and experimentally demonstrate the viability of such technology (presented here for the ethyl lactate production).

The SMBR exhibits high conversion (far beyond equilibrium value), high productivity, and operates at moderate temperatures. However, it requires further separation units for desorbent recovery from both extract and raffinate streams. Different methodologies to minimize the desorbent consumption should be evaluated, as the use of different materials as catalyst/adsorbent with less affinity toward water. Contrarily to the SMB, the SMBR is not a mature technology; till the moment, as far as our knowledge goes, no known commercial application uses the SMBR.

## NOMENCLATURE

| | |
|---|---|
| $a$ | Liquid-phase activity |
| $Da$ | Damköhler number |
| $C$ | Liquid-phase concentration (mol/m$^3$) |
| $\overline{C_p}$ | Average liquid-phase concentration inside the particle (mol/m$^3$) |
| $C_T$ | Total liquid-phase concentration (mol/m$^3$) |
| $D_{ax}$ | Axial dispersion coefficient (m$^2$/min) |
| $DC$ | Desorbent consumption (m$^3$/kg) |
| $d_p$ | Particle diameter (m) |
| $K$ | Adsorption equilibrium parameter Langmuir isotherm (m$^3$/mol) |
| $k_c$ | Reaction kinetic constant (mol/g$^{-1}$.s$^{-1}$) |
| $k_e$ | External mass-transfer coefficient (m/s) |
| $K_{eq}$ | Equilibrium reaction constant |
| $k_i$ | Internal mass-transfer coefficient (m/s) |
| $k_L$ | Global mass-transfer coefficient (m/s) |
| $k_p$ | Mass-transfer coefficient (s$^{-1}$) |
| $K_S$ | Adsorption constant of Langmuir—Hinshelwood reaction rate |
| $L_c$ | Column length (m) |
| $M_{W,C}$ | Molecular weight of component C (kg/mol) |
| $n$ | Total number of components |
| $PR$ | Raffinate productivity (kg$_C$/(m$^3_{resin}$. day)) |
| $PUR$ | Raffinate purity (%) |
| $PUX$ | Extract purity (%) |
| $\overline{q}$ | Average solid-phase concentration (mol/m$^3$) |
| $Q$ | Volumetric flow rate (m$^3$/s) |
| $Q_{Ads}$ | Adsorption monolayer capacity (mol/m$^3$) |
| $Q_V$ | Volumetric monolayer capacity (m$^3$/m$^3_{solid}$) |
| $r$ | Rate of reaction (mol/kg$^{-1}$.s$^{-1}$) |
| $R_p$ | Particle radius (m) |
| $t$ | Time variable (s) |
| $t^*$ | Switching time (s) |
| $u_s$ | Solid interstitial velocity (m/s) |
| $v$ | Fluid interstitial velocity (m/s) |

| | |
|---|---|
| $V_{mol,i}$ | Molar volume of species i (m$^3$/mol) |
| $x$ | Dimensionless axial position |
| $X$ | Conversion |
| $y_i$ | Molar fraction of component $i$ |
| $z$ | Axial coordinate |

## Greek Letters

| | |
|---|---|
| $\alpha_p$ | Number of mass-transfer units |
| $\gamma$ | Dimensionless velocity ratio |
| $\gamma^*$ | Activity coefficient |
| $\varepsilon$ | Bulk porosity |
| $\varepsilon_P$ | Particle porosity |
| $\upsilon$ | Solid/liquid ratio |
| $\sigma$ | Stoichiometric coefficient |
| $\rho_b$ | Bulk density (kg/m$^3$) |
| $\rho_p$ | Particle density (kg/m$^3$) |
| $\mu$ | Viscosity (cP) |
| $\eta$ | Effectiveness factor of the catalyst |

## Subscripts

| | |
|---|---|
| $D$ | Desorbent |
| $F$ | Feed |
| $i$ | Component $i$ ($i = A, B, C, D$) |
| $j$ | Section ($j = 1, 2, 3, 4$) |
| $k$ | Column |
| $P$ | Product |
| $R$ | Raffinate |
| $Rec$ | Recycle |
| $X$ | Extract |

## REFERENCES

Agreda, V.H., Partin, L.R., 1984. Reactive Distillation Process for the Production of Methyl Acetate. U.S. Patent 4435595.

Borges da Silva, E.A., Ulson de Souza, A.A., de Souza, S.G.U., Rodrigues, A.E., 2006. Analysis of the high-fructose syrup production using reactive SMB technology. Chem. Eng. J. 118, 167–181.

Catchpoole, H.J., Andrew Shalliker, R., Dennis, G.R., Guiochon, G., 2006. Visualising the onset of viscous fingering in chromatography columns. J. Chromatogr. A 1117, 137–145.

Faria, R.P.V., 2014. Glycerol Valorisation as Biofuels: Glycerol Acetals Production by Simulated Moving Bed Reactor (Ph.D. thesis). University of Porto.

Faria, R.P.V., Pereira, C.S.M., Silva, V.M.T.M., Loureiro, J.M., Rodrigues, A.E., 2013. Glycerol valorisation as biofuels: selection of a suitable solvent for an innovative process for the synthesis of GEA. Chem. Eng. J. 233, 159–167.

Faria, R.P.V., Pereira, C.S.M., Silva, V.M.T.M., Loureiro, J.M., Rodrigues, A.E., 2014. Sorption enhanced reactive process for the synthesis of glycerol ethyl acetal. Chem. Eng. J. 258, 229–239.

Fricke, J., Meurer, M., Dreisörner, J., Schmidt-Traub, H., 1999. Effect of process parameters on the performance of a simulated moving bed chromatographic reactor. Chem. Eng. Sci. 54, 1487–1492.

Fricke, J., Schmidt-Traub, H., 2003. A new method supporting the design of simulated moving bed chromatographic reactors. Chem. Eng. Process. 42, 237–248.

Graça, N.S., Pais, L.S., Silva, V.M.T.M., Rodrigues, A.E., 2011. Analysis of the synthesis of 1,1-dibutoxyethane in a simulated moving-bed adsorptive reactor. Chem. Eng. Process. 50, 1214–1225.

Krishna, R., 2002. Reactive separations: more ways to skin a cat. Chem. Eng. Sci. 57, 1491–1504.

Lode, F., Francesconi, G., Mazzotti, M., Morbidelli, M., 2003. Synthesis of methylacetate in a simulated moving-bed reactor: experiments and modeling. AIChE J. 49, 1516–1524.

Mallmann, T., Burris, B.D., Ma, Z., Wang, N.H.L., 1998. Standing wave design of nonlinear SMB systems for fructose purification. AIChE J. 44, 2628–2646.

Mazzotti, M., Neri, B., Gelosa, D., Kruglov, A., Morbidelli, M., 1997. Kinetics of liquid-phase esterification catalyzed by acidic resins. Ind. Eng. Chem. Res. 36, 3–10.

Minceva, M., Rodrigues, A.E., 2005. Simulated moving-bed reactor: reactive-separation regions. AIChE J. 51, 2737–2751.

Minceva, M., Silva, V.M.T., Rodrigues, A.E., 2005. Analytical solution for reactive simulated moving bed in the presence of mass transfer resistance. Ind. Eng. Chem. Res. 44, 5246–5255.

Pereira, C.S.M., Gomes, P.S., Gandi, G.K., Silva, V.M.T.M., Rodrigues, A.E., 2008a. Multifunctional reactor for the synthesis of dimethylacetal. Ind. Eng. Chem. Res. 47, 3515–3524.

Pereira, C.S.M., Pinho, S.P., Silva, V.M.T.M., Rodrigues, A.E., 2008b. Thermodynamic equilibrium and reaction kinetics for the esterification of lactic acid with ethanol catalyzed by acid ion exchange resin. Ind. Eng. Chem. Res. 47, 1453–1463.

Pereira, C.S.M., Silva, V.M.T.M., Rodrigues, A.E., 2009a. Fixed bed adsorptive reactor for ethyl lactate synthesis: experiments, modelling, and simulation. Sep. Sci. Technol. 44, 2721–2749.

Pereira, C.S.M., Silva, V.M.T.M., Rodrigues, A.E., 2011. Ethyl lactate as a solvent: properties, applications and production processes - a review. Green Chem. 13, 2658–2671.

Pereira, C.S.M., Zabka, M., Silva, V.M.T.M., Rodrigues, A.E., 2009b. A novel process for the ethyl lactate synthesis in a simulated moving bed reactor (SMBR). Chem. Eng. Sci. 64, 3301–3310.

Pöpken, T., Götze, L., Gmehling, J., 2000. Reaction kinetics and chemical equilibrium of homogeneously and heterogeneously catalyzed acetic acid esterification with methanol and methyl acetate hydrolysis. Ind. Eng. Chem. Res. 39, 2601–2611.

Rodrigues, A.E., Silva, V.M.T.M., 2004. Industrial process for acetals production in a simulated moving bed reactor. PT Patent 103123; WO Patent 2005/113476A1; US Patent 2008/0287714; EP Patent 1748974, 2004.

Sanz, M.T., Murga, R., Beltrán, S., Cabezas, J.L., Coca, J., 2002. Autocatalyzed and ion-exchange-resin-catalyzed esterification kinetics of lactic acid with methanol. Ind. Eng. Chem. Res. 41, 512–517.

Silva, V.M.T.M., Rodrigues, A.E., 2002. Dynamics of a fixed-bed adsorptive reactor for synthesis of diethylacetal. AIChE J. 48, 625–634.

Silva, V.M.T.M., Rodrigues, A.E., 2005. Novel process for diethylacetal synthesis. AIChE J. 51, 2752–2768.

Silva, V.T., Silva, R., Rodrigues, A.E., 2009. Green diesel additive synthesis: elimination of azeotropic distillation by coupling simulated moving bed reactor with solvent recovery units. 2009 AIChE Annual Meeting, Nashville.

Stankiewicz, A.I., Moulijn, J.A., 2000. Process intensification: transforming chemical engineering. Chem. Eng. Prog. 96, 22–33.

Taylor, R., Krishna, R., 2000. Modelling reactive distillation. Chem. Eng. Sci. 55, 5183–5229.

Zhang, Y., Hidajat, K., Ray, A.K., 2004. Optimal design and operation of SMB bioreactor: production of high fructose syrup by isomerization of glucose. Biochem. Eng. J. 21, 111–121.

Zhang, Z., Hidajat, K., Ray, A.K., 2001. Application of simulated countercurrent moving-bed chromatographic reactor for MTBE synthesis. Ind. Eng. Chem. Res. 40, 5305–5316.

# CHAPTER 9

# Process Reintensification: PermSMBR

## 9.1 COMBINING THE SIMULATED MOVING BED REACTOR WITH PERMEABLE MEMBRANES: THE CONCEPT OF PermSMBR

The simulated moving bed membrane reactor (PermSMBR) is a newly developed technology that combines the simulated moving bed reactor (SMBR) with membranes (Silva et al., 2009). Its principle of operation is very similar to that of the SMBR technology; however, in the PermSMBR, each SMBR column is replaced by a column containing a set of tubular membranes (integrated PermSMBR) (Figure 9.1) or each SMBR column is followed by a membrane module (coupled PermSMBR) (Figure 9.2). In both integrated and coupled configurations the membranes applied are permeable to one of the reaction products. Traditionally, as in the SMBR, there are two liquid inlet streams (the feed and the desorbent) and two liquid outlet streams (the extract and the raffinate), but the countercurrent solid movement is simulated by a synchronous shift of these streams by one column (integrated PermSMBR) or a column plus a membrane module (coupled PermSMBR) in the direction of the fluid, at regular time intervals (switching time). Moreover, an additional stream is collected (permeate) that combines all the flows removed through the membranes and that comprises the permeative species.

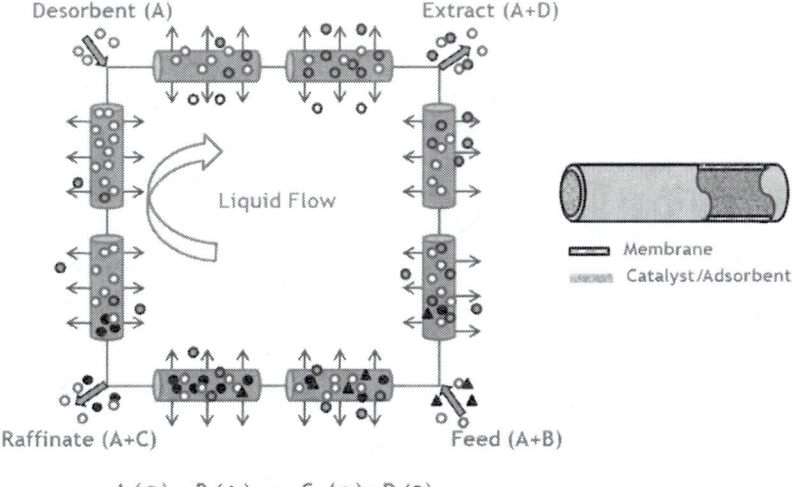

**Figure 9.1** Schematic representation of the integrated PermSMBR unit with four sections and two columns per section considering a reaction of type A + B ↔ C + D, where C is the least-adsorbed product and D is the most-adsorbed product and the one for which the membranes are selective.

*Simulated Moving Bed Technology*
ISBN 978-0-12-802024-1

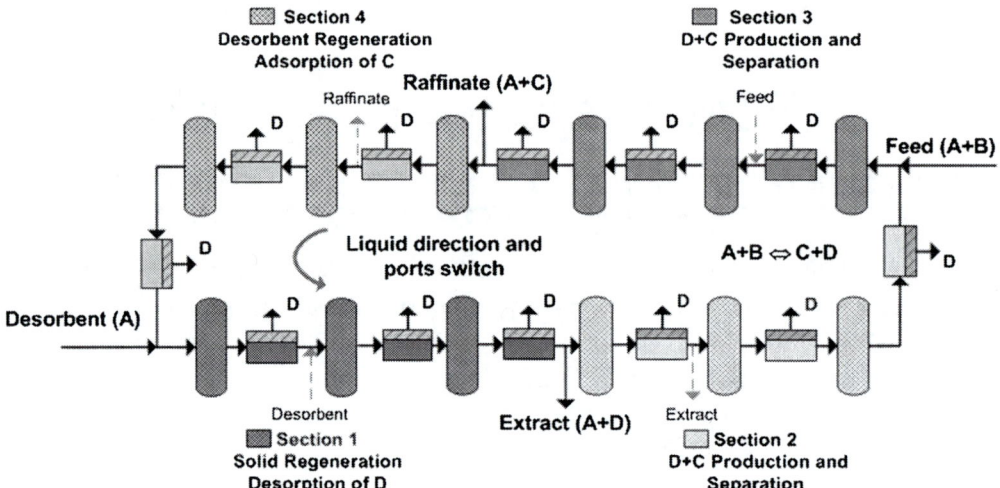

**Figure 9.2** Schematic representation of the coupled PermSMBR unit with four sections and three columns plus three membrane modules per section, considering a reaction of type A + B ↔ C + D, where C is the least-adsorbed product and D is the most-adsorbed product and the one for which the membranes are selective. The dashed arrows represent the port switches.

The typical PermSMBR also contains four sections defined by the liquid inlet/outlet streams, as represented in Figure 9.2, and having the same functions as those of the SMBR described in the previous chapter (Chapter 8, Section 8.1.2). Because in all sections one of the products is continuously removed by the membranes, both solid/desorbent regeneration (sections I and IV, respectively) and reaction conversion (sections II and III) are improved compared with the SMBR.

The PermSMBR can have different numbers of sections, depending on the total number of streams fed/removed from the unit, without accounting for the permeate stream. The possible configurations are conveniently addressed in the PermSMBR patent (Silva et al., 2009). Among all of them, a particularly interesting configuration for the production of compounds involving equilibrium-limited reactions of the type A + B ↔ C + D, is the PermSMBR with just three sections (called PermSMBR-3s), with which, depending on the product for which the membrane is selective, the most-adsorbed or the least-retained product, the extract or the raffinate stream is eliminated, respectively.

In the PermSMBR-3s, in which the membranes are selective for the most-adsorbed product and therefore the extract stream is eliminated, the reactants are continuously fed between sections I (located between the desorbent and the feed ports) and II (located between the feed and the raffinate ports), where the reaction takes place and the formed products are separated. The least-retained product is collected at the raffinate stream, which is positioned at the outlet of section II, whereas the most-adsorbed product is removed through the membranes. The desorbent is continuously fed into section I and it is regenerated in section III, the region between the raffinate and the desorbent streams, before being recycled to section I. In this case, it is not mandatory to completely regenerate the adsorbent in section I before going to section III, because the most-adsorbed product can also be removed at the end of section III through the membranes, avoiding the contamination of the raffinate stream. On the other hand, when the membranes are selective for the least-adsorbed product and the raffinate stream is eliminated, the reactants are fed between sections II (now located between the extract and the feed ports) and III (located between the feed and the desorbent ports), where the reaction takes place and the products are separated as they are formed. The product is collected at the outlet of section I in the extract stream, whereas the least-adsorbed product is collected at the permeate streams. Once again, the desorbent is introduced into section I, and it is regenerated in section III, before being recycled to section I. In this operation mode, it is not imperative to completely regenerate the desorbent on section III, because the less-retained component can also be removed at the beginning of section I, through the selective membranes, without contamination of the extract stream. Nevertheless, it is necessary to ensure the complete regeneration of the adsorbent in section I.

## 9.2 PermSMBR APPLICATIONS

The PermSMBR has the potential to produce with high yield and purity a product C and a by-product D obtained by

1. Equilibrium-limited reactions (e.g., A + B ↔ C + D), by which 100% conversion can be achieved because of the equilibrium displacement through the continuous removal of the products by adsorption and selective membranes (for instance, C is collected in the raffinate, D in the extract, and additionally one of them is also collected in the permeate);
2. Parallel and sequential reactions (e.g., A + B → C; B + C → D), by which component C selectivity is improved by using a membrane permeable to this compound; therefore, C is collected in the permeate and also in the raffinate or extract stream depending on if it is the less- or the more-adsorbed component, respectively.

The PermSMBR equipment can also be used to perform difficult separations, for instance, the separation of a mixture comprising three compounds (A, B, and C) with similar boiling points. They can be separated based on their different affinities toward the adsorbent and the membrane selectivity (A, being the more-adsorbed species, is collected in the extract; B, being the less-adsorbed component, is collected in the raffinate; and C, being the permeative component, is collected in the permeate).

The PermSMBR concept has not yet been experimentally proven. However, its integrated configuration was theoretically assessed for the production of ethyl lactate (EL), diethyl acetal, and dibutyl acetal (DBE), using mathematical models that strongly rely on kinetic, adsorption, and permeation experimental data (Silva et al., 2010; Pereira et al., 2012). Moreover, as mentioned in the previous chapter, all these compounds have already been produced using the SMBR technology, which is a significant step in the proof of concept of the new equipment. In all the studied systems, the integrated PermSMBR reveals a high productivity and a low solvent consumption; for instance, for an EL productivity of 16 $kg_{EL}/(L_{resin}$ day), at 50 °C, the SMBR ethanol consumption is 165% higher than the ethanol (Eth) consumption of the PermSMBR (1.96 $L_{Eth}/kg_{EL}$). A reactive distillation process at 128 °C (bottom temperature) requires a larger amount of ethanol, by 152% compared to the PermSMBR at 70 °C (0.86 $L_{Eth}/kg_{EL}$), for a productivity of about 41 $kg_{EL}/(L_{resin}$ day) (Silva et al., 2010). These results demonstrate the potential of this new technology for the sustainable production of oxygenated compounds derived from equilibrium-limited reactions, as is the case of acetals, esters, and ketones, among others.

In the following sections, the PermSMBR process development is illustrated using, as an example, the production of DBE, and the performance of this hybrid reactor is compared with the performance of the SMBR.

The DBE production by PermSMBR was explored under the scope of a European project, EuroBioref (www.eurobioref.org/), motivated by the fact that this compound can be produced from common biorefinery building blocks (n-butanol and acetaldehyde) and used as a green fuel additive: it enhances the renewable fraction in diesel, which positively contributes to the important European environmental commitment that aims to replace 10% of total transport fuels with biofuels by 2020 (UE Directive 2009/28/EC); moreover, it improves the diesel cetane number (Boennhoff and Obenaus, 1980). Therefore, the development of a sustainable, eco-friendly process for DBE synthesis is of significant importance.

## 9.3 CASE STUDY: DIBUTYL ACETAL PRODUCTION PROCESS DEVELOPMENT BY PermSMBR

To develop a PermSMBR process, independent of the adopted configuration (integrated or coupled), in addition to the basic data required in the SMBR technology, information related to the membranes flux and selectivity is also required for all the species involved in the synthesis of the desired product.

DBE is produced through a liquid-phase reversible reaction between n-butanol and acetaldehyde, in acid medium, having water as a by-product. For the implementation of the PermSMBR for DBE production, the following was considered: the strong acid ion-exchange resin Amberlyst-15 (A15) is used as catalyst and water-selective adsorbent. Commercial microporous silica membranes from Pervatech (The Netherlands) are used to enhance the removal of water; these were selected because they revealed higher flux and selectivity in the dehydration of aqueous mixtures than other membranes (e.g., Pervap SMS from Sulzer Chemtech) (Casado et al., 2005; Sommer and Melin, 2005). n-Butanol is used as reactant and also as desorbent, and so water (in addition to being removed through the membranes) is collected in the extract stream and DBE is collected in the raffinate stream, both diluted in n-butanol.

Because, at the time of the beginning of the EuroBioref project, DBE production by SMBR using A15 as catalyst/adsorbent had been already assessed (Graça et al., 2011), kinetic and adsorption data were available (see Table 9.1), and so, the PermSMBR process development for DBE production started with the determination of pervaporation data.

## 9.3.1 Pervaporation Data

The pervaporation process, for the DBE system, was evaluated using two commercial tubular hydrophilic silica membranes (Pervatech) following the experimental procedure described in a previous study (Pereira et al., 2010).

The influence of the boundary layer mass transfer resistance (concentration polarization) on the pervaporation process was evaluated by performing experiments with water/n-butanol mixtures, at different feed flow rates, keeping constant the feed concentration and the operating temperature. The temperature was set to the highest value, because the flux through the membrane is higher and the mass transfer effects in the boundary layer are more significant. As can be observed in Figure 9.3, at a feed flow rate above 25 l/h the

**Table 9.1** Kinetic and Adsorption Data for the DBE System (Graça et al., 2010a,b)

| Langmuir–Hinshelwood Rate Expression | Extended Langmuir Adsorption Isotherms | | |
|---|---|---|---|
| $$r = k_c \frac{a_A a_B - a_C a_D/(a_A K_{eq})}{(1 + K_{s,D} a_D)^2} \quad (9.1)$$ | $$\overline{q}_i = \frac{Q_{ads,i} K_i \overline{C}_{p,i}}{1 + \sum_{j=1}^{NC} K_j \overline{C}_{p,j}} \quad (9.5)$$ | | |
| with | **Component** | $Q_{ads}$ **(mol/ $L_{wet\ solid}$)** | **K (L/ mol)** |
| $$K_{eq} = 0.00959 \exp\left[\frac{1755.3}{T(K)}\right] \quad (9.2)$$ | n-Butanol | 8.5 | 7.5 |
| | Acetaldehyde | 15.1 | 0.5 |
| | Water | 44.9 | 12.1 |
| $$k_c = 2.39 \times 10^9 \exp\left[\frac{-6200.9}{T(K)}\right]\left(\frac{mol}{g_{cat} min}\right) \quad (9.3)$$ | DBE | 5.8 | 0.4 |
| $$K_{S,D} = 2.25 \times 10^{-4} \exp\left[\frac{3303.1}{T(K)}\right] \quad (9.4)$$ | | | |
| A = n-Butanol, B = acetaldehyde, C = DBE, D = water | | | |

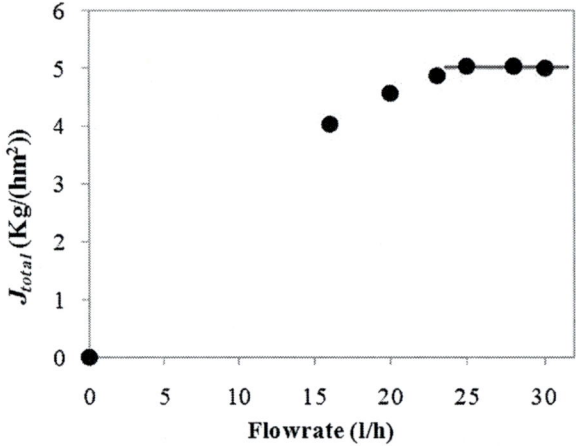

**Figure 9.3** Total permeation flux as a function of feed flow rate ($T = 70°C$, $x_{W,F} = 0.686 \pm 0.004$, $P_{perm} = 53$ mbar). *(Reprinted with permission from Pereira, C.S.M., Silva, V.M.T.M., Rodrigues, A.E., 2012. Green Fuel Production Using the PermSMBR Technology. Industrial and Engineering Chemistry Research 51, 8928–8938. Copyright (2012) American Chemical Society.)*

total flux remains constant, which indicates the absence of mass transfer resistance from the bulk liquid phase to the feed−membrane interface. As a result, the remaining perva-poration experiments were performed using a feed flow rate of about 28 l/h.

The performance of the silica membranes was evaluated experimentally for binary (water and $n$-butanol) and quaternary (water, $n$-butanol, acetaldehyde, and DBE) mix-tures, at different feed compositions, and at two temperatures, 50 and 70 °C, measuring the total flux and the permeate composition.

For each component, the permeation molar flux through the membrane ($J_i$) was described by the solution-diffusion model (Wijmans and Baker, 1995):

$$J_i = Q_{memb,i}\left(a_i p_i^0 - y_i P_{perm}\right) \tag{9.6}$$

where $Q_{memb,i}$ is the permeativity of component $i$ through the membrane, $a_i$ is the activity of component $i$ in bulk (calculated using the UNIFAC model); $p_i^0$ is the saturation pressure of component $i$; $P_{perm}$ is the total pressure on the permeate side; and $y_i$ is the molar fraction of component $i$ in the vapor phase (permeate side).

In Figure 9.4, the water molar flux as a function of the water driving force $\left(a_W p_W^0 - y_W P_{perm}\right)$, considering all the binary and quaternary experiments performed at 50 and 70 °C, is shown. A linear pattern is observed, indicating a water permeativity of $(4.43 \pm 0.66) \times 10^{-6}$ mol/(Pa s m$^2$) at 50 °C and $(3.93 \pm 0.94) \times 10^{-6}$ mol/

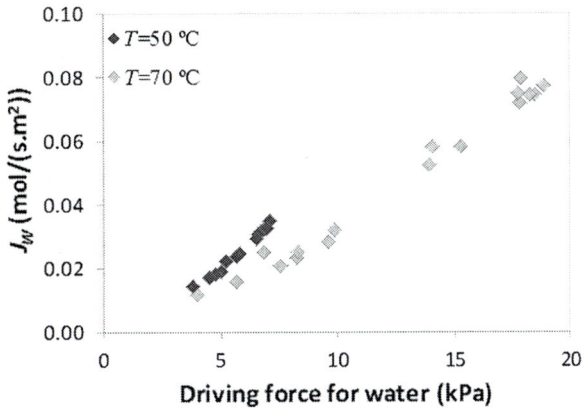

**Figure 9.4** Water flux as a function of water driving force for various water contents on binary (water/ *n*-butanol) and quaternary (acetaldehyde, *n*-butanol, DBE, and water) mixtures. *(Reprinted with permission from Pereira, C.S.M., Silva, V.M.T.M., Rodrigues, A.E., 2012. Green fuel production using the PermSMBR technology. Industrial and Engineering Chemistry Research 51, 8928–8938. Copyright (2012) American Chemical Society.)*

(Pa s m$^2$) at 70 °C, in the range of 0.08–0.38 of water molar fraction. For the other species, it was found that the permeativity of *n*-butanol is $(9.84 \pm 2.21) \times 10^{-9}$ mol/ (Pa s m$^2$) at 50 °C and $(6.48 \pm 2.07) \times 10^{-9}$ mol/(Pa s m$^2$) at 70 °C; the acetaldehyde flux is negligible, with a permeativity of $(7.93 \pm 4.24) \times 10^{-10}$ mol/(Pa s m$^2$) at 70 °C; and that DBE does not permeate the membrane.

## 9.3.2 PermSMBR Mathematical Models

The mathematical models used to describe the performance of the integrated and coupled PermSMBR for DBE synthesis consider axial dispersion flow for the bulk fluid phase; linear driving force approximation for the inter- and intraparticle mass transfer rates; multicomponent adsorption equilibrium at the adsorbent phase described by the extended Langmuir isotherm model (Graça et al., 2010a); kinetic law based on Langmuir—Hinshelwood rate expression (Graça et al., 2010b); liquid velocity variations due to reaction, adsorption/desorption, and species permeation; constant porosity and length of the packed bed; membrane concentration polarization; and isothermal operation.

The integrated and coupled PermSMBR model equations are summarized in Tables 9.2 and 9.3, respectively.

In both integrated and coupled PermSMBR mathematical models:

The global mass transfer coefficient, $k_L$, was determined by Eqn (9.37).

**Table 9.2** Model Equations of the Integrated PermSMBR

---

Bulk fluid mass balance to component $i$, in column $k$

$$\frac{\partial C_{ik}}{\partial t} + \frac{\partial(C_{ik}u_k)}{\partial z} + \frac{(1-\varepsilon)}{\varepsilon}\frac{3}{r_p}k_{L,ik}\left(C_{ik}-\overline{C}_{p,ik}\right) = D_{ax,k}\frac{\partial}{\partial z}\left(C_T\frac{\partial x_{ik}}{\partial z}\right) - \frac{A_m}{\varepsilon}J_{ik} \tag{9.7}$$

Pellet mass balance to component $i$, in column $k$

$$\varepsilon_p\frac{\partial\overline{C}_{p,ik}}{\partial t} + \left(1-\varepsilon_p\right)\frac{\partial q_{ik}}{\partial t} = \frac{3}{r_p}k_{L,ik}\left(C_{ik}-\overline{C}_{p,ik}\right) + v_i\rho_p\eta r\left(\overline{C}_{p,ik}\right) \tag{9.8}$$

Interstitial fluid velocity variation

$$\frac{du_k}{dz} = -\frac{(1-\varepsilon)}{\varepsilon}\frac{3}{r_p}\sum_{i=1}^{n}k_{L,ik}V_{mol,i}\left(C_{ik}-\overline{C}_{p,ik}\right) - \frac{A_m}{\varepsilon}\sum_{i=1}^{n}J_{iK} \tag{9.9}$$

Initial and Danckwerts boundary conditions

$$t=0:\quad C_{ik}=\overline{C}_{p,ik}=C_{ik,0};\quad q_{ik}=q_{ik,0} \tag{9.10}$$

$$z=0:\quad u_kC_{ik}-D_{ax,k}C_T\frac{\partial x_{ik}}{\partial z}\bigg|_{z=0}=u_kC_{ik,F};\quad u_k=u_{k,0} \tag{9.11a}$$

$$z=L_C:\quad \frac{\partial x_{ik}}{\partial z}\bigg|_{z=L_c}=0 \tag{9.11b}$$

Node mass balances

*Desorbent node*

$$C_{i(j=4,z=L_c)}=\frac{u_1}{u_4}C_{i(j=1,z=0)}-\frac{u_D}{u_4}C_i^D \tag{9.12}$$

*Feed node*

$$C_{i(2,z=L_c)}=\frac{u_3}{u_2}C_{i(3,z=0)}-\frac{u_F}{u_2}C_i^F \tag{9.13}$$

*Extract node*

$$C_{i(j=1,z=L_c)}=C_{i(j=2,z=0)} \tag{9.14}$$

*Raffinate node*

$$C_{i(j=3,z=L_c)}=C_{i(j=4,z=0)} \tag{9.15}$$

Axial dispersion coefficient $(D_{ax})$ (Butt, 1980)

$$\varepsilon Pe_p=0.2+0.011Re_p^{0.48} \tag{9.16}$$

$$Pe_p=d_pu/D_{ax} \tag{9.17}$$

$$Re_p-\rho d_pu/\mu \tag{9.18}$$

**Table 9.3** Model Equations of the Coupled PermSMBR

Bulk fluid mass balance to component $i$ in column $k$

$$\frac{\partial C_{ik}}{\partial t} + \frac{\partial (C_{ik}u_k)}{\partial z} + \frac{(1-\varepsilon)}{\varepsilon}\frac{3}{r_p}k_{L,ik}\left(C_{ik} - \overline{C}_{p,ik}\right) = D_{ax,k}\frac{\partial}{\partial z}\left(C_T\frac{\partial x_{ik}}{\partial z}\right) \tag{9.19}$$

Pellet mass balance to component $i$ in column $k$

$$\varepsilon_p\frac{\partial \overline{C}_{p,ik}}{\partial t} + \left(1 - \varepsilon_p\right)\frac{\partial q_{ik}}{\partial t} = \frac{3}{r_p}k_{L,ik}\left(C_{ik} - \overline{C}_{p,ik}\right) + v_i\rho_p r\left(\overline{C}_{p,ik}\right) \tag{9.20}$$

Interstitial fluid velocity variation in column $k$

$$\frac{du_k}{dz} = -\frac{(1-\varepsilon)}{\varepsilon}\frac{3}{r_p}\sum_{i=1}^{n}k_{L,ik}V_{mol,i}\left(C_{ik} - \overline{C}_{p,ik}\right) \tag{9.21}$$

Mass balance to component $i$ in the retentate side of membrane $m$

$$\frac{\partial C_{im}}{\partial t} + \frac{\partial (C_{im}v_m)}{\partial z} + A_m J_i = D_{ax,m}\frac{\partial}{\partial z}\left(C_{Tm}\frac{\partial x_{im}}{\partial z}\right) \tag{9.22}$$

Fluid velocity variation in membrane $m$

$$\frac{dv_m}{dz} = -A_m\sum_{i=1}^{n}J_{im}V_{mol,i} \tag{9.23}$$

Initial and Danckwerts boundary conditions

$$t = 0: \quad C_{im} = C_{ik} = \overline{C}_{p,ik} = C_{ik,0} \text{ and } q_{ik} = q_{ik,0} \tag{9.24}$$

$$z = 0: \quad u_k C_{ik}|_{z=0} - D_{ax,k}C_T\frac{\partial x_{ik}}{\partial z}\bigg|_{z=0} = u_k C_{ik,F} \text{ and}$$

$$v_m C_{im}|_{z=0} - D_{ax,m}C_{Tm}\frac{\partial x_{im}}{\partial z}\bigg|_{z=0} = v_m C_{ik}|_{z=L} \tag{9.25a}$$

$$z = L: \quad \frac{\partial x_{ik}}{\partial z}\bigg|_{z=L_k} = 0 \text{ and } \frac{\partial x_{im}}{\partial z}\bigg|_{z=L_m} = 0 \tag{9.25b}$$

*Continued*

**Table 9.3** Model Equations of the Coupled PermSMBR—cont'd

Node mass balances
*Desorbent node* $(j = 1)$

$$C_{i(j=4)}\Big|_{z=L_4} = \frac{Q_1|_{z=0}}{Q_4|_{z=L_4}} C_{i(j=1)}\Big|_{z=0}$$
$$- \frac{Q_D}{Q_4|_{z=L_4}} C_i^D \qquad (9.26)$$

$$Q_1|_{z=0} = Q_4|_{z=L_4} + Q_{Ds} \qquad (9.28)$$

*Extract node* $(j = 2)$

$$C_{i(j-1)}\Big|_{z=L_{(j-1)}} = C_{i(j)}\Big|_{z=0} \qquad (9.30)$$

$$Q_2|_{z=0} = Q_1|_{z=L_1} - Q_X \qquad (9.32)$$

*Feed node* $(j = 3)$

$$C_{i(j=2)}\Big|_{z=L_2} = \frac{Q_3|_{z=0}}{Q_2|_{z=L_2}} C_{i(j=3)}\Big|_{z=0}$$
$$- \frac{Q_F}{Q_2|_{z=L_2}} C_i^F \qquad (9.27)$$

$$Q_3|_{z=0} = Q_2|_{z=L_2} + Q_F \qquad (9.29)$$

*Raffinate node* $(j = 4)$

$$C_{i(j-1)}\Big|_{z=L_{(j-1)}} = C_{i(j)}\Big|_{z=0} \qquad (9.31)$$

$$Q_4|_{z=0} = Q_3|_{z=L_3} - Q_R \qquad (9.33)$$

For the remaining columns:

$$Q_k|_{z=L_k} = Q_m|_{z=0} \qquad (9.34)$$

$$Q_m|_{z=L_m} = Q_{k+1}|_{z=0} \qquad (9.35)$$

Axial dispersion coefficient in the retentate side of the membranes ($D_{ax,m}$) (Levenspiel, 1999)

$$\frac{D_{ax,m}}{v L_m} = \frac{1}{Re.Sc} + \frac{Re.Sc}{192} \qquad (9.36)$$

The axial dispersion coefficient of the packed beds was determined experimentally from the Peclet number (Graça et al., 2010a).

$$\frac{1}{k_L} = \frac{1}{k_e} + \frac{1}{\varepsilon_p k_i} \tag{9.37}$$

where $k_e$ and $k_i$ are, respectively, the external and internal mass transfer coefficients to the liquid phase. The correlations used to obtain these values are presented in detail in the literature (Silva and Rodrigues, 2002).

The permeate flux of species $i$ was defined as

$$J_i = k_{ov,i}\left(a_i p_i^0 - y_i P_{perm}\right) \tag{9.38}$$

where $k_{ov,i}$ is the global membrane mass transfer coefficient, which combines the resistance due to the diffusive transport in the boundary layer with the membrane resistance (Wijmans et al., 1996):

$$\frac{1}{k_{ov,i}} = \frac{1}{Q_{memb,i}} + \frac{\gamma_i^* p_i^0 V_{mol,i}}{k_{bl,i}} \tag{9.39}$$

in which $k_{bl}$ is the boundary layer mass transfer coefficient that was determined by the Lévêque correlation (Lévêque, 1928):

$$Sh = 1.62 Re^{0.33} Sc^{0.33} \left(\frac{d_{int}}{L_m}\right)^{0.33} \quad (Re < 2300) \tag{9.40}$$

where

$$Sh = \frac{k_{bl} d_{int}}{D_m} \tag{9.41}$$

$$Re = \frac{\rho d_{int} u}{\mu} \tag{9.42}$$

$$Sc = \frac{\mu}{\rho D_m} \tag{9.43}$$

The Perkins and Geankoplis method (Perkins and Geankoplis, 1969) was used to estimate the molecular diffusivity of the compounds. Further details with regard to these calculations can be found in the literature (Graça et al., 2011).

Similar to the SMBR, the PermSMBR performance parameters (extract and raffinate purity, conversion, productivity, and desorbent consumption) are important indicators used to evaluate the process feasibility under different operating conditions. These parameters are determined using the expressions defined in the previous chapter (Chapter 8) for the SMBR.

### 9.3.2.1 Numerical Solution

The system of partial differential equations, ordinary differential equations, and algebraic equations previously presented for the integrated PermSMBR were numerically solved by using the gPROMS-general Process Modelling System, version 3.1.3 (www.psenterprise.com). The axial discretization step was performed using third-order orthogonal collocation in finite element method (OCFEM), using 10 finite elements per column with two collocation points in each element. For the coupled PermSMBR, the model equations were solved numerically by using the gPROMS-general Process Modelling System, version 3.5.3. The axial domain was discretized using second-order OCFEM over 20 finite elements. In both cases, the system of ordinary differential and algebraic equations was integrated over time using the DASOLV integrator implementation in gPROMS. For all simulations a tolerance value of $10^{-5}$ was set.

### 9.3.3 PermSMBR versus SMBR

The PermSMBR performance was assessed and compared with that of the SMBR using their reactive/separation regions and/or optimal operating points, to check the improvement caused by the membrane integration for the production of DBE.

#### 9.3.3.1 Geometrical Specifications

The SMBR unit considered in this study has the same geometrical specifications of one of the SMB units available at LSRE (the Licosep 12-26 SMB Pilot Unit (Novasep, France)) in which the proof of concept for DBE production was performed (Graça et al., 2011). This unit comprises 12 columns (230 × 26, length × i.d., mm) packed with the commercial ion-exchange resin A15.

The integrated PermSMBR unit considered consists in 12 columns; each column with 13 silica tubular membranes from Pervatech (255 × 7, length × i.d., mm) packed in the lumen side (inside the membrane tube) with the A15 resin. The length of the bed and number of membranes were set by imposing the same mass of catalyst and effective cross-sectional area ($\varepsilon A$) as those of the SMBR unit.

The coupled PermSMBR consists in 12 fixed-bed columns, each of which is followed by a membrane module comprising 13 parallel hydrophilic tubular membranes (Pervatech). The geometry and number of the fixed-bed columns and the geometry and number of the membranes were set to have the same amount of catalyst/adsorbent and the same membrane area, respectively, as in the integrated PermSMBR.

A summary of the column characteristics for each of the technologies is presented in Table 9.4.

#### 9.3.3.2 Reactive/Separation Regions

The reactive/separation regions for 95% criteria for both extract and raffinate purities as well as for acetaldehyde conversion, for the SMBR, the integrated PermSMBR, and the coupled PermSMBR, defined by the liquid flow rates at the beginning of sections II and III, because, along with the columns of each section, there will be fluctuations in the flow rate caused by adsorption/desorption and permeation (integrated and coupled

**Table 9.4** Characteristics of the SMBR and PermSMBR Columns (Pereira et al., 2014)

| | SMBR | PermSMBR | Coupled PermSMBR | |
| --- | --- | --- | --- | --- |
| | | | Fixed-bed column | Pervaporation Module |
| Solid weight (A15) | 47.6 g | 47.6 g | 47.6 g | — |
| Length of the bed ($L$) | 23 cm | 25.45 cm | 23 cm | 25.45 cm |
| Internal diameter (i.d.) | 2.6 cm | 0.7 cm[a] | 2.6 cm | 0.7 cm[a] |
| Bed porosity ($\varepsilon$) | 0.4 | 0.424 | 0.4 | — |
| Bulk density ($\rho_b$) | 390 kg/m$^3$ | 374 kg/m$^3$ | 390 kg/m$^3$ | — |
| Number of membranes | — | 13 | — | 13 |

[a]Pervatech membrane internal diameter.

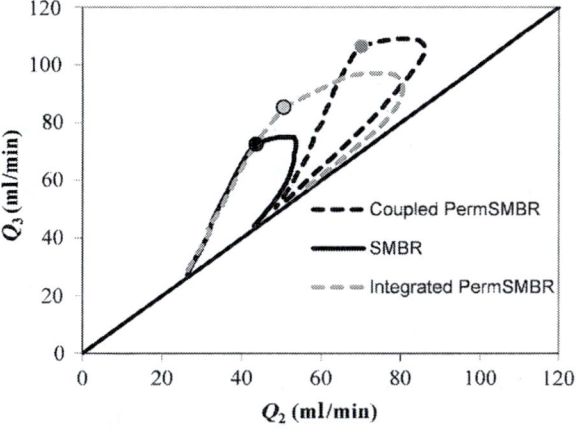

**Figure 9.5** Reactive/separation region for the SMBR and PermSMBR processes. ● SMBR optimal operating point; ◎ integrated PermSMBR optimal operating point; ◍ coupled PermSMBR optimal operating point (Pereira et al., 2014).

PermSMBR) of the species, are shown in Figure 9.5. The following operating conditions were set: feed of 51% acetaldehyde in $n$-butanol, configuration of 3—3—3—3, temperature of 50 °C, and desorbent and recycle flow rates of 105 and 21 ml/min, respectively. For the SMBR the switching time was 3.1 min, leading to $\gamma_1 = 7.997$ and $\gamma_4 = 1.333$. These $\gamma$ values were also imposed for the integrated PermSMBR by considering a switching time of 3.4 min (Pereira et al., 2012). For the coupled PermSMBR, this equivalence is not straightforward and therefore the same switching time as in the SMBR was assumed. Additionally, for the integrated and the coupled PermSMBR, the permeate pressure was considered equal to 5 mbar. The adopted configuration was 3—3—3—3 for the SMBR and integrated PermSMBR and 3(3)—3(3)—3(3)—3(3) for the coupled PermSMBR, meaning three packed bed columns plus three membrane modules per section (the number of membrane modules is indicated in parentheses).

As can be observed in Figure 9.5, the larger reactive/separation region is that of the integrated PermSMBR, which indicates that a wider range of operating conditions can be applied in sections II and III to achieve the same purities and conversion requirements as the SMBR and the coupled PermSMBR. This was expected because, compared to the SMBR, water is also removed by the membranes, enhancing the overall process performance. Relative to the coupled PermSMBR configuration, usually integrated processes, at suitable conditions for reaction and separation steps, lead to the best results.

At first glance, what is surprising is that, at the optimal operating point (vertex of the reactive/separation regions), the coupled PermSMBR has a slightly better performance than the integrated configuration (see Table 9.5). Nevertheless, this can be justified by the fact that these two configurations of the PermSMBR technology (integrated and coupled) are not being compared under identical conditions.

In the integrated PermSMBR and SMBR technologies, at the switching time, all the inlet and outlet streams are switched one column ahead in the direction of the liquid flow to simulate the countercurrent movement of the solid, and so, the simulated solid velocity is given by the length of the column divided by the switching time. Therefore, as previously mentioned, to have the same $\gamma_1$ and $\gamma_4$ (by having the same solid velocities) to compare the SMBR and integrated PermSMBR technologies under similar conditions, for the integrated PermSMBR a switching time of 3.4 min was considered. Nevertheless, in the coupled PermSMBR, at the switching time, all the inlet and outlet streams are switched one column plus one membrane module ahead, also, in the direction of the liquid (see Figure 9.2). Therefore, the solid velocity cannot be determined in the same way as for the SMBR and integrated PermSMBR; the switching time should be corrected to compensate for the residence time in the membrane modules to have the same solid velocity as in the SMBR and integrated PermSMBR technologies. This compensation is not straightforward because the residence time in the membrane modules is different in each section owing to the different flow rates. For instance in section I, the residence time in the membrane module (13 membranes in parallel) is about 1 min, whereas in section IV the residence time is 6 min.

The strategy adopted, to compare the SMBR and the integrated and coupled PermSMBR technologies in a fair way, was the determination of the optimal switching time (the one that leads to the best performance at the optimal operating point) for each of them. The values obtained are presented in Table 9.6 as well as the corresponding performance parameters at the optimal operating point (vertex of the reactive/separation region). As can be observed, the PermSMBR leads to better results than the SMBR,

**Table 9.5** Performance Parameters at the Optimal Operating Points

|  | SMBR | Integrated PermSMBR | Coupled PermSMBR |
|---|---|---|---|
| PR ($kg_{DBE}$ $L_{resin}^{-1}$ $day^{-1}$) | 53.15 | 64.15 | 69.14 |
| DC ($L_{n\text{-}Butanol}$/$kg_{DBE}$) | 2.69 | 2.15 | 1.97 |

Adapted from Pereira et al. (2014).

**Table 9.6** Performance Parameters for the Optimal Switching Time

| | $t^*$ (min) | PR ($kg_{DBE}\ L_{resin}^{-1}\ day^{-1}$) | DC ($L_{n\text{-Butanol}}/kg_{DBE}$) |
|---|---|---|---|
| SMBR | 3.1 | 53.39 | 2.83 |
| Integrated PermSMBR | 2.5 | 69.70 | 2.04 |
| Coupled PermSMBR | 3.0 | 69.42 | 2.07 |

but the integrated solution presents just a slight improvement compared with the coupled one. Nevertheless, the integrated PermSMBR allows the use of a wider range of operating conditions (in sections II and III) in order to obey the predefined requirements of purities and conversion compared to the coupled PermSMBR, which is proven by the size of the reactive separation regions shown in Figure 9.6. It should be mentioned that these reactive separation regions are narrower, for low feed flow rates, owing to the incomplete regeneration of the A15 resin in section I when the optimal switching time values are set.

### 9.3.4 PermSMBR-3s

The PermSMBR technology can be reduced to just three sections, for both coupled and integrated setups, as explained in Section 9.1. In this case, hydrophilic membranes and water-selective adsorbent (A15) were selected and so this simplification was made by the elimination of the extract stream. In this operation mode, DBE is still collected in the raffinate stream, whereas water is removed only through the membranes.

The coupled PermSMBR with three sections (coupled PermSMBR-3s) was evaluated for the DBE synthesis considering the operating conditions that previously led to the highest productivity: feed of 51% acetaldehyde in $n$-butanol, temperature of 50 °C, permeate pressure of 5 mbar, switching time of 3.0 min, and recycle flow rate

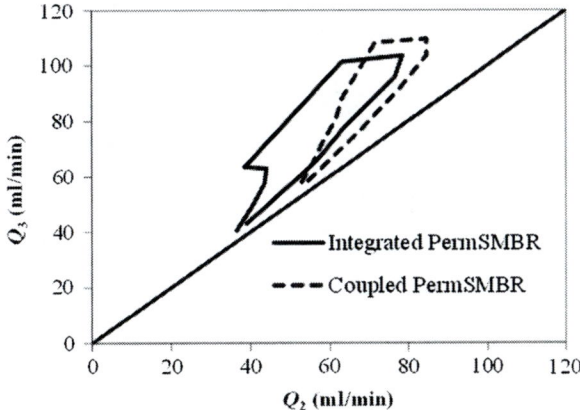

**Figure 9.6** Reactive separation regions of the integrated and coupled PermSMBR at the optimal switching times (2.5 min for the integrated PermSMBR; 3.0 min for the coupled PermSMBR) (Pereira et al., 2014).

of 18 ml/min. The optimal operating point that satisfies the preset purity and conversion criteria was determined by changing the desorbent flow rate for each feed flow rate. As in the previous studies, a minimum of 95% DBE purity and acetaldehyde conversion was set. However, the amount of water removed through the membranes was not enough to satisfy the 95% purity criterion for any flow rate. To increase the membranes' water flux, the temperature of the coupled PermSMBR-3s was increased from 50 to 70 °C. At this new temperature, it was possible to obey the purity and conversion criteria, and the highest productivity obtained was 73.44 $kg_{DBE} L_{resin}^{-1} day^{-1}$ with a desorbent consumption of 0.89 $L_{n\text{-Butanol}}/kg_{DBE}$.

Compared with the coupled PermSMBR with four sections, the coupled PermSMBR-3s leads to the formation of more DBE (6% increase on productivity), using significantly less $n$-butanol as desorbent (57% reduction on solvent consumption). Moreover, in this configuration mode just one stream has to be treated (the raffinate) and as a consequence, the global costs associated with the downstream separation units will be lower, which might compensate for the temperature increase from 50 to 70 °C.

The integrated PermSMBR with three sections (integrated PermSMBR-3s) was also assessed, and similar to the coupled PermSMBR, the integrated setup operated with just three sections seems to be the most attractive solution (Pereira et al., 2012).

When comparing the integrated and the coupled PermSMBR-3s, similar productivities are achieved, but in the integrated solution the solvent savings are even more significant (integrated PermSMBR-3s: DC = 0.42 $L_{n\text{-Butanol}}/kg_{DBE}$ (Pereira et al., 2012)) under the operational conditions evaluated. In Figure 9.7, a schematic representation of a DBE production plant using the integrated technology with just three sections is shown.

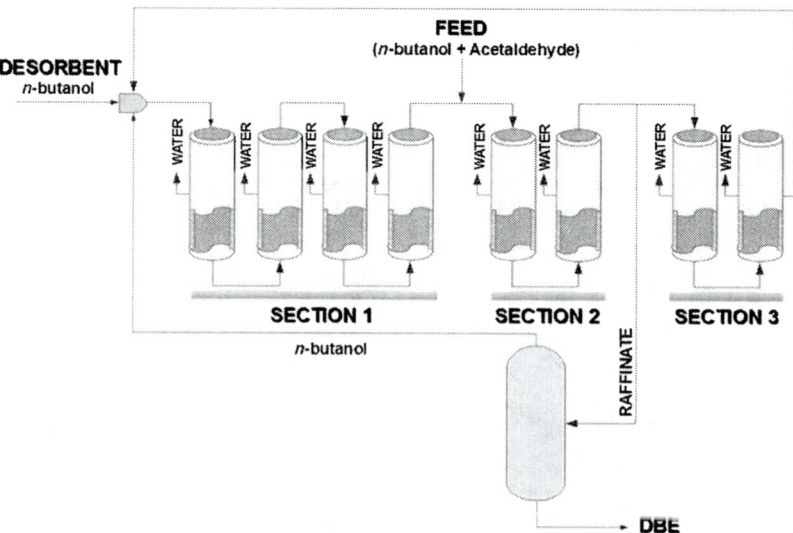

**Figure 9.7** Integrated PermSMBR-3s process scheme for DBE production (Pereira and Rodrigues, 2013).

## 9.4 PermSMBR CONCLUDING REMARKS

The PermSMBR is a newly developed technology that integrates the SMBR with permeant-selective membranes to enhance product separation.

Two possible PermSMBR setups were evaluated for DBE production: the integrated PermSMBR and the coupled PermSMBR. The main difference between them is that, in the integrated setup, the tubular membranes are packed with the catalyst/adsorbent, whereas in the coupled solution, the membranes follow a fixed-bed column packed with the solid and, as a result, they are not in direct contact with the catalyst/adsorbent. The coupled PermSMBR is, therefore, easier to implement than the integrated mode, because the membrane installation/replacement procedures and cleanup are simpler. Moreover, in the coupled configuration, the membranes' mechanical and chemical stability is not compromised by the solid presence.

In terms of productivity and desorbent consumption, the integrated and coupled PermSMBR have similar performance at the optimal operating points, but the integrated setup allows the use of a wider range of operating conditions to fulfill the same purity and conversion requirements. In both setups (coupled and integrated), the operation with just three sections leads to better results than when operating with four. Similar productivities are achieved using the coupled or the integrated PermSMBR-3s; nevertheless the solvent savings are significantly higher when the integrated PermSMBR-3s is used (integrated PermSMBR-3s, $DC = 0.42$ $L_{n\text{-Butanol}}/kg_{DBE}$; coupled PermSMBR-3s, $DC = 0.89$ $L_{n\text{-Butanol}}/kg_{DBE}$).

Compared with the SMBR, both PermSMBR setups are more efficient, having higher productivity and lower desorbent consumption for the same purity and conversion criteria. Moreover, when the PermSMBR is operated with just three sections (PermSMBR-3s), one less separation unit is needed for desorbent recovery, which is not possible when using the SMBR technology. However, in addition to requiring additional capital cost (membranes and vacuum pumps), the new technology is in the process development phase; its proof of concept has not been performed yet. The SMBR is a mature technology in the sense that the concept is already proven and has been tested through a lab prototype that is producing efficiently different oxygenates.

The decision on the most sustainable solution for the production of DBE (as for the production of other products) must pass for an economical evaluation of the whole production plant: SMBR/PermSMBR plus separation units.

## NOMENCLATURE

| | |
|---|---|
| $a_i$ | Liquid-phase activity of component $i$ in bulk side |
| $A_m$ | Membrane area per unit reactor volume ($m^2_{membrane}/m^3_{bulk}$) |
| $C$ | Liquid-phase concentration ($mol/m^3$) |
| $\overline{C}_p$ | Average liquid-phase concentration inside the particle ($mol/m^3$) |
| $C_T$ | Total liquid-phase concentration ($mol/m^3$) |

| $D_{ax}$ | Axial dispersion coefficient (m$^2$/min) |
|---|---|
| DC | Desorbent consumption (m$^3$/mol) |
| $d_{int}$ | Membrane internal diameter (m) |
| $d_p$ | Particle diameter (m) |
| $J_i$ | Permeate flux of species $i$ (mol/(m$^2$s)) |
| $k_{bl}$ | Boundary layer mass transfer coefficient (m/s) |
| $k_e$ | External mass transfer coefficient (m/s) |
| $k_i$ | Internal mass transfer coefficient (m/s) |
| $K_L$ | Global mass transfer coefficient (m/s) |
| $k_{ov}$ | Global membrane mass transfer coefficient (mol/(m$^2$ s Pa)) |
| $L$ | Column length (m) |
| $L_m$ | Membrane length (m) |
| $n$ | Total number of components |
| $N_c$ | Total number of columns |
| $Pe_p$ | Peclet number relative to particle |
| $p_i^0$ | Saturation pressure of component $i$ (Pa) |
| $P_{perm}$ | Total pressure on the permeate side (Pa) |
| PR | Raffinate productivity (kg$_C$/(m$^3_{resin}$ day)) |
| PUR | Raffinate purity (%) |
| PUX | Extract purity (%) |
| $q$ | Solid-phase concentration in equilibrium with the fluid concentration inside the particle (mol/l) |
| $Q$ | Volumetric flow rate (m$^3$/s) |
| $Q_{memb}$ | Permeance (mol s$^{-1}$ m$^{-2}$ Pa$^{-1}$) |
| $r$ | Rate of reaction (mol kg$^{-1}$ s$^1$) |
| $Re$ | Reynolds number |
| $Re_p$ | Reynolds number relative to particle |
| $r_p$ | Particle radius (m) |
| $Sc$ | Schmidt number |
| $Sh$ | Sherwood number |
| $t$ | Time variable (s) |
| $t^*$ | Switching time (s) |
| $u$ | Interstitial velocity (m/s) |
| $v$ | Superficial velocity (m/s) |
| $V_{mol,i}$ | Molar volume of species $i$ (m$^3$/mol) |
| $X$ | Acetaldehyde conversion |
| $x_i$ | Molar fraction in the liquid phase of component $i$ |
| $y_i$ | Molar fraction in the vapor phase of component $i$ |
| $z$ | Axial coordinate (m) |

## Greek Letters

| $\gamma$ | Dimensionless velocity ratio |
|---|---|
| $\gamma^*$ | Activity coefficient |
| $\varepsilon$ | Bulk porosity |
| $\varepsilon_p$ | Particle porosity |
| $v_i$ | Stoichiometric coefficient of component $i$ |
| $\rho_b$ | Bulk density (kg/m$^3$) |
| $\rho_p$ | Particle density (kg/m$^3$) |

| $\mu$ | Viscosity (cP) |
|---|---|
| $\eta$ | Effectiveness factor of the catalyst |
| $\Delta P$ | Pressure drop (Pa) |
| $\Phi$ | Sphericity of the particles |

## Subscripts

| $i$ | Component $i$ ($I = A, B, C, D$) |
|---|---|
| $j$ | Section ($j = 1, 2, 3, 4$ or I, II, III, IV) |
| $k$ | Column |
| $m$ | Membrane module in coupled PermSMBR |
| $0$ | Initial conditions |
| $A$ | $n$-butanol |
| $B$ | Acetaldehyde |
| $C$ | 1,1-dibutoxyethane |
| $D$ | Water |
| $D_s$ | Desorbent |
| $F$ | Feed |
| $p$ | Particle |
| $R$ | Raffinate |
| $Rec$ | Recycle |
| $X$ | Extract |

## REFERENCES

Boennhoff, K., Obenaus, F., 1980. 1,1-Diethoxyethane as Diesel Fuel. DE Patent 2 911.

Butt, J.B., 1980. Reaction Kinetics and Reactor Design. Prentice-Hall, Englewood Cliffs, NJ.

Casado, C., Urtiaga, A., Gorri, D., Ortiz, I., 2005. Pervaporative dehydration of organic mixtures using a commercial silica membrane: determination of kinetic parameters. Sep. Purif. Technol. 42, 39–45.

Graça, N.S., Pais, L.S., Silva, V.M.T.M., Rodrigues, A.E., 2010a. Dynamic study of the synthesis of 1,1-dibutoxyethane in a fixed-bed adsorptive reactor. Sep. Sci. Technol. 46, 631–640.

Graça, N.S., Pais, L.S., Silva, V.M.T.M., Rodrigues, A.E., 2010b. Oxygenated biofuels from butanol for diesel blends: synthesis of the acetal 1,1-dibutoxyethane catalyzed by amberlyst-15 ion-exchange resin. Ind. Eng. Chem. Res. 49, 6763–6771.

Graça, N.S., Pais, L.S., Silva, V.M.T.M., Rodrigues, A.E., 2011. Analysis of the synthesis of 1,1-dibutoxyethane in a simulated moving-bed adsorptive reactor. Chem. Eng. Process. 50, 1214–1225.

Levenspiel, O., 1999. Chemical Reaction Engineering, third ed. John Wiley & Sons, New York.

Lévêque, M.A., 1928. Les lois de transmission de chaleur par convection. Ann. Mines 13, 201.

Pereira, C.S.M., Rodrigues, A.E., May/June 2013. New sorption enhanced reaction technologies (SMBR and PermSMBR) for the production of diesel blends and green solvents. Chim. Oggi 31 (3), 64–67.

Pereira, C.S.M., Silva, V.M.T.M., Pinho, S.P., Rodrigues, A.E., 2010. Batch and continuous studies for ethyl lactate synthesis in a pervaporation membrane reactor. J. Membr. Sci. 361, 43–55.

Pereira, C.S.M., Silva, V.M.T.M., Rodrigues, A.E., 2012. Green fuel production using the PermSMBR technology. Ind. Eng. Chem. Res. 51, 8928–8938.

Pereira, C.S.M., Silva, V.M.T.M., Rodrigues, A.E., 2014. Coupled PermSMBR — process design and development for 1,1-dibutoxyethane production. Chem. Eng. Res. Des. 92, 2017–2026. http://dx.doi.org/10.1016/j.cherd.2013.11.015.

Perkins, L.R., Geankoplis, C.J., 1969. Molecular diffusion in a ternary liquid system with the diffusing component dilute. Chem. Eng. Sci. 24, 1035–1042.

Silva, V.M.T.M., Pereira, C.S.M., Rodrigues, A.E., 2009. Simulated Moving Bed Membrane Reactor, New Hybrid Separation Process and Used Thereof. PT Patent 104496; WO Patent 2010/116335.

Silva, V.M.T.M., Pereira, C.S.M., Rodrigues, A.E., 2010. PermSMBR—A new hybrid technology: application on green solvent and biofuel production. AIChE J. 57, 1840–1851.

Silva, V.M.T.M., Rodrigues, A.E., 2002. Dynamics of a fixed-bed adsorptive reactor for synthesis of diethylacetal. AIChE J. 48, 625–634.

Sommer, S., Melin, T., 2005. Performance evaluation of microporous inorganic membranes in the dehydration of industrial solvents. Chem. Eng. Process 44, 1138–1156.

Wijmans, J.G., Athayde, A.L., Daniels, R., Ly, J.H., Kamaruddin, H.D., Pinnau, I., 1996. The role of boundary layers in the removal of volatile organic compounds from water by pervaporation. J. Membr. Sci. 109, 135–146.

Wijmans, J.G., Baker, R.W., 1995. The solution-diffusion model: a review. J. Membr. Sci. 107, 1–21.

# CHAPTER 10

# Sequential Centrifugal Partition Chromatography

## 10.1 SOLID SUPPORT-FREE LIQUID CHROMATOGRAPHY

### 10.1.1 Introduction

Solid support-free liquid—liquid chromatography was developed by Ito in the 1960s (Ito et al., 1966), based on the idea of Craig's countercurrent distribution liquid—liquid apparatus (Craig and Post, 1949). In support-free liquid—liquid chromatography, also known as countercurrent chromatography (CCC) and centrifugal partition chromatography (CPC), the mobile and the stationary phases are two phases of a biphasic liquid system. One of the liquid phases is kept stationary with the help of a centrifugal force in a specially designed housing ("column") mounted on the axis of a centrifuge, whereas the other one, the mobile phase, is pumped through (Berthod, 2002; Foucault, 1995; Ito, 2005a; Mandava and Ito, 1988; Pauli et al., 2008). As a result of different partitioning between the mobile and the stationary phase, solutes present in a mixture injected into the mobile phase entering the column travel at different velocities along the column and can be collected in a purified form at the outlet.

The support-free liquid—liquid chromatography is the only chromatographic technique in which the user prepares not only the mobile phase, but also the stationary phase. This is done, simply, by mixing different portions of two or more solvents that form a biphasic liquid system (see Figure 10.1). The equilibrium composition of the resulting liquid phases corresponds to the composition of the mobile and stationary phase. Hence, the two phases cannot be selected independently because any change in the composition of either phase affects the composition of the other phase. The choice of biphasic liquid systems is almost limitless, which is what makes this technique extremely versatile and allows for a tailor-made system.

The key features of CCC/CPC are related to the liquid nature of the stationary phase, which occupies 60—80% of the column volume. The entire volume of the stationary phase is accessible to the solutes, making high sample loading feasible. Moreover, either of the phases of the biphasic liquid system (upper or lower phase) can be used as a stationary phase (see Figure 10.2). Not only the flow direction of the mobile phase but also the roles of the phases can be switched during the separation run, that is, the phase used as a mobile phase becomes stationary and vice versa. This creates an opportunity for several original operating modes (not possible when working with solid stationary phases), which can be used to improve the separation resolution, increase productivity, and reduce solvent consumption

*Simulated Moving Bed Technology*
ISBN 978-0-12-802024-1

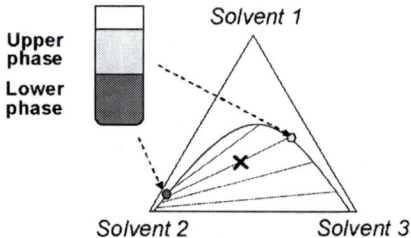

**Figure 10.1** Illustration of a preparation of mobile and stationary phases in support-free liquid—liquid chromatography. The upper or lower phase can be used as a mobile phase. The other phase is used as the stationary phase.

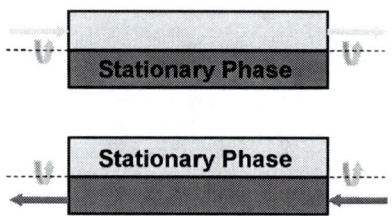

**Figure 10.2** Illustration of the possible uses of two liquid phases: upper phase as mobile phase and lower phase as mobile phase, with change of pumping direction.

(Agnely and Thiebaut, 1997; Berthod et al., 2003; Couillard et al., 2005; Delannay et al., 2006; Hewitson et al., 2009; Lu et al., 2008; Mekaoui et al., 2012; Foucault, 1995; Hopmann et al., 2012a). All above-listed characteristics distinguish support-free liquid—liquid chromatography (CCC/CPC) from conventional chromatography with a solid stationary phase, that is, high-performance liquid chromatography (HPLC).

Since the early years of CCC/CPC, natural products isolation and purification has been the leading area of application (Berthod et al., 2009; Pauli et al., 2008). However, the vast range of applications, including separation of pharmaceuticals, vitamins, dyes, herbicides, pesticides, inorganic elements, amino acids, and peptides and proteins, undoubtedly demonstrates the versatility of this technology. An extensive overview of the CCC/CPC application can be found in the several review articles and books (Marston and Hostettmann, 1994; Foucault, 1995; Ito, 2005a; Menet and Thiebaut, 1999; Berthod et al., 2009; Pauli et al., 2008).

For a long time support-free liquid—liquid chromatography was regarded as a niche lab-scale chromatographic technique. Despite its potential, the technology never reached the popularity of liquid—solid chromatography. The engineering progress in the design and scale-up of the support-free liquid—liquid chromatography equipment achieved since 2005, mainly by Professor Ito (2005b) and the companies Dynamics Extractions (UK), Armen Instruments (France), and Kromaton Technologies (France), has turned the technology from a slow to a rapid, robust, and scalable separation tool

(Sutherland, 2007). Currently the technology is making its breakthrough into industry (Sutherland et al., 2014).

## 10.1.2 Column Design

The "column" in support-free liquid–liquid chromatography is in a specially designed housing mounted on the axis of a centrifuge. This assembly is commonly referred to as the machine and replaces the classical cylindrical column used in HPLC. There are basically two types of commercially available machines, so-called hydrodynamic and hydrostatic machines, or hydrodynamic and hydrostatic columns.

Hydrodynamic machines (J-type coil–planet centrifuge) have two axes of rotation and generate a variable centrifugal force field. The column is a continuous piece of tubing wound on a bobbin, which rotates around its own axis and revolves around the axis of the centrifuge (see Figure 10.3(a)). In this way, mixing zones are created in the column in

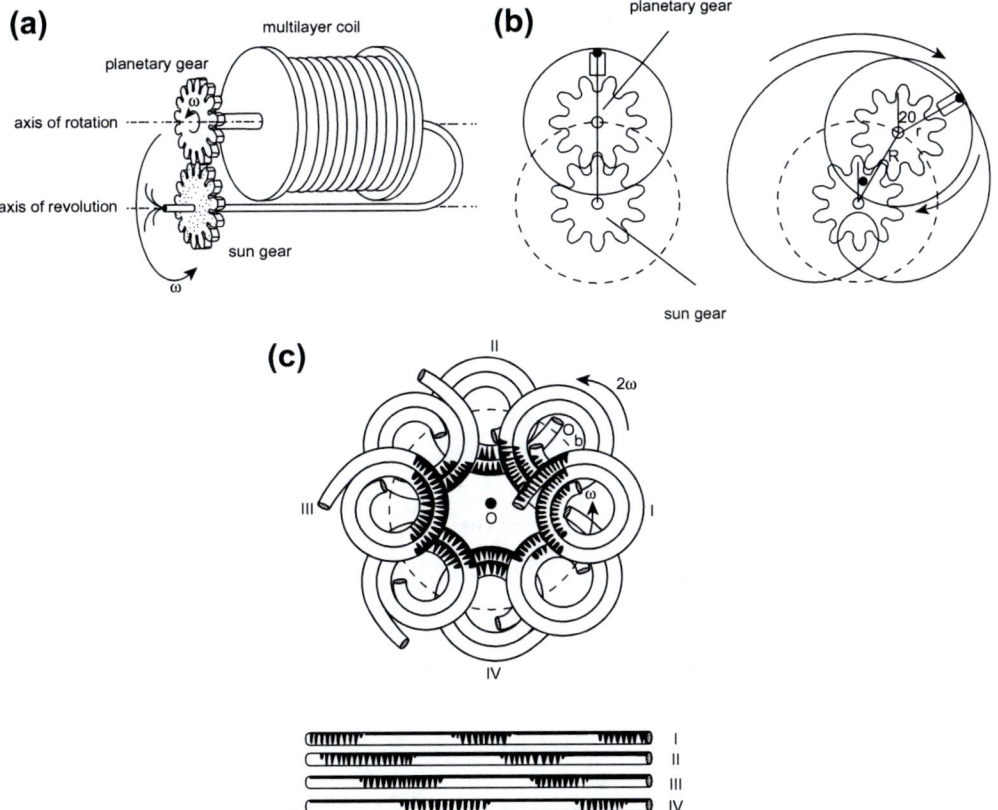

**Figure 10.3** Hydrodynamic machine. (a) Design, (b) planetary motion, and (c) schematic presentation of the mixing and settling zones in positions I, II, III, and IV during one full rotation of the bobbin around the central axis in a wound and unwound column (Ito, 2005a; Sutherland et al., 1998).

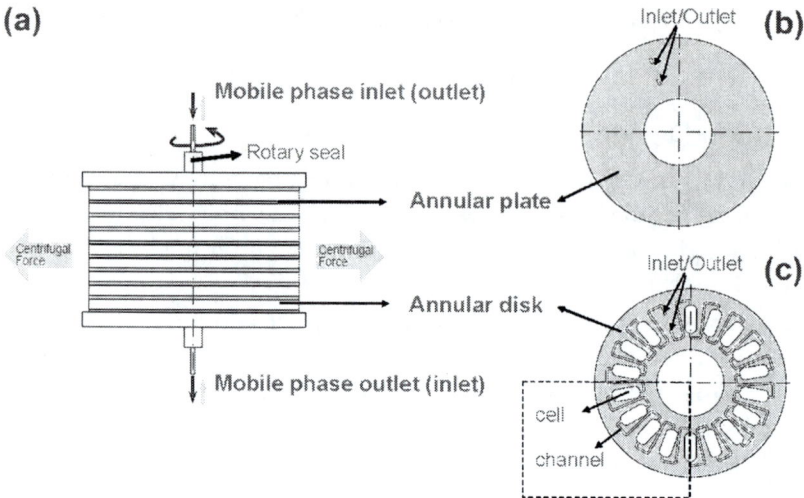

**Figure 10.4** (a) CPC column built by stacked disks and Teflon plates in between. (b) Annular Teflon plate. (c) Annular disk with engraved cells interconnected by channels.

the region closest to the central axis and settling zones in the region farthest from the central axis (Figure 10.3(b)). The mixing and settling zones are successively distributed along the whole column length.

In hydrostatic machines, the "column" is made of several identical disks, which are laid one above the other (see Figure 10.4(a)). Each disk has a series of circumferentially engraved cells, which are interconnected by channels (see Figure 10.4(c)). Between each two adjacent disks an annular Teflon plate is placed. Through the openings of the annular Teflon plate (see Figure 10.4(b)), the last cell of one disk is connected with the first cell of the next disk. The column is placed on the rotor of a centrifuge. The mobile phase enters and leaves the column via rotary seals (see Figure 10.4(a)). The main difference between machines from different producers is in the geometry of the cells engraved in the disk (see Figure 10.5).

In both types of columns, the upper or the lower phase can be used as a stationary phase. This is achieved with the right selection of the flow direction relative to the centrifugal force field (Berthod, 2002).

In general, a chromatographic separation performed in hydrostatic columns is called CPC and that with hydrodynamic columns is called CCC. The volume of the columns ranges between 20 ml and 18 l (Sutherland et al., 2008, 2009).

### 10.1.3 Stationary Phase Retention, Efficiency, and Resolution

Batch CPC and CCC separations are performed using a conventional HPLC setup, only the HPLC column is replaced by a hydrostatic or hydrodynamic machine,

**(a)**                    **(b)**

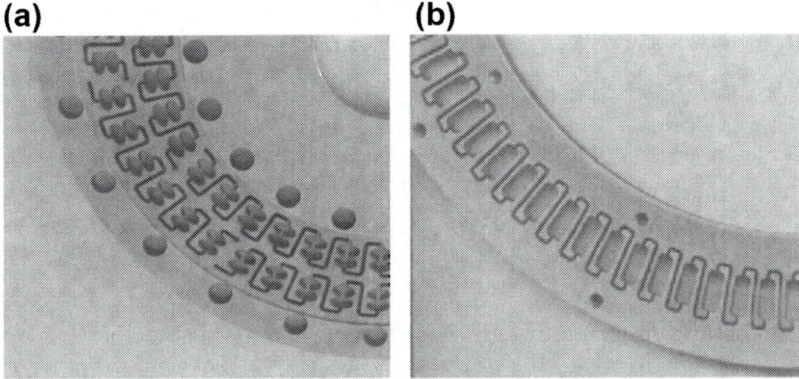

**Figure 10.5** Design of disk cells in hydrostatic columns from different producers: (a) Armen Instruments (Model TCPC250) and (b) Kromaton Technologies (FCPC250).

respectively. The two liquid phases can be preprepared or prepared online by using the pumps of the setup.

The machine, that is, the column, is first completely filled with the phase intended to be used as a stationary phase. Then the rotation is started, and after the set rotational speed is achieved the mobile phase is pumped into the column using a preselected flow rate. The mobile phase starts replacing the stationary phase and this occurs until the hydrodynamic equilibrium is achieved in the column. From this moment on, no more stationary phase leaves the column and the sample can be injected in a same way as in the classical HPLC columns.

The column volume fraction occupied by the stationary phase is called the stationary-phase retention, $S_F$, calculated as

$$S_F = \frac{V_S}{V_C} \tag{10.1}$$

where $V_S$ is the volume of the stationary phase and $V_C$ is the volume of the column.

The stationary phase retention is influenced by several factors, including the column geometry (type and dimensions), the physical properties of the two liquid phases (density, viscosity, and interfacial tension), the rotational speed (the created centrifugal force), and the mobile phase flow rate.

Unlike in liquid–solid chromatography, in which the volume of the stationary phase and the volume of the mobile phase accessible to the solutes are fixed once the column is packed, in CCC/CPC the volumes of the mobile and stationary phases change with the change of the mobile phase flow rate and the rotational speed. The stationary phase retention decreases with the increase in the mobile phase flow rate and to some extent with the increase in the rotational speed (centrifugal force) (Adelmann and Schembecker, 2011; Foucault et al., 1992a,b, 1994; Ignatova et al., 2001; Ignatova and Sutherland, 2003; Marchal et al., 2000). Likewise, for different biphasic systems, the volumes of the two phases in the same CCC/CPC column under the same operating conditions (mobile phase flow rate and rotational speed) differ as a result of the different physical properties of the phases (Adelmann and Schembecker, 2011; Foucault et al., 1994; Ignatova et al., 2001).

The above-listed parameters determine also the flow regime of the mobile phase in the "column," the contact area between the two phases, and consecutively the mass transfer rate. The columns of hydrostatic type have been investigated by direct visual observation using a transparent disk and camera by three different groups (Adelmann and Schembecker, 2011; Marchal et al., 2002; van Buel et al., 1998). The result of these studies show that the contact area between the mobile and the stationary phase and thus mass transfer rate increases with increasing mobile phase flow rate and/or rotational speed. This explains the increase in the column efficiency with the increase in the mobile phase flow rate, a tendency opposite that of columns packed with solid stationary phase. In Figure 10.6, the hydrostatic column efficiency is presented as a function of the mobile phase flow rate for three different stationary phase retentions, obtained with upper phase as mobile phase (Figure 10.6(a)) and lower phase as mobile phase (Figure 10.6(b)).

For concentrations in the linear part of the distribution (partition) equilibrium, the retention volume of solute $i$, $V_{R_i}$, can be calculated using the equation

$$V_{R_i} = V_M + P_i^{SM} V_S \tag{10.2}$$

where $V_M$ is the volume of the mobile phase and $V_S$ is the volume of the stationary phase.

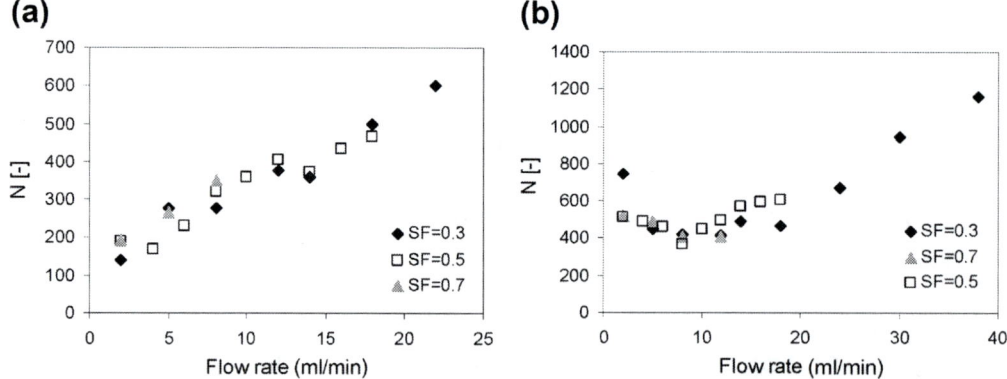

**Figure 10.6** The column efficiency (number of theoretical plates, $N$) as a function of mobile phase flow rate calculated from pulse injection experiments of hydroquinone at different stationary phase retentions ($S_F$). (a) Upper phase used as mobile phase; (b) lower phase used as mobile phase. Hydrostatic column Model TCPC 250 (Armen Instruments); column volume 250 ml; biphasic system, heptane/ethyl acetate/methanol/water 1/2/1/2 (v/v/v/v); injection volume 0.5 ml; concentration 5 g/l; rotation speed 1700 rpm. Note: The experiments were performed in the range of the mobile phase flow rate in which the selected value of $S_F$ stays unchanged.

The partition coefficient, $P_i^{SM}$, is defined as the ratio of the solute concentration in the stationary $(c_i^S)$ and the mobile $(c_i^M)$ phase at equilibrium:

$$P_i^{SM} = \frac{c_i^S}{c_i^M} \tag{10.3}$$

For a symmetrical elution peak, a Gaussian distribution can be assumed and the number of theoretical plates, that is, the column efficiency, for solute $i$ can be calculated using Eqn (10.4):

$$N_i = \left(\frac{V_{R_i}}{\sigma_i}\right)^2 = 16\left(\frac{V_{R_i}}{w_i}\right)^2 \tag{10.4}$$

The resolution of two eluting components is defined as

$$R_s = \frac{V_{R_2} - V_{R_1}}{(w_2 + w_1)/2} \tag{10.5}$$

where $V_{R_1}$ and $V_{R_2}$ are the retention volumes of the first and second eluting components, respectively, and $w_1$ and $w_2$ are their peak widths at baseline.

Assuming that the number of theoretical plates is the same for both components $(N_1 = N_2 = N)$ and combining Eqns (10.2), (10.4), and (10.5), the resolution can be written as follows:

$$R_s = \frac{1}{4}\sqrt{N} \; \frac{\left(P_2^{SM} - P_1^{SM}\right)}{\left(\dfrac{V_M}{V_S}\right) + \left(\dfrac{P_2^{SM} + P_1^{SM}}{2}\right)} \tag{10.6}$$

The number of theoretical plates of typical hydrodynamic and hydrostatic columns ranges between several hundreds and few thousands.

In Figure 10.7(a), the separation resolution of a binary mixture with a separation factor of 1.3 ($\alpha = P_2^{SM}/P_1^{SM} = 1.3$) in a column with 1000 theoretical plates, calculated using Eqn (10.6), is presented as a function of the partition coefficient of the first eluting component $(P_1^{SM})$ for different phase ratios $(V_M/V_S)$. As expected, the resolution increases with the decrease in the $V_M/V_S$ ratio, that is, the increase of the volume of the stationary phase $(V_S)$. In addition, two other clear trends can be observed. Namely, the resolution drops considerably with the decrease of the partition coefficients below approximately 0.5 and does not increase significantly for partition coefficients above approximately 3. Furthermore, it should be taken into account that the high partition coefficients lead to increased separation time, solvent consumption, and decreased concentration of collected products as a result of the extensive peak broadening.

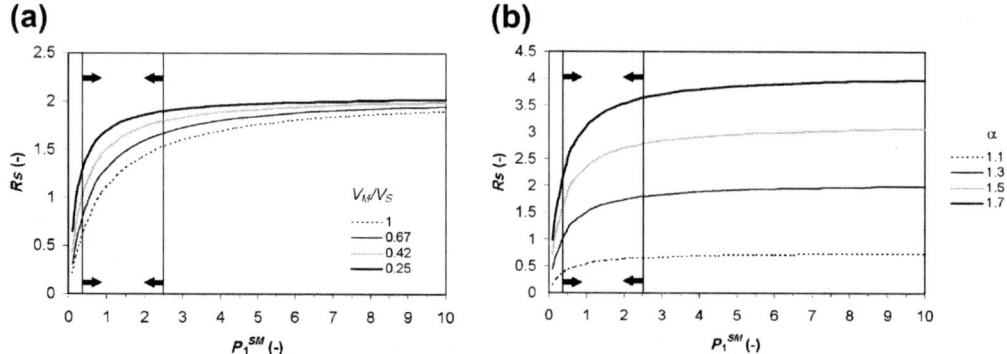

**Figure 10.7** Resolution of a binary mixture as a function of the partition coefficient of the first eluting component ($P_1^{SM}$), (a) for different ratios of the volumes of the mobile and stationary phases ($V_M/V_S$), $N = 1000$, $\alpha = P_2^{SM}/P_1^{SM} = 1.3$, and (b) for different separation factors $\alpha$, $N = 1000$, $V_M/V_S = 0.43$ ($S_F = 0.7$). The vertical lines present the preferred range in CCC/CPC.

These trends are also observable in Figure 10.7(b), in which the separation resolution of binary mixtures with different separation factors ($\alpha$) is presented for a fixed value of $V_M/V_S$.

In conclusion, a satisfactory separation resolution with balanced mobile phase consumption is possible only for high stationary phase volumes and for partition coefficients from a certain range of values (Berthod, 2002).

The conventional wisdom in the CCC/CPC community is that the value of the partition coefficient of the solute of interest should be ideally in a range between 0.4 and 2.5 (Friesen and Pauli, 2005). This range is practical, because according to Eqn (10.3) it stays the same independent of which phase, upper or lower, is used as the mobile phase. Indeed, this range provides a good balance between the separation resolution, productivity, and mobile phase consumption. However, it should not be regarded as a strict but rather as an orientation range, because the resolution is a function not only of the absolute value of the partition coefficient, but also of the number of theoretical plates and selectivity (see Eqn (10.6)).

## 10.1.4 Selection of Mobile and Stationary Phase

The selection of mobile and stationary phase in solid support-free liquid–liquid chromatography is equivalent to the selection of a biphasic solvent system and its global composition. The partition coefficient of the target component is used as a screening parameter for this selection.

A miscibility gap is normally observed between two solvents with different chemical functionality (polarity, solvation). Taking into consideration that the preferred range of the partition coefficients is between 0.4 and 2.5, it becomes clear that in most of the binary biphasic liquid systems the partition coefficients will be out of the preferred range, because the components would partition preferably in one of the phases. Thus, in most of the cases, the biphasic systems used in CCC/CPC are composed of three or more solvents.

The conventional procedure for the selection of a biphasic system for a particular separation task includes experimental determination of the partition coefficients of the target components in a series of preselected biphasic systems with the aim of finding one with the partition coefficient of the target component in the preferred range. Owing to the immense choice of biphasic liquid systems, the selection of the biphasic system is the most time-consuming step during the development of CCC/CPC separations. Hence, the selection is mostly done from previously published literature data for similar separation problems or by experimental screening of predefined biphasic multisolvent system compositions organized in tables according to overall polarity of the system, so-called solvent system families (Camacho-Frias and Foucault, 1996; Foucault, 1995; Jean-Hugues Renault, 2002). "Solvent system family" is a mixture of three to five solvents that form two phases when mixed in certain proportions. One family consists of several systems with different global compositions organized in a table, in which each system (composition) is marked with a number or letter (see Figure 10.8).

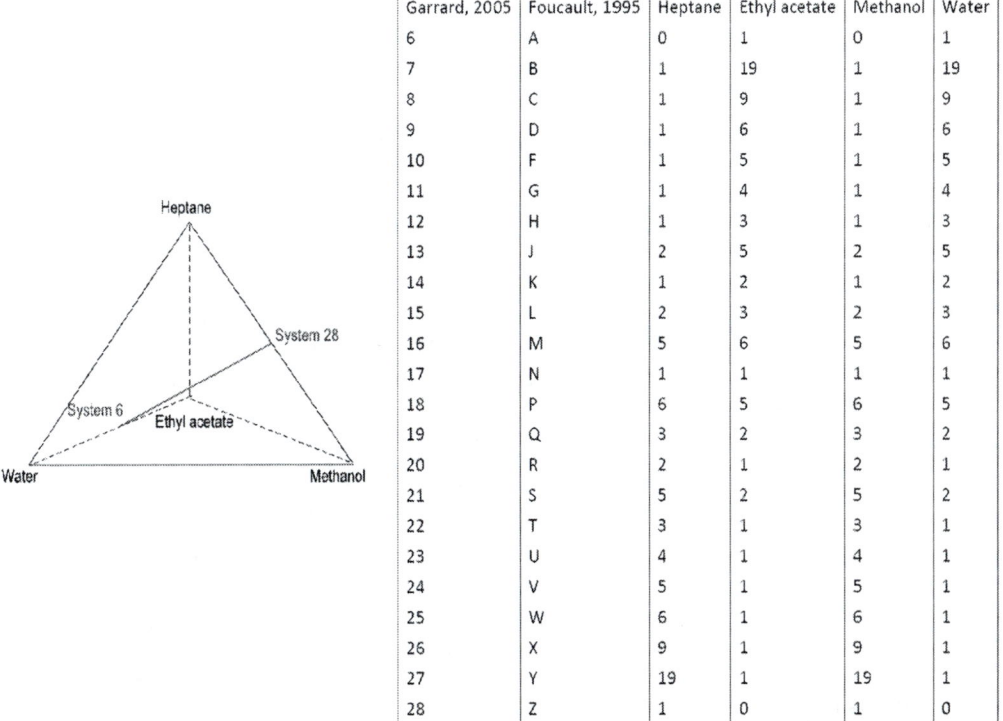

| Garrard, 2005 | Foucault, 1995 | Heptane | Ethyl acetate | Methanol | Water |
|---|---|---|---|---|---|
| 6 | A | 0 | 1 | 0 | 1 |
| 7 | B | 1 | 19 | 1 | 19 |
| 8 | C | 1 | 9 | 1 | 9 |
| 9 | D | 1 | 6 | 1 | 6 |
| 10 | F | 1 | 5 | 1 | 5 |
| 11 | G | 1 | 4 | 1 | 4 |
| 12 | H | 1 | 3 | 1 | 3 |
| 13 | J | 2 | 5 | 2 | 5 |
| 14 | K | 1 | 2 | 1 | 2 |
| 15 | L | 2 | 3 | 2 | 3 |
| 16 | M | 5 | 6 | 5 | 6 |
| 17 | N | 1 | 1 | 1 | 1 |
| 18 | P | 6 | 5 | 6 | 5 |
| 19 | Q | 3 | 2 | 3 | 2 |
| 20 | R | 2 | 1 | 2 | 1 |
| 21 | S | 5 | 2 | 5 | 2 |
| 22 | T | 3 | 1 | 3 | 1 |
| 23 | U | 4 | 1 | 4 | 1 |
| 24 | V | 5 | 1 | 5 | 1 |
| 25 | W | 6 | 1 | 6 | 1 |
| 26 | X | 9 | 1 | 9 | 1 |
| 27 | Y | 19 | 1 | 19 | 1 |
| 28 | Z | 1 | 0 | 1 | 0 |

**Figure 10.8** Heptane/ethyl acetate/methanol/water (ARIZONA) systems global composition (v/v/v/v) (Foucault, 1995; Garrard, 2005).

The most frequently used biphasic CCC solvent system families are heptane/ethyl acetate/methanol/water (the so-called ARIZONA family) (Camacho-Frias and Foucault, 1996), hexane/ethyl acetate/methanol/water (the so-called HEMWat family) (Oka et al., 1998), heptane/butanol/acetonitrile/water, and chloroform/methanol/water (Foucault, 1995; Oka et al., 1998; Berthod, 1991). The solvent systems used for a wide range of compounds are compiled in Foucault (1995), Ito (2005a), and Menet and Thiebaut (1999).

At present, 90% of all CCC/CPC separations are performed using systems from the ARIZONA and HEMWat solvent system families (Pauli et al., 2008). These systems include four typically used solvents to prepare the mobile phase in HPLC. Namely, the standard solvents in normal-phase chromatography, hexane (heptane) and ethyl acetate, and the standard solvents in reversed-phase chromatography. By combining different portions of these four solvents (see Figure 10.8), two liquid phases, that is, mobile and stationary phases, with different polarities can be obtained.

The experimental effort needed to select a biphasic solvent system and its global composition could be significantly reduced, if not eliminated, by using thermodynamic models for the prediction of solute partition coefficient. This is demonstrated in Hopmann et al. (2011, 2012b).

Using the definition of the thermodynamic equilibrium in liquid–liquid systems:

$$x_i^S \gamma_i^S = x_i^M \gamma_i^M \tag{10.7}$$

The partition coefficient defined by Eqn (10.3) can be rewritten as follows:

$$P_i^{SM} = \frac{c_i^S}{c_i^M} = \frac{\frac{n_i^S}{V^S}}{\frac{n_i^M}{V^M}} = \frac{x_i^S \frac{n^S}{V^S}}{x_i^M \frac{n^M}{V^M}} = \frac{\gamma_i^M}{\gamma_i^S} \frac{v^M}{v^S} \tag{10.8}$$

where $n_i^M$ and $n_i^S$ are the numbers of moles of solute $i$ in the mobile and stationary phases, respectively; $V^M$ and $V^S$ are the volumes of the mobile and stationary phases, respectively; $\gamma_i^M$ and $\gamma_i^S$ are the activity coefficients of solute $i$ in the mobile and stationary phases, respectively; and $v^M$ and $v^S$ are the molar volumes of the mobile and stationary phases, respectively.

For diluted solutions at constant temperature and pressure, the partition coefficient can be approximated by the activity coefficients at infinite dilution ($x_i^S, x_i^M \to 0$):

$$P_i^{SM} = \frac{\gamma_i^{M\infty}}{\gamma_i^{S\infty}} \frac{v^M}{v^S} \tag{10.9}$$

The molar volume of the phases is a sum of the ideal molar volume $v_{ideal}$ (=molar volume of an ideal mixture) and the excess volume of mixing $v^E$ (Eqn (10.10)):

$$v = v_{ideal} + v^E = \sum x_i v_{0i} + v^E \tag{10.10}$$

Under the assumption of an ideal mixture, the molar volume of the phases can be calculated as a weighted sum of the molar volumes of pure substances, $v_{0i}$.

Thus, the partition coefficient of a solute in a biphasic liquid system with a given global composition can be calculated according to Eqn (10.9) with the help of the activity coefficient model, once the compositions of the upper and lower phases are known.

The most commonly used activity coefficient models are the correlation methods, such as NRTL model (non-random two-liquid model) (Renon and Prausnitz, 1968) and UNIQUAC model (UNIversal QUAsi-Chemical model) (Abrams and Prausnitz, 1975), and the predictive group-contribution model UNIFAC (UNIversal QUAsi-Chemical Functional-Group Activity Coefficient) (Fredenslund et al., 1975). The disadvantage of the correlation methods is the limitation in scope. The group contribution models are dependent on group interaction parameters, which are often limited or missing.

In Hopmann et al. (2011, 2012b), the activity coefficient was calculated using the conductor-like screening model for real solvents (COSMO-RS) (Klamt et al., 2001). The model is based on unimolecular quantum mechanics calculations combined with methods of statistical thermodynamics. COSMO-RS can be used to predict all thermo-dynamic properties that can be derived from the chemical potential, including activity coefficients, based on the molecular structure of the involved components. COSMO-RS has been successfully applied for a priori predictions of solute partition co-efficients in octanol/water and micelle/water systems (Buggert et al., 2009; Mokrushina et al., 2007) and used to screen solvents for extraction and biocatalytic biphasic reaction systems (Burghoff et al., 2009; Spiess et al., 2008).

In Figure 10.9, the experimental and COSMO-RS predicted partition coefficients of six solutes in different biphasic solvent systems are presented.

The accuracy of the a priori prediction is system dependent and, as shown in Hopmann et al. (2011), sufficient for screening and selection of biphasic solvent systems (mobile and stationary phase) in CCC/CPC. This is also illustrated in Figure 10.10, which shows the calculated and experimentally determined partition coefficients in different global compositions of a given quaternary biphasic system. Namely, using the screening criteria ($0.5 < P_i^{SM} < 2.5$, meaning $-0.4 < \log P_i^{SM} < 0.4$), the same system global composition will be selected based on the COSMO-RS prediction and solute partitioning experiment.

In conclusion, COSMO-RS is a promising tool with which the laborious exper-imental work normally needed for screening and selection of a suitable solvent system in CCC/CPC can be significantly reduced. For the calculation of the parti-tion coefficient with Eqn (10.9) using COSMO-RS, the only input information needed is the molecular structure of the solute and the solvents of the biphasic system and the composition of the two liquid phases (in absence of the solute). The latter can be taken from the experimental data available in the literature, for example, Sørensen et al. (1979), or calculated from the liquid—liquid thermodynamic

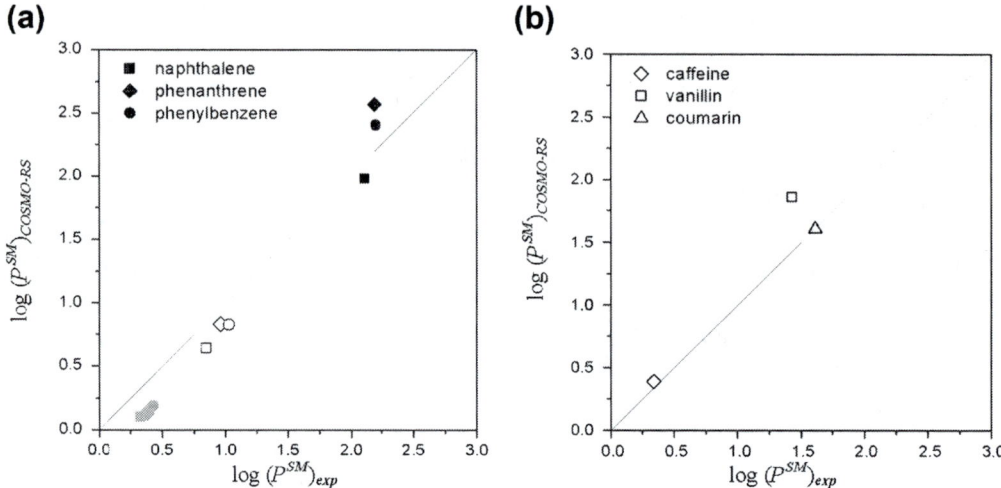

**Figure 10.9** Comparison of COSMO-RS predicted and experimental partition coefficients (a) of polyphenols in various heptane/ethyl acetate/methanol/water systems (black symbols, 1/2/1/2 v/v/v/v; open symbols, 2/1/2/1 v/v/v/v; gray symbols, 9/1/9/1 v/v/v/v; $T = 20\,°C$ *(experimental data from Ignatova et al., 2011)*) and (b) of caffeine, vanillin, and coumarin in ethyl acetate/*n*-butanol/water (4/6/10; $T = 22\,°C$ *(experimental data from Friesen and Pauli, 2007)*).

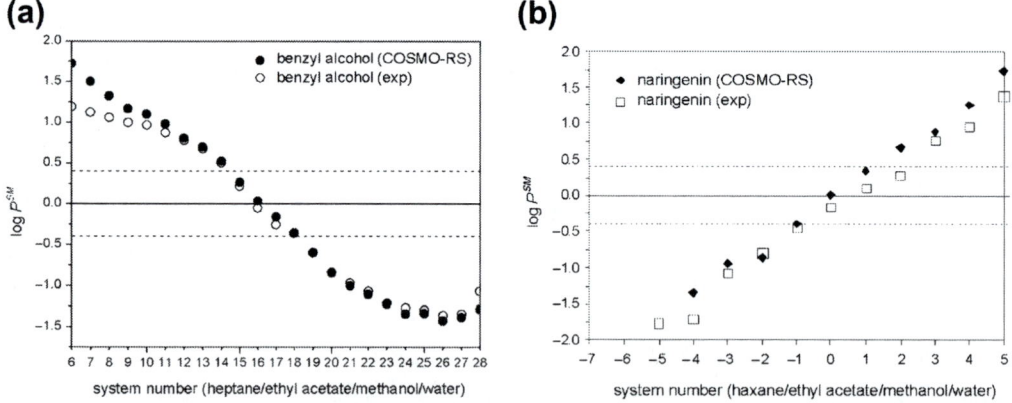

**Figure 10.10** Comparison of experimental and predicted partition coefficients of (a) benzyl alcohol in different global compositions of heptane/ethyl acetate/methanol/water systems (system numbers from Figure 10.8) and (b) naringenin in different global compositions of hexane/ethyl acetate/methanol/water systems (system number from Foucault, 1995). The dashed horizontal lines indicate the preferred range of the partition coefficient.

equilibrium condition (Eqn (10.7)) using COSMO-RS or any of the above-mentioned activity coefficient models (Frey, Hopmann, Minceva, 2014).

## 10.2 SEQUENTIAL CENTRIFUGAL PARTITION CHROMATOGRAPHY

### 10.2.1 Principle of Continuous Solid Support-Free Liquid Chromatography

In support-free liquid chromatography, a feed mixture can be continuously separated into two products using the process design concept patented by Couillard et al. (2005). The principle of operation is schematically presented in Figure 10.11 for a separation of a binary mixture of components $A$ and $B$. It is assumed here that component $A$ distributes preferentially into the lower phase, whereas component $B$ distributes preferentially into the upper phase. The process is realized in two columns connected in series. The columns are filled with predefined volumes of the upper and lower phases from a selected biphasic liquid system. The feed is introduced continuously between the two columns and the two products are collected sequentially at the opposite ends of each column. The process is cyclic. One cycle has two steps, called the descending step (*Des*) and the ascending step (*As*).

In the descending step, the lower phase is used as a mobile phase and it is introduced into the unit through column 2. Components $A$ and $B$ dissolved in the lower phase are used as a feed, which is introduced between columns 1 and 2. During the descending step component $A$ travels faster with the mobile phase toward the left end of column 1 and can be collected in a pure form in product 1. Just before component $B$ reaches the product collection point the unit is switched to the ascending step.

In the ascending step, the upper phase is used as a mobile phase and it is introduced into the unit through column 1. Components $A$ and $B$ dissolved in the upper phase are now used as a feed stream. In this step, component $B$ travels faster with the mobile phase

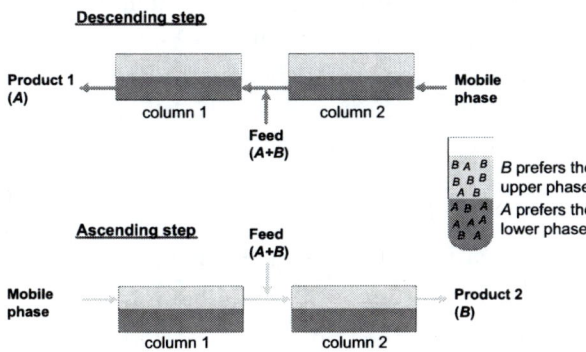

**Figure 10.11** Schematic presentation of the principle of operation of sequential centrifugal partition chromatography, showing the descending step and the ascending step. The lower phase is presented in dark gray, upper phase in light gray.

**Table 10.1** Overview of the Separations Performed with sCPC/ICcE, Including the Global Compositions of the Solvent Systems Used and the Corresponding Separation Factors of the Separated Components

| Separated Components | Solvent System (v/v/v/v) | Separation Factor, $\alpha$ (−) | Publication |
|---|---|---|---|
| Magnolol–honokiol | Hexane/ethyl acetate/ methanol/water (2/1/2/1) | 1.96 | Peng et al. (2010) |
| Naringenin–vanillin | Hexane/ethyl acetate/ methanol/water (2/3/2/3) | 3.17 | Ignatova et al. (2011) |
| Hydroquinone– pyrocatechol | Heptane/ethyl acetate/ methanol/water (1/2/1/2) | 3.50 | Hopmann et al. (2012a) |
| Aspirin–salicylic acid | Hexane/ethyl acetate/ methanol/water (1/1/1/1) | 5.33 | Hewitson et al. (2011) |
| Polar impurities–dye– nonpolar impurities | Heptane/1-butanol//water (2/3/4) | $\geq 8$ | Mekaoui et al. (2012) |
| Capsaicin– dihydrocapsaicin | Heptane/ethyl acetate/ methanol/water (1/2/1/2) | 1.36 | Goll et al. (2013) |

toward the right end of column 2 and can be collected in product 2. Shortly before component $A$ elutes out of column 2, one cycle is completed and the unit is switched back to the descending step.

The inventors have named the process true moving bed CPC, which actually does not reflect the real nature of the process, because in each step of the cycle one of the phases is kept stationary. Two alternative names for the same concept have been suggested in the literature: intermittent countercurrent extraction (Hewitson et al., 2009) and sequential CPC (sCPC), suggested by the Minceva group (Völkl et al., 2011), both suggesting the intermittent/sequential nature of the process. In the following sections, sCPC is used.

The few experimental demonstrations of the process have shown its potential (Audo and Le Quemeneur, 2012; Hewitson et al., 2009, 2011; Hopmann and Minceva, 2012; Peng et al., 2010; Hopmann et al., 2012a,b; Goll et al., 2013). The separations were performed in units with hydrostatic (Goll et al., 2013; Hopmann et al., 2012a,b; Hopmann and Minceva, 2012) and hydrodynamic (Hewitson et al., 2009, 2011; Peng et al., 2010) columns and are summarized in Table 10.1.

## 10.2.2 Shortcut Method for Selecting sCPC Operating Parameters

Owing to the cyclic nature of the sCPC process, the selection of the unit operating parameters is not straightforward. For fixed volumes of the upper and lower phases in the columns, six operating parameters (see Figure 10.11), four flow rates and two durations of the process steps, should be selected. Namely, the feed flow rate during the descending

($F_{L,F}$) and ascending ($F_{U,F}$) steps, the mobile phase flow rate in the ascending ($F_U$) and descending ($F_L$) steps, and the duration of the ascending ($t_{As}$) and descending ($t_{Des}$) steps.

In Völkl et al. (2011) a shortcut design method for the selection of the sCPC operating parameters for a complete separation of a binary feed mixture of components $A$ and $B$ is developed based on the assumption of instantaneous equilibrium between the two phases. Further, it is assumed that:

- The off-column volume is negligible.
- The partition coefficients the solutes present in the feed mixture are constant (independent of the solute concentration). They describe the linear part of the partition equilibrium; hence, the partition coefficient of solute $i$ is equal to the distribution constant of solute $i$, $P_i^{SM} = K_i$.
- The distribution constant of solute $i$ in each step is defined as the ratio of its concentration in the stationary phase to that in the mobile phase, consequently:

$$K_{i,Des} = \frac{1}{K_{i,As}}.$$

- $K_{A,Des} < K_{B,Des}$ (consequently $K_{A,As} > K_{B,As}$), that is, component $A$ is collected during the descending step in product 1 and component $B$ is collected during the ascending step in product 2.

To achieve a complete separation of $A$ from $B$ the requirements listed in the first column in Table 10.2 should be fulfilled.

In the second column in Table 10.2, the requirements are written in terms of the distance traveled by each component $i$ in each column $j$ during the ascending and descending steps, $x_{i,j,step}$.

The distance traveled by component $i$ in column $j$ during the descending step is

$$x_{i,j,Des} = v_{i,j,Des} t_{Des} \tag{10.15}$$

and during the ascending step is

$$x_{i,j,As} = v_{i,j,As} t_{as} \tag{10.16}$$

The velocity of component $i$ in column $j$ during the descending step and the ascending step is, respectively:

$$v_{i,j,Des} = \frac{F_{L,j} L_C}{V_L + K_i^{Des} V_U} \tag{10.17}$$

$$v_{i,j,As} = \frac{F_{U,j} L_C}{V_U + \frac{1}{K_i^{As}} V_L} \tag{10.18}$$

By replacing Eqns (10.11)−(10.14) with Eqns (10.15)−(10.18) the constraints on the sCPC operating parameters (Eqns (10.19)−(10.22)), presented in the third column in Table 10.2, are obtained.

**Table 10.2** Sequential Centrifugal Partition Chromatography (sCPC) Unit Operation Restrictions and Operating Parameter Constraints for a Complete Separation of a Binary Feed Mixture

| Restrictions | Restrictions on Distance Traveled by Components A and B during the Ascending and Descending Steps | Operating Parameter Constraints |
|---|---|---|
| At the end of the descending step component B should stay in column 1 | $x_{B,1,Des} < L_c$ (10.11) | $t_{Des} < \dfrac{V_L + K_B^{Des} V_U}{F_L + F_{L,F}}$ (10.19) |
| At the end of the ascending step component B should be completely eluted from column 1 | $x_{B,1,As} - x_{B,1,Des} > 0$ (10.12) | $\dfrac{t_{Des}}{t_{As}} < \dfrac{F_U}{F_L + F_{L,F}} K_B^{Des}$ (10.20) |
| At the end of the ascending step component A should stay in column 2 | $x_{A,2,As} < L_c$ (10.13) | $t_{As} < \dfrac{v_U + \dfrac{1}{K_A^{Des}} V_L}{F_U + F_{U,F}}$ (10.21) |
| At the end of the descending step component A should be completely eluted from column 2 | $x_{A,2,Des} - x_{A,2,As} > 0$ (10.14) | $\dfrac{t_{Des}}{t_{As}} > \dfrac{F_U + F_{U,F} K_A^{Des}}{F_L}$ (10.22) |

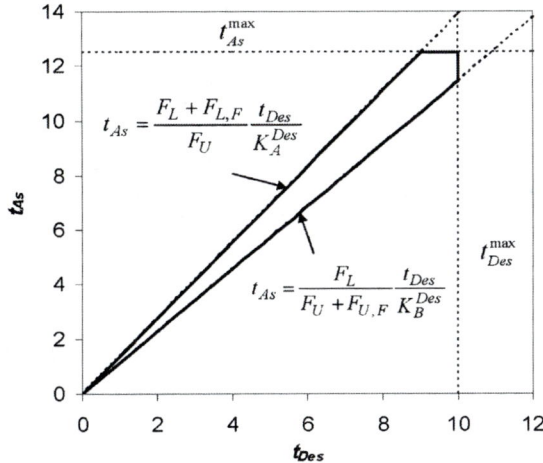

**Figure 10.12** Region of complete separation presented in a $t_{As}$ versus $t_{Des}$ plot for fixed values of the feed and mobile phase flow rates in the ascending and descending steps.

In Eqns (10.19)–(10.22), the flow rates and step durations are not explicitly defined and hence cannot be selected independent of each other. There are two possible options: (1) first the feed and mobile phase flow rates in both steps ($F_{L,F}$, $F_{U,F}$, $F_U$, $F_L$) are selected and the durations of the step times ($t_{As}$ and $t_{Des}$) are calculated using Eqns (10.19)–(10.22) or (2) the durations of the two steps ($t_{As}$ and $t_{Des}$) are selected and then the feed and mobile phase flow rates in the ascending and descending steps ($F_{L,F}$, $F_{U,F}$, $F_U$, $F_L$) are calculated using the same equations. Because there are no restrictions on the duration of the sCPC cycle steps, switching from one step to the other is achieved simply by switching the valves in the unit, and the first option is more reasonable.

For preselected values of $F_{L,F}$, $F_{U,F}$, $F_U$, and $F_L$ the region of complete separation, defined by Eqns (10.19)–(10.22), can be easily visualized in a $t_{As} = f(t_{Des})$ plot, presented in Figure 10.12.

Under the assumption of negligible dispersion effects, for any pair of step times ($t_{As}, t_{Des}$) within this region a complete separation of components $A$ and $B$ is achieved (100% purity and 100% recovery). Furthermore, for any point inside this region the productivity and solvent consumptions in cyclic steady state (CSS) are equal. The only difference between different ($t_{As}, t_{Des}$) pairs is the time (number of cycles) and consequently the solvent consumed until the unit reaches CSS (Völkl et al., 2011).

It is recommended to select the operating point ($t_{As}, t_{Des}$) as far as possible from the borders of the separation region to account for the influence of the dispersive effects on the product purity and to use longer step times (i.e., near the apex of the separation region) to ensure a system has hydrodynamic stability and to extend the valves' life time.

The proposed shortcut design method was validated experimentally for separation of binary mixtures containing components with similar physical properties, namely: (1) two isomers, hydroquinone and pyrocatechol (Hopmann et al., 2012a; Hopmann and Minceva, 2012), and (2) two molecules with molecular structures that differ by only one double bond, capsaicin and dihydrocapsaicin (Goll et al., 2013). In both cases, a complete separation was achieved experimentally using operating parameters selected with the proposed shortcut design method.

### 10.2.3 Modeling of sCPC Unit Operation

The hydrostatic columns consist of a series of cells with defined geometry connected by channels (see Figures 10.4(c) and 10.5). The cells are filled with the mobile and stationary phases, whereas in the channels only the mobile phase is present. The chromatographic models that were derived starting from a mass balance of differential volume of columns with defined cross-sectional area and length are thus not directly applicable to hydrostatic columns.

In Völkl et al. (2011), a model of an sCPC unit with hydrostatic columns is presented. The columns of the sCPC unit are modeled using the stage model of Martin and Synge (1941). In this model, the column is presented as a series of cells (stages) of equal volume filled with a constant volume of mobile and stationary phase; the mobile phase flows continuously from cell to cell with a constant flow rate. Further, it is assumed that the mobile phase leaving one cell is in equilibrium with the stationary phase in the cell. The number of cells (stages) is a parameter related to the axial dispersion and mass transfer resistance. The model is quite simple and easy to solve and most importantly does not require information about the cell geometry.

The sCPC unit model equations are the following.

*In descending mode*, the upper phase is the stationary phase and the mobile phase is introduced into column 2 (Figure 10.13(a)).

Mass balance of component $i$ in cell $k$ of column $j$:

$$\frac{V_U}{N} \frac{dc_{i,k,j}^U}{dt} + \frac{V_L}{N} \frac{dc_{i,k,j}^L}{dt} = F_{L,j}\left(c_{i,k+1,j}^L - c_{i,k,j}^L\right) \tag{10.23}$$

The flow rate and mass balance at the feed node are

$$F_{L,1} = F_{L,2} + F_{L,F} \tag{10.24}$$

$$F_{L,1}c_{i,in,1}^L = F_{L,2}c_{i,1,2}^L + F_{L,F}c_{i,F}^L \tag{10.25}$$

*In ascending mode*, the lower phase is the stationary phase and the mobile phase is introduced into column 1 (see Figure 10.13(b)).

**(a)**    **Descending step (Mobile phase=lower phase; Feed=lower phase + A +B)**

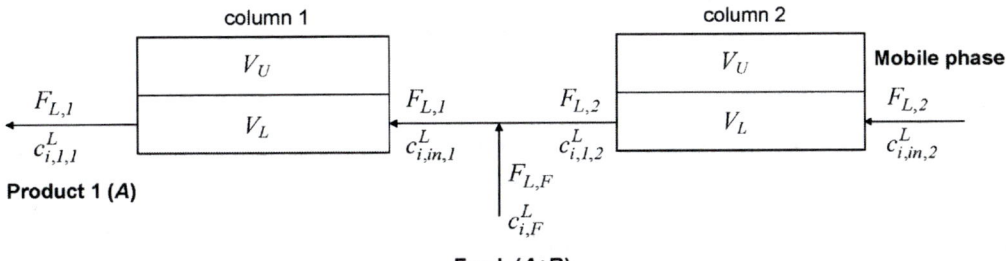

Product 1 (A)

Feed (A+B)

**(b)**    **Ascending step (Mobile phase= upper phase; Feed= upper phase + A +B)**

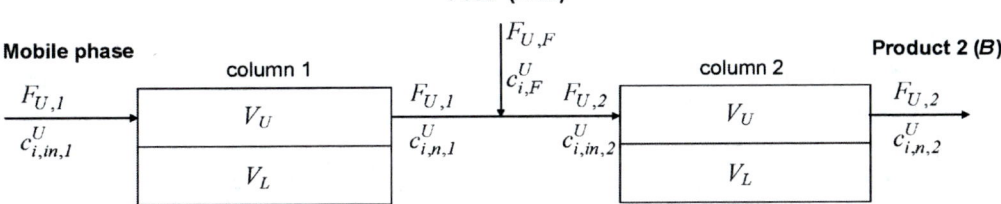

**Figure 10.13** Stream flow rates and concentrations in an sCPC unit during the (a) descending step and (b) ascending step.

Mass balance of component $i$ in cell $k$ of column $j$:

$$\frac{V_U}{N}\frac{dc_{i,k,j}^U}{dt} + \frac{V_L}{N}\frac{dc_{i,k,j}^L}{dt} = F_{U,j}\left(c_{i,k-1,j}^U - c_{i,k,j}^U\right) \qquad (10.26)$$

The flow rate and mass balance at the feed node are

$$F_{U,2} = F_{U,1} + F_{U,F} \qquad (10.27)$$

$$F_{U,2}c_{i,in,2}^U = F_{U,1}c_{i,n,1}^U + F_{U,F}c_{i,F}^U \qquad (10.28)$$

where $i$ refers to the solute ($i = A, B$); $k$ is the cell number ($k = 1, 2,\ldots,N$); $j$ is the column number ($j = 1,2$); $N$ is the number of cells (stages); $V_U$ and $V_L$ are the volumes of the upper and lower phases, respectively; $F_{U,j}$ and $F_{L,j}$ are the volumetric flow rates of the upper phase (during the ascending step) and the lower phase (during the descending step) in column $j$, respectively; $F_{L,F}$ and $F_{U,F}$ are the volumetric flow rates of the feed in the ascending and descending steps, respectively; $c_{i,k,j}^U$ is the concentration of component $i$ in the upper phase in cell $k$ of column $j$; $c_{i,k,j}^L$ is the concentration of component $i$ in the lower phase in cell $k$ of column $j$; $c_{i,in,j}^U$ is the inlet concentration of component $i$ in the upper phase in column $j$ during the ascending step; and $c_{i,in,j}^L$ is the inlet concentration of component $i$ in the lower phase in column $j$ during the descending step.

The distribution constant of component $i$ is defined as the ratio of its concentration in the stationary phase to that in the mobile phase, hence:

$$K_{i,Des} = \frac{1}{K_{i,As}} = \frac{c_{i,k,j}^{U}}{c_{i,k,j}^{L}} \quad i = A, B, \quad k = 1, 2 \ldots N, \quad j = 1, 2 \tag{10.29}$$

The number of cells (stages) at given mobile phase flow rates can be determined from pulse–injection experiments performed with each component (or the mixture) in ascending and descending modes, using the method of moments. For the simulation of the sCPC operation in the linear range of distribution (partition) equilibrium a different number of cells can be used for each component (i.e., no competitive effects between the components present in the feed mixture). For each component, the number of the cells is calculated as the average of the number of the cells obtained in ascending and descending steps at a given mobile phase flow rate. The distribution constants of the components in the feed mixture, the second model parameter, can be determined experimentally by a shake–flask experiment or from a pulse injection experiment according to Eqn (10.2).

### 10.2.4 Selection of the sCPC Unit Operating Parameters

In Figure 10.14 a general procedure for the selection of the sCPC unit operating parameters is presented.

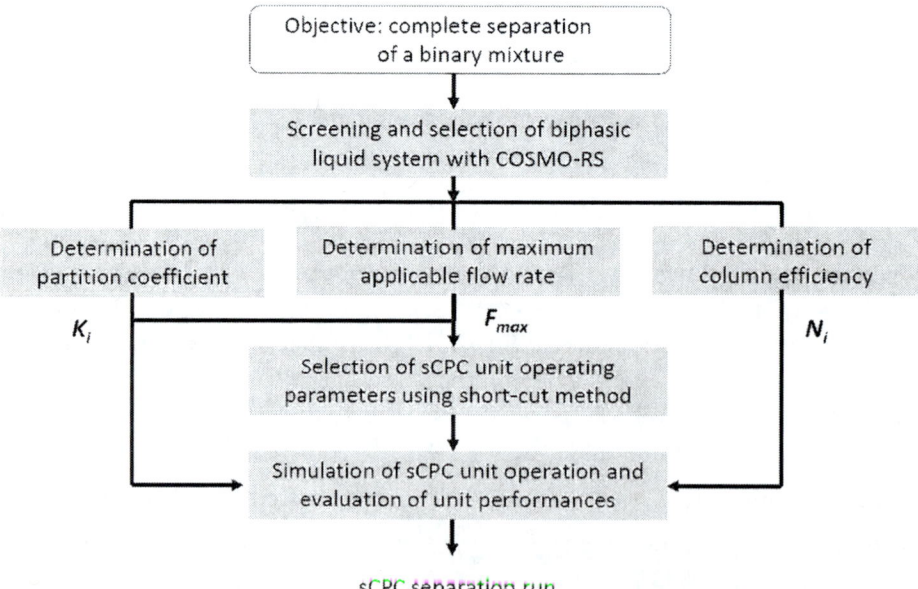

**Figure 10.14** Procedure for the selection of the sCPC unit operating parameters.

The first step involves screening for possible biphasic solvent systems using the predicted values of the distribution constants (partition coefficients) of the feed mixture components as a screening parameter.

Once a promising system is found, the distribution constants are verified experimentally. Next, the influence of the mobile phase flow rate on the volumes of the phases (upper and lower phase) in the sCPC columns in the ascending and descending step is determined. The goal is to find the maximum applicable mobile phase flow rate for a preset phase volume ratio. The column efficiency at different mobile phase flow rates, in ascending and descending modes, is determined from pulse injection experiments with the feed mixture.

In the next step, the region of complete separation for the preselected feed and mobile-phase flow rates is constructed using the shortcut design method presented in Section 10.2.2. Before the sCPC separation is performed, the separation process is simulated for the selected operating parameters using the model presented in Section 10.2.3 and the product purity is evaluated. At the end, an sCPC separation with the selected operating parameters is performed and the obtained experimental data are compared with the simulated ones. Once the model predictions are validated the model can be further used for an optimization of the unit operating parameters.

## 10.3 EXAMPLE

In Goll et al. (2013), the concepts described in the Section 10.2 are demonstrated for an sCPC unit design for the separation of a binary mixture of capsaicin and dihydrocapsaicin. The molecular structure of these two components differs in only one double bond (see Figure 10.15), and hence it is a good representative of a "difficult" separation task.

For the selection of a biphasic liquid system the heptane/ethyl acetate/methanol/water system was screened using COSMO-RS. The screening procedure includes

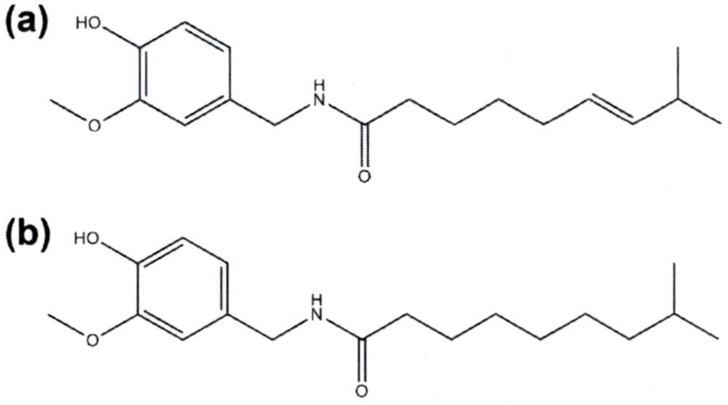

**Figure 10.15** Chemical structures of (a) capsaicin and (b) dihydrocapsaicin.

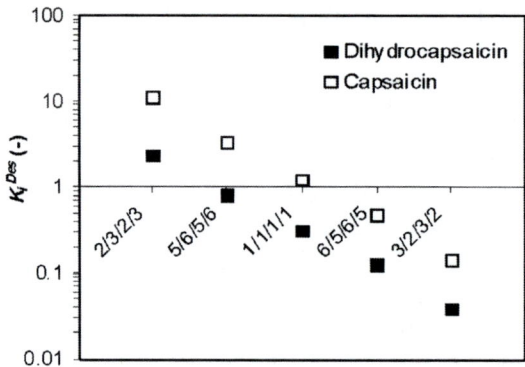

**Figure 10.16** COSMO-RS predicted values of the distribution constants of capsaicin and dihydrocapsaicin in heptane/ethyl acetate/methanol/water systems with different global compositions.

prediction of the distribution constants (partition coefficients) of capsaicin and dihydrocapsaicin in systems with different predefined global compositions of heptane/ethyl acetate/methanol/water. The results of the screening are presented in Figure 10.16.

The distribution constants of capsaicin and dihydrocapsaicin decrease from the system heptane/ethyl acetate/methanol/water 2/3/2/3 (v/v/v/v) to the system heptane/ethyl acetate/methanol/water 3/2/3/2 (v/v/v/v), whereas the separation factor (defined as the ratio of the distribution constants of dihydrocapsaicin and capsaicin) only slightly decreases (see Figure 10.16).

In the heptane/ethyl acetate/methanol/water systems with global compositions 2/3/2/3, 6/5/6/5, and 3/2/3/2 (v/v/v/v) both components distribute preferably in the same phase. The use of these systems in sCPC will result in a very low retention of the solutes in one sCPC step and a very high retention in the other sCPC step. As discussed in Section 10.1.3, low retention (very low values of partition coefficients) leads to low resolution, whereas high retention (very high values of partition coefficients) leads to an excessive solvent consumption and results in diluted products (dispersed elution profiles).

Hence, the most appropriate systems for carrying out an sCPC separation are the systems heptane/ethyl acetate/methanol/water 5/6/5/6 and 1/1/1/1 (v/v/v/v). In Goll et al. (2013) the system with lower separation factor, that is, heptane/ethyl/acetate/methanol/water 1/1/1/1 (v/v/v/v), was selected because the authors wanted to explore the separation limits of sCPC technology.

Next, the distribution constants of capsaicin and dihydrocapsaicin in the selected biphasic system were determined experimentally, by performing pulse injections in batch mode. The values of the distribution constants obtained are presented in Table 10.3.

To achieve similar separation conditions during the ascending and descending steps, the columns were filled with 50/50 (vol%/vol%) upper and lower phases of the biphasic system, resulting in a stationary phase retention of 0.5 in both modes (see Eqn (10.1)).

**Table 10.3** Sequential Centrifugal Partition Chromatography (sCPC) Unit Operating Parameters and System Parameters

| Operating Parameter | Value | Model Parameter | Value |
|---|---|---|---|
| $F_U$, $F_L$ (ml/min) | 18 | $V_U$, $V_L$ (ml) | 123.5 |
| $F_{U,F}$, $F_{L,F}$ (ml/min) | 2 | $K_{Cs}^{Des}$ (−) | 1.26 |
| $t_{Des}$ (min) | 7.00 | $K_{Dhs}^{Des}$ (−) | 1.66 |
| $t_{As}$ (min) | 4.83 | $N_{Cs}$ (−) | 450 |
| $c_{Cs,F}^{U}$, $c_{Cs,F}^{L}$ (mg/ml) | 3.50 | $N_{Dhs}$ (−) | 529 |
| $c_{Dhs,F}^{U}$, $c_{Dhs,F}^{L}$ (mg/ml) | 1.63 | | |

These conditions will lead to comparable hydrodynamics in the cells of both columns, and the maximum allowed mobile phase flow rate for the pre-set phase volume ratio of 0.5 will be in a similar range in both operating modes. In Figure 10.17 the stationary phase retention as a function of the mobile phase flow rate, in the ascending and descending steps, is presented.

The maximum applicable mobile phase flow rates in the ascending and descending steps are 20 and 22 ml/min, respectively. This means that the sum of the selected mobile phase and feed flow rates in each step should be lower than the maximal applicable flow rate. In addition, to ensure similar hydrodynamics and mass transfer conditions, the same mobile phase and feed flow rates in the ascending and descending steps were selected, that is, $F_U = F_L$ and $F_{U,F} = F_{L,F}$ (see Table 10.3).

The region of complete separation constructed using the selected flow rates and above described shortcut method is presented in Figure 10.18.

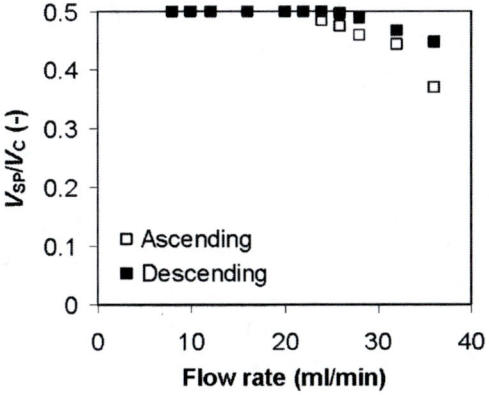

**Figure 10.17** Stationary phase retention in the ascending and descending modes as a function of the mobile phase flow rate. Biphasic system: heptane/ethyl acetate/methanol/water 1/1/1/1 (v/v/v/v), columns filled with 50/50 (vol%/vol%) upper and lower phase, 1700 rpm.

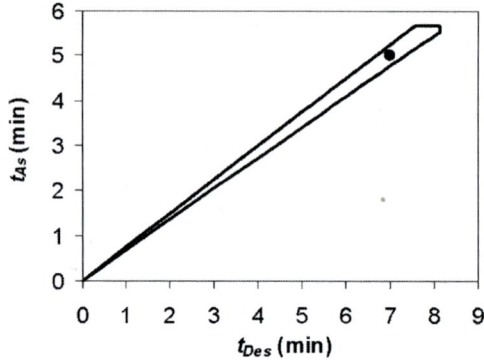

**Figure 10.18** sCPC separation region and selected operating point. sCPC unit operating conditions given in Table 10.3.

Under ideal conditions (instantaneous partition equilibrium) a complete separation of a binary mixture is always possible, as long as a separation region exists. In reality, owing to the dispersion and mass transfer effects, the separation region might be smaller than the one defined using the constraints of the shortcut design method, given in Table 10.2. Therefore, the sCPC operating point should be selected not too close to the borders of the separation region. As recommended in Völkl et al. (2011), long step times near the maximum allowed value are preferable, because of the time needed for the hydrodynamic equilibration inside the cells of the CPC columns after the switch from one step to another. Considering the above, the sCPC operating point presented in Figure 10.18 was selected ($t_{Des} = 7$ min and $t_{As} = 4$ min 50 s).

For the selected operating point the sCPC unit operation was simulated with the cell model (Eqns (10.23)−(10.29)). As an input for the model, the values of the sCPC unit operating parameters, the distribution constants, and the number of theoretical stages are required. The last was determined by performing pulse-injection experiments in the ascending and descending modes with a mobile phase flow rate of 20 ml/min (=feed + mobile phase flow rate, see Table 10.3).

According to the simulation results, the cyclic steady state is achieved after approximately 13 cycles. The purity of dihydrocapsaicin collected in the ascending step and of capsaicin collected in the descending step is 97% and 99%, respectively. Even though the sCPC operating point lies in the separation region, the purity of the products is lower than 100%. The difference between the expected and the predicted purities can be traced back to the fact that the sCPC cell model takes the mass transfer effects and hydrodynamics of the real system into account, whereas these parameters are neglected in the shortcut approach used to construct the region of complete separation.

In Figure 10.19, the experimental and simulated mean concentrations of the collected products in ascending and descending modes are plotted against the cycle number. There is a satisfactory agreement between the experimental and the simulated data.

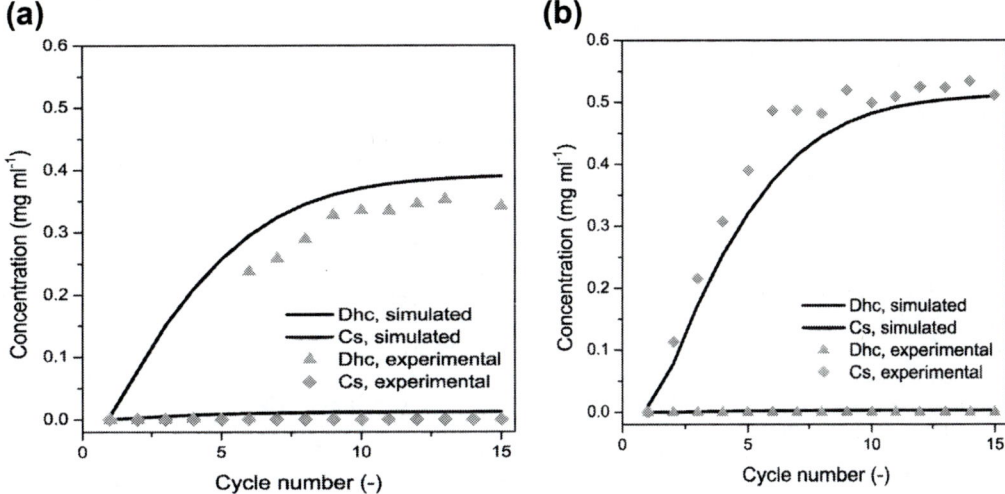

**Figure 10.19** Comparison of the experimental and simulated mean concentrations of the collected products in (a) ascending and (b) descending mode. sCPC unit operating conditions given in Table 10.3. Dhc, dihydrocapsaicin; Cs, capsaicin.

This example demonstrates the usefulness of theoretical models for design of sCPC separations. More details about applied theoretical models and methods are available in Goll et al. (2013), Hopmann et al. (2011, 2012a,b), Hopmann and Minceva (2012), and Völkl et al. (2011).

## 10.4  CONCLUDING REMARKS

Solid support-free liquid—liquid chromatography is a versatile technology that combines the principles of extraction (partitioning between two liquid phases) and chromatography (one of the liquid phases is stationary).

The described method for screening and selection of biphasic systems provides a way to design tailor-made stationary and mobile phases for a specific separation task. The main advantage of the activity coefficient predictive method used in this approach, COSMO-RS, is that the molecular structures of the solutes and solvents are the only input information needed.

The sCPC technology is at its very beginning, very similar to the early years of the now successful SMB technology. The presented shortcut design method and mathematical model for process simulation are the first engineering inputs to this new technology. The shortcut design method is very useful for the fast design and preliminary selection of the sCPC unit operating parameters. The method is equivalent to the state-of-the-art shortcut method used for designing SMB units, the so-called triangle theory. The relatively simple sCPC mathematical model gives a satisfactory prediction of experimentally acquired data and hence is a useful tool for process simulation studies.

# REFERENCES

Abrams, D.S., Prausnitz, J.M., 1975. Statistical thermodynamics of liquid mixtures: a new expression for the excess Gibbs energy of partly or completely miscible systems. AIChE J. 21, 116.

Adelmann, S., Schembecker, G., 2011. Influence of physical properties and operating parameters on hydrodynamics in centrifugal partition chromatography. J. Chromatogr. A 1218, 5401.

Agnely, M., Thiebaut, D., 1997. Dual-mode high-speed counter-current chromatography: retention, resolution and examples. J. Chromatogr. A 790, 17.

Audo, G., Le Quemeneur, C., 2012. True moving bed CPC for continuous purification of natural substances. In: Seventh International Conference on Countercurrent Chromatography, August 6–8, Hangzhou, China.

Berthod, A., 1991. Practical approach to high-speed counter-current chromatography. J. Chromatogr. A 550, 677.

Berthod, A., 2002. Countercurrent Chromatography: The Support-Free Liquid Stationary Phase. In: Comprehensive Analytical Chemistry, vol. 38. Elsevier Science & Technology Books, Amsterdam.

Berthod, A., Ruiz-Angel, M.J., Carda-Broch, S., 2009. Countercurrent chromatography: people and applications. J. Chromatogr. A 1216, 4206.

Berthod, A., Ruiz-Angel, M.J., Carda-Broch, S., 2003. Elution-extrusion countercurrent chromatography. Use of the liquid nature of the stationary phase to extend the hydrophobicity window. Anal. Chem. 75, 5886.

Buggert, M., Cadena, C., Mokrushina, L., Smirnova, I., Maginn, E.J., Arlt, W., 2009. COSMO-RS calculations of partition coefficients: different tools for conformation search. Chem. Eng. Technol. 32, 977.

Burghoff, B., Schiferli, J., Sousa Marques, J., de Haan, A.B., 2009. Extractant screening and selection for methyl tert-butyl ether removal from aqueous streams. Chem. Eng. Sci. 64, 2887.

van Buel, M.J., Van Halsema, F.E.D., Van der Wielen, L.A.M., Luyben, K.C., 1998. Flow regimes in centrifugal partition chromatography. AIChE J. 44, 1356.

Camacho-Frias, E., Foucault, A., 1996. Solvent systems in centrifugal partition chromatography. Analysis 24, 159.

Couillard, F., Foucault, A., Durand, A., 2005. Method and Device for Separating Constituents of a Liquid Charge by Means of Liquid-Liquid Centrifuge Chromatography. WO 2005/011835.

Craig, L.C., Post, O., 1949. Apparatus for countercurrent distribution. Anal. Chem. 21, 500.

Delannay, E., Toribio, A., Boudesocque, L., Nuzillard, J.M., Zeches-Hanrot, M., Dardennes, E., Le Dour, G., Sapi, J., Renault, J.H., 2006. Multiple dual-mode centrifugal partition chromatography, a semi-continuous development mode for routine laboratory-scale purifications. J. Chromatogr. A 1127, 45.

Foucault, A., 1995. Centrifugal Partition Chromatography. M. Dekker, New York.

Foucault, A.P., Bousquet, O., Le Goffic, F., 1992a. Importance of the parameters Vm/Vc in countercurrent chromatography: tentative comparison between instrument designs. J. Liq. Chromatogr. 15, 2691.

Foucault, A.P., Bousquet, O., Le Goffic, F., Cazes, J., 1992b. Countercurrent chromatography with a new centrifugal partition chromatographic system. J. Liq. Chromatogr. 15, 2721.

Foucault, A.P., Camacho Frias, E., Bordier, C.G., Le Goffic, F., 1994. Centrifugal partition chromatography: stability of various biphasic systems and pertinence of the "Stokes model" to describe the influence of the centrifugal field upon the efficiency. J. Liq. Chromatogr. 17, 1.

Fredenslund, A., Jones, R.L., Prausnitz, J.M., 1975. Group-contribution estimation of activity coefficients in nonideal liquid mixtures. AIChE J. 21, 1086.

Frey, A., Hopmann, E., Minceva, M., 2014. Selection of biphasic liquid systems in liquid-liquid chromatography using predictive thermodynamic models. Chem. Eng. Technol. 37, 1663.

Friesen, J.B., Pauli, G.F., 2005. G.U.E.S.S.—a generally useful estimate of solvent systems for CCC. J. Liq. Chromatogr. Relat. Technol. 28, 2777.

Friesen, J.B., Pauli, G.F., 2007. Rational development of solvent system families in counter-current chromatography. J. Chromatogr. A 1151, 51.

Garrard, I.J., 2005. Simple approach to the development of a CCC solvent selection protocol suitable for automation. J. Liq. Chromatogr. Relat. Technol. 28, 1923.

Goll, J., Frey, A., Minceva, M., 2013. Study of the separation limits of continuous solid support free liquid-liquid chromatography: separation of capsaicin and dihydrocapsaicin by centrifugal partition chromatography. J. Chromatogr. A 1284, 59.

Hewitson, P., Ignatova, S., Sutherland, I., 2011. Intermittent counter-current extraction-effect of the key operating parameters on selectivity and throughput. J. Chromatogr. A 1218, 6078.

Hewitson, P., Ignatova, S., Ye, H., Chen, L., Sutherland, I., 2009. Intermittent counter-current extraction as an alternative approach to purification of Chinese herbal medicine. J. Chromatogr. A 1216, 4187.

Hopmann, E., Goll, J., Minceva, M., 2012a. Sequential centrifugal partition chromatography: a new continuous chromatographic technology. Chem. Eng. Technol. 35, 72.

Hopmann, E., Arlt, W., Minceva, M., 2011. Solvent system selection in counter-current chromatography using conductor-like screening model for real solvents. J. Chromatogr. A 1218, 242.

Hopmann, E., Frey, A., Minceva, M., 2012b. A priori selection of the mobile and stationary phase in centrifugal partition chromatography and counter-current chromatography. J. Chromatogr. A 1238, 68.

Hopmann, E., Minceva, M., 2012. Separation of a binary mixture by sequential centrifugal partition chromatography. J. Chromatogr. A 1229, 140.

Ignatova, S.N., Maryutina, T.A., Spivakov, B.Y., 2001. Effect of physicochemical properties of two-phase liquid systems on the retention of stationary phase in a CCC column. J. Liq. Chromatogr. Relat. Technol. 24, 1655.

Ignatova, S.N., Sutherland, I.A., 2003. A fast, effective method of characterizing new phase systems in CCC. J. Liq. Chromatogr. Relat. Technol. 26, 1551.

Ignatova, S., Sumner, N., Colclough, N., Sutherland, I., 2011. Gradient elution in counter-current chromatography: a new layout for an old path. J. Chromatogr. A 1218, 6053.

Ito, Y., 2005a. Golden rules and pitfalls in selecting optimum conditions for high-speed counter-current chromatography. J. Chromatogr. A 1065, 145.

Ito, Y., 2005b. Origin and evolution of the coil planet centrifuge: a personal reflection of my 40 years of CCC research and development. Sep. Purif. Rev. 34, 131.

Ito, Y., Weinstein, M., Aoki, I., Harada, R., Kimura, E., Nunogaki, K., 1966. Coil planet centrifuge. Nature (London) 212, 985.

Klamt, A., Eckert, F., Hornig, M., 2001. COSMO-RS: a novel view to physiological solvation and partition questions. J. Comput. Aided Mol. Des. 15, 355.

Lu, Y., Pan, Y., Berthod, A., 2008. Using the liquid nature of the stationary phase in counter-current chromatography. V. The back-extrusion method. J. Chromatogr. A 1189, 10.

Mandava, N.B., Ito, Y., 1988. Countercurrent chromatography: theory and practice. In: Chromatographic Science Series, vol. 44. Marcel Dekker, Inc., New York.

Marchal, L., Foucault, A., Patissier, G., Rosant, J.M., Legrand, J., 2000. Influence of flow patterns on chromatographic efficiency in centrifugal partition chromatography. J. Chromatogr. A 869, 339.

Marchal, L., Legrand, J., Foucault, A., 2002. Mass transport and flow regimes in centrifugal partition chromatography. AIChE J. 48, 1692.

Marston, A., Hostettmann, K., 1994. Counter-current chromatography as a preparative tool—applications and perspectives. J. Chromatogr. A 658, 315.

Martin, A.J., Synge, R.L., 1941. A new form of chromatogram employing two liquid phases: a theory of chromatography. 2. Application to the micro-determination of the higher monoamino-acids in proteins. Biochem. J. 35, 1358.

Mekaoui, N., Chamieh, J., Dugas, V., Demesmay, C., Berthod, A., 2012. Purification of Coomassie Brilliant Blue G-250 by multiple dual mode countercurrent chromatography. J. Chromatogr. A 1232, 134.

Menet, J.M., Thiebaut, D. (Eds.), 1999. Countercurrent chromatography. Chromatographic Science Series, vol. 82. Marcel Dekker, New York.

Mokrushina, L., Buggert, M., Smirnova, I., Arlt, W., Schomaecker, R., 2007. COSMO-RS and UNIFAC in prediction of micelle/water partition coefficients. Ind. Eng. Chem. Res. 46, 6501.

Oka, H., Harada, K.I., Ito, Y., Ito, Y., 1998. Separation of antibiotics by counter-current chromatography. J. Chromatogr. A 812, 35.

Pauli, G.F., Pro, S.M., Friesen, J.B., 2008. Countercurrent separation of natural products. J. Nat. Prod. 71, 1489.

Peng, A., Ye, H., Shi, J., He, S., Zhong, S., Li, S., Chen, L., 2010. Separation of honokiol and magnolol by intermittent counter-current extraction. J. Chromatogr. A 1217, 5935.

Renault, Jean-Hugues, J.-M.N.O.I.A.M, 2002. Solvent systems. Countercurrent chromatography: the support-free liquid stationary phase. In: Comprehensive Analytical Chemistry, vol. 38. Elsevier Science & Technology Books, Amsterdam.

Renon, H., Prausnitz, J.M., 1968. Local compositions in thermodynamic excess functions for liquid mixtures. AIChE J. 14, 135.

Sørensen, J.M., Arlt, W., Macedo, E., Rasmussen, P., 1979. Liquid-Liquid Equilibrium Data Collection. In: DECHEMA Chemistry Data Series, vol. V. DECHEMA, Frankfurt/Main.

Spiess, A.C., Eberhard, W., Peters, M., Eckstein, M.F., Greiner, L., Buechs, J., 2008. Prediction of partition coefficients using COSMO-RS: solvent screening for maximum conversion in biocatalytic two-phase reaction systems. Chem. Eng. Process. 47, 1034.

Sutherland, I., Garrard, I., Hewitson, P., Ignatova, S., Thickitt, C., Douillet, N., Freebairn, K., Johns, D., Mountain, C., Wood, P., Edwards, N., Rooke, D., Harris, G., Keay, D., Mathews, B., Brown, R., 2014. Scalable technology for the extraction of pharmaceutics: outcomes from a 3 year collaborative industry/academia research programme. J. Chromatogr. A 1282, 84.

Sutherland, I.A., Audo, G., Bourton, E., Couillard, F., Fisher, D., Garrard, I., Hewitson, P., Intes, O., 2008. Rapid linear scale-up of a protein separation by centrifugal partition chromatography. J. Chromatogr. A 1190, 57.

Sutherland, I.A., Brown, L., Forbes, S., Games, G., Hawes, D., Hostettmann, K., Mckerrell, E.H., Marston, A., Wheatley, D., Wood, P., 1998. Countercurrent chromatography (CCC) and its versatile application as an industrial purification and production process. J. Liq. Chromatogr. Relat. Technol. 21, 279.

Sutherland, I., Hewitson, P., Ignatova, S., 2009. New 18-l process-scale counter-current chromatography centrifuge. J. Chromatogr. A 1216, 4201.

Sutherland, I.A., 2007. Recent progress on the industrial scale-up of counter-current chromatography. J. Chromatogr. A 1151, 6.

Völkl, J., Arlt, W., Minceva, M., 2011. Method for selection of operating parameters in sequential centrifugal partition chromatography. AIChE J. 59, 241.

# CHAPTER 11

# Conclusions and Perspectives

A modern view of chemical engineering (ChE) can be summarized in the ChE diamond (Rodrigues, 2012) shown in Figure 11.1 with focus on process and product engineering and molecular and materials engineering or, in short, ChE = $M^2P^2E$. This is not only about how "to make" but also "to service" and "to care," in the words of Solke Bruin (2004), to stress the whole life cycle of a product.

The simulated moving bed (SMB) was initially used in a class of *processes* (Sorbex processes from UOP) to make *products* (commodities such as *p*-xylene) using *materials* (adsorbents). SMB technology is now a mature technology spread over various areas such as the petrochemical, pharmaceutical, sugar/food, and bioprocessing industries. The realization of the technology may differ depending on the application (rotary valve vs set of valves). The separation problems involved liquid-phase bulk separations from pseudo-binary mixtures as in the PAREX process, to binary mixtures as in chiral separations, up to the recovery of one particular component out of a complex mixture as is the case in many bioindustries. In this last case two-column SMB may be a good solution because of its simplicity and robustness (Nestola et al., 2014). Further developments are expected from the materials corner by improving adsorbents (e.g., binderless zeolites), making new adsorbents (metal organic frameworks), or new-shaped adsorbents (monoliths).

Future developments will be certainly coming from the *molecular engineering* corner leading to faster process development, less experimentation to acquire basic data, and better understanding of adsorption or catalytic processes at the molecular level (Granato et al., 2014).

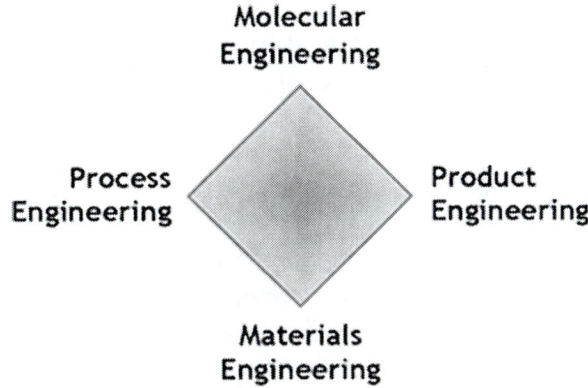

**Figure 11.1** The chemical engineering diamond: ChE = $M^2P^2E$.

*Simulated Moving Bed Technology*
ISBN 978-0-12-802024-1

Today one can find in the market SMB equipment for the laboratory scale-up to the industrial scale from various vendors as well as home-made equipment in various laboratories, including the FlexSMB® from our group. However, in industrial practice it is important to get the product within specifications even if process disturbances occur. SMB process control is still one area of future research with solutions adapted for various separation problems (Suvarov et al., 2014).

We should look at SMB and SMB-related processes (continuous ion-exchange or sequential multicolumn processes in general) not as a panacea but simply as one more available technology to be considered, in particular in the case of difficult separations. One should keep in mind that SMB does not deliver the pure components we want; instead the products are in streams (extract and raffinate in the case of binary separations) together with the desorbent (or eluent). This means that further separation units (distillation, membranes, crystallization) must be considered to optimize the overall separation train. There is a window for research on hybrid processes combining different separation processes in order to develop criteria to decide what is the best cut for the first separation in combination with the second unit to reach at the end of the required specifications in terms of purity and recovery (Ayotte-Sauve et al., 2010).

Current developments in biorefineries (de Jong et al., 2012) also open up many opportunities to implement SMB or SMB-based processes in various platforms (lignin, C6 sugars, C5 and C6 sugars, pulp, and biogas, to name a few) for the production of bio-based products such as $n$-butanol, ethanol, lactic acid, ethyl lactate, xylitol, sorbitol, succinic acid, vanillin, syringaldehyde, galacto-oligosaccharides, isomalto-oligosaccharides, fructo-oligosaccharides, etc. In particular, supercritical fluid SMB is a technology that is ready to use but has not yet reached the industrial scale. Nevertheless there is a potential for such technology in areas of separation of extracts from natural products (Cristancho et al., 2012).

Industrial SMB separations are in liquid phase but there is room for new gas-phase SMB separations, as illustrated in this book with propane/propylene separations; in this case the downstream separation units could be simple pressure-swing adsorption processes for $C_3/C_4$ separations using $C_4$ as the desorbent. Further development is expected not only in the area of olefins/paraffins separation but also in energy-related applications (biogas, syngas).

The coupling of adsorption and reaction processes under the umbrella of SMB technology led to the development of the simulated moving bed reactor (SMBR). This should be of interest for equilibrium-limited reactions and has increasing value when going from reactants to products. The concept has been tested at the laboratory scale and patented for acetals (diethyl acetal, dimethyl acetal, dibutyl acetal, diethoxybutane, glycerol acetal). It has also been tested for the production of ethyl lactate and butyl acrylate. However, as far as we know, no industrial plant exists using SMBR technology. Further process intensification combining SMBR with membrane permeation

(PermSMBR) would improve the process performance by reducing desorbent consumption.

The acceptance of a new technology takes time and one should be aware that combined technologies require a deeper understanding of the systems characteristics (reaction and adsorption equilibrium and kinetics, catalyst deactivation, adsorbent aging). In fact, systems with separate units for reaction and adsorption (separation) are more robust than integrated systems. Moreover, process intensification through the combination of reaction and adsorption processes requires the matching of operating conditions for reaction and adsorption, and also the rate of reaction should be similar to the rate of removal of one product of the reversible reaction. This concept is also known in gas-phase processes as sorption-enhanced reaction processes, for hydrogen production by steam reforming of methane (Hufton et al., 1999; Oliveira et al., 2011) or ethanol (Wu et al., 2013), and sorption-enhanced water gas shift, demonstrated in the European project CAESAR (Jansen et al., 2013).

Last, but not the least, we expect a growing development of sequential centrifugal partition chromatography (sCPC), presented in Chapter 10 of this book; sCPC really combines two unit operations: extraction and chromatography.

Astarita once said: "The amount of information available grows continuously, but the amount of information that any one of us can digest does not grow." By writing this book we intended to facilitate the digestion of various aspects of SMB technology and help shape the future use of cyclic adsorption/reaction processes.

## REFERENCES

Ayotte-Sauve, E., Sorin, M., Rheault, F., 2010. Energy requirement of a distillation/membrane parallel hybrid: a thermodynamic approach. Industrial Eng. Chem. Res. 49, 2295–2305.

Bruin, S., 2004. Product-driven Process Engineering. The Eternal Triangle Molecules, Product, Process. Inaugural Lecture given at TU Eindhoven Jan 9.

Cristancho, C.A.M., Peper, S., Johannsen, M., 2012. Supercritical fluid simulated moving bed chromatography for the separation of ethyl linoleate and ethyl oleate. J. Supercrit. Fluids 66, 129–136.

de Jong, E., Higson, A., Walsh, P., Wellisch, M., 2012. Bio-based Chemicals. Value Added Products from Biorefinery. IEA Bioenergy Task 42 Biorefinery.

Granato, M.A., Martins, V.D., Santos, J.C., Jorge, M., Rodrigues, A.E., 2014. From molecules to processes: molecular simulations applied to the design of simulated moving bed for ethane/ethylene separation. Can. J. Chem. Eng. 92, 148–155.

Hufton, J.R., Mayorga, S., Sircar, S., 1999. Sorption-enhanced reaction process for hydrogen production. Aiche J. 45, 248–256.

Jansen, D., Selow, E.V., Cobden, P., Manzolini, G., Macchi, E., Gazzani, M., Blom, R., Henriksen, P., Beavis, R., Wright, A., 2013. SEWGS technology is now ready for scale-up! Energy Procedia 37, 2265–2273.

Nestola, P., Silva, R.J.S., Peixoto, C., Alves, P.M., Carrondo, M.J.T., Mota, J.P.B., 2014. Adenovirus purification by two-column, size-exclusion, simulated countercurrent chromatography. J. Chromatogr. A 1347, 111–121.

Oliveira, E.L.G., Grande, C.A., Rodrigues, A.E., 2011. Effect of catalyst activity in SMR-SERP for hydrogen production: commercial vs. large-pore catalyst. Chem. Eng. Sci. 66, 342–354.

Rodrigues, A.E., 2012. Adsorption. What else? Mater. Adsorción Catálisis 4, 6—16.

Suvarov, P., Kienle, A., Nobre, C., De Weireld, G., Wouwer, A.V., 2014. Cycle to cycle adaptive control of simulated moving bed chromatographic separation processes. J. Process Control 24, 357—367.

Wu, Y.J., Li, P., Yu, J.G., Cunha, A.F., Rodrigues, A.E., 2013. Sorption-enhanced steam reforming of ethanol on NiMgAl multifunctional materials: experimental and numerical investigation. Chem. Eng. J. 231, 36—48.

# INDEX

*Note*: Page numbers followed by "f" and "t" indicate figures and tables, respectively.

CPSIA information can be obtained at www.ICGtesting.com
Printed in the USA
LVOW09*0954120715

445927LV00010B/189/P